Advances in Asian Human-Environmental Research

Aims and Scope

The series aims at fostering the discussion on the complex relationships between physical landscapes, natural resources, and their modification by human land use in various environments of Asia. It is widely acknowledged that human-environment interactions become increasingly important in area studies and development research, taking into account regional differences as well as bio-physical, socio-economic and cultural particularities.

The book series seeks to explore theoretic and conceptual reflection on dynamic human-environment systems applying advanced methodology and innovative research perspectives. The main themes of the series cover urban and rural landscapes in Asia. Examples include topics such as land and forest degradation, glaciers in Asia, mountain environments, dams in Asia, medical geography, vulnerability and mitigation strategies, natural hazards and risk management concepts, environmental change, impacts studies and consequences for local communities. The relevant themes of the series are mainly focused on geographical research perspectives of area studies, however there is scope for interdisciplinary contributions.

More information about this series at http://www.springer.com/series/8560

Rais Akhtar
Editor

Climate Change and Human Health Scenario in South and Southeast Asia

 Springer

Editor
Rais Akhtar
International Institute of Health Management and Research (IIHMR)
New Delhi, India

ISSN 1879-7180   ISSN 1879-7199   (electronic)
Advances in Asian Human-Environmental Research
ISBN 978-3-319-23683-4   ISBN 978-3-319-23684-1   (eBook)
DOI 10.1007/978-3-319-23684-1

Library of Congress Control Number: 2016936378

Springer Cham Heidelberg New York Dordrecht London

Cover image: Nomads near Nanga Parbat, 1995. Copyright © Marcus Nüsser (used with permission)

Printed on acid-free paper

Springer International Publishing AG Switzerland is part of Springer Science+Business Media (www.
springer.com).

Foreword 1

Climate change is increasingly recognised as a major threat to public health with the potential to affect many health outcomes including vector-borne and diarrhoeal diseases, under-nutrition, air pollution-related outcomes, allergies, heat- and cold-related mortality and a range of outcomes related to extreme events such as floods, droughts and intense storms. Despite growing evidence of a range of adverse effects, progress in addressing the emissions of greenhouse gases from fossil fuel combustion and land use change is too slow. The agreement of 195 nations in the Conference of the Parties (COP21) in Paris in December 2015 represents a considerable achievement which aims to hold a global temperature rise this century to well below 2 °C above pre-industrial levels and to catalyse efforts to limit the temperature increase even further to 1.5 °C. However, the Intended Nationally Determined Contributions (INDCs) that were communicated to the Treaty secretariat prior to the conference fell substantially short of these aspirations and would result in warming of around 2.7 °C (http://climateactiontracker.org/global.html). Thus greater efforts to curb emissions are needed to protect human health and reduce the risk of wide ranging and potentially catastrophic effects.

Disadvantaged populations are generally likely to be most vulnerable to the adverse effects of climate change despite having contributed a little to the problem. There is a danger that climate change could reverse some of the hard-won gains in health that have been achieved in many parts of the world in recent decades.

In order to protect public health, it will be essential to develop and implement effective policies to reduce vulnerability and increase the resilience of populations to climate change as far as possible. Strengthening the capacity of health systems to detect and respond to threats to health posed by climate change is a vital component of effective adaptation strategies. Health facilities and staff need to be equipped to deal with climate-related health outcomes. However, there are limits to adaptation that need to be recognised, including for example to intense thermal stress, which will impair labour productivity particularly in those who work outdoors in tropical and sub-tropical regions or to sea-level rise in low lying islands. This makes further action to reduce emissions of greenhouse gases imperative in order to reduce the risk of 'dangerous climate change', which is the focus of the UN Framework

Convention on Climate Change. There is also growing recognition that many policies to reduce greenhouse gas emissions can also have ancillary (co-)benefits for
health; for example, reduced combustion of coal for electricity generation can
reduce both carbon dioxide emissions and fine particulate air pollution. Awareness
of these near-term benefits can make policies to reduce GHG emissions more attrac-
tive to decision makers.

This book provides a rich source of evidence of many of the potential threats to
public health posed by climate change but also demonstrates how well-designed
policies can help protect against the risks posed by climate change. It covers a
diverse range of countries and health outcomes covering issues such as the vulnerability of older people and the implications of demographic change together with
the impact of climate change on the distribution of major vector-borne diseases such
as malaria and dengue. It also addresses the health challenges facing coastal populations in the face of sea level rise often accompanied by other local environmental
changes such as loss of mangroves, depletion of aquifers and damming of rivers. Its
focus on South and Southeast Asia makes it a valuable source of evidence for public
health professionals, clinicians and those who are interested to learn more about the
potential regional impacts of climate change. This book deserves to be widely read
and acted upon by all those concerned to protect vulnerable populations against the
burgeoning threats to health from climate change.

London School of Hygiene and Tropical Medicine   Andy Haines
London, UK

Foreword 2

Where and on whom will humandriven climate change have most of its early and serious impacts? It has long been recognised that, for two broad reasons, most of that climateattributable trauma, disease and death will occur in the regions of the world that are geographically vulnerable and burdened by poverty, crowding and, often, weak, misguided or corrupt government.

The first general reason is that the effects of climate change on human health are mostly those of a risk multiplier. Climate change comprises a complex of altered environmental conditions, ecological processes and social relationships; it is not a classical 'exposure'. Hence, populations that already have high preexisting rates of child diarrhoeal disease, undernutrition and stunting and various vector-borne infectious diseases, when exposed to increases in temperature, flooding or wild weather, are likely to experience the greatest *absolute* increases in those and related adverse health impacts.

Second, many such populations exist within the lowerlatitude world, with high agricultural dependence on seasonal monsoonal rainfall, heavily populated low-lying coastal regions and various endemic infectious diseases that thrive in warm and moist conditions. Many of those populations have high and persistent levels of poverty, crowded periurban living, limited understanding of risks to health and a low level of economic, technical and human resources available to help avert adverse health impacts of climate change.

For those reasons this book is welcome. It focuses on the circumstances and future health risks from climate change in South and Southeast Asia, where more than two billion of the world's population lives. The conditions and risks within the region are many and diverse, and adaptive strategies will vary greatly across place and over time.

So too will the ways in which different countries will need to curtail their greenhouse gas emissions – whether in relation to wetrice agriculture (methane), increased livestock production (predominantly methane; also carbon dioxide), electricity generation (carbon dioxide and black carbon), industrial production and transport systems (carbon dioxide, methane precursors, etc.). Alternative technologies, not curbed national development, must provide much of that resolution.

The book provides broad geographic coverage, including India, Bangladesh, Taiwan, Thailand, Indonesia and Malaysia. That should provide comparative opportunities to learn about and learn from one another and to understand more clearly the essential transboundary nature of the climate change problem. Pleasingly, issues to do with mental health, livelihoods and adaptation strategies are integral parts of this volume.

There is much more that could be written about if space and current knowledge permitted. Climate change knows no boundaries in the diversity of ways that it influences human health, wellbeing and physical survival. Those ways will assume different intensities and forms as the climate and world around us continue changing.

So we should learn from what is already known, including from the contents of this book, and expand our efforts in research, public communication and discussion, policy advocacy and progressive intersectoral action to both abate and insulate against climate change.

The Australian National University   Tony (A.J.) McMichael
Canberra, ACT, Australia

Preface

Globally, climate change-induced natural disasters are occurring nearly five times as often as they were in the 1970s. But some disasters – such as floods and storms – pose a bigger threat than others. Flooding and storms are also impacting the economy. Heatwaves are also an emerging killer.

Thus climate change represents a range of environmental hazards and will affect populations in the regions such as South and Southeast Asia, which already suffer from a number of natural hazards in general and climate-sensitive diseases in particular.

According to WHO report about 44 % of all disasters, globally during 1996–2005: 57 % of people killed globally in natural disasters were from South East Asia Region. The WHO report further adds that by 2025, Asia will be home to 21 of the world's 37 global megacities, making multibillion dollar disasters even more common (WHO 2008).

The climate change scenario in the South and Southeast Asia has been a major focus in the recent assessment of IPCC that indicates warming trends and increasing temperature extremes which have been observed across most of the Asian region over the past century (IPCC 2014).

Earlier, the IPCC 4th Assessment Report had stated that climate change, in particular increased risk of floods and droughts, is expected to have severe impact on South Asian countries, since their economies rely heavily on agriculture, natural resources, forestry and fisheries sectors. About 65 % of the population of South Asia live in rural areas and account for about 75 % of the poor who are most vulnerable and are impacted by climate change. Changes in the intensity of rainfall events and the overall impact of El Niño on the Monsoon pattern and increased risk of critical temperature being exceeded more frequently exist.

Researchers in Southeast Asia used data on spatial distribution of various climaterelated hazards (cyclones/typhoons), drought risks, landslides, sea level rise, population density and adaptive capacity, in areas of Indonesia, Thailand, Vietnam Lao PDR, Cambodia, Malaysia and the Philippines in the Southeast Asian region. Thus both South and Southeast Asian countries are vulnerable due to heatwaves and floods, and the region experienced the largest number of flood events.

South Asian countries – India, Pakistan and Bangladesh – and Thailand in the
Southeast Asian region suffer from heatwaves. A World Bank report warned that
'Countries in the South East Asia region are particularly vulnerable to the sealevel
rise, increases in heat extremes, increased intensity of tropical cyclones, and ocean
warming and acidification because many are archipelagoes located within a tropical
cyclone belt and have relatively high coastal population densities' (The World Bank
2013).

Natural Disaster-Affected Population

An analysis of data from South and Southeast Asia reveals the pattern of access to
safe drinking water to the population. Inequalities are not widespread, though
improved water access is better in countries such as Bhutan, Maldives, Malaysia,
Thailand, India and Pakistan (Fig. 1).

However, there are wide disparities in using improved sanitation, with Maldives,
Malaysia, Thailand and Sri Lanka better off in comparison with other countries.
India, Nepal and Cambodia fared poorly in the provision of improved sanitation.

In Southeast Asia, many more people died as a result of natural disasters from 2001 to 2010
than during the previous decade, mainly because of two extreme events: the Indian Ocean
earthquake and tsunami of 2004 and Cyclone Nargis in Myanmar in 2008. While the floods
in Pakistan raised deaths due to natural disaster to over 2,100 with over 18 million people
affected. The record floods in Pakistan contributed to the large economic damages and
losses experienced by Pakistan in 2010 (US$7.4 billion) making it the most costly year with

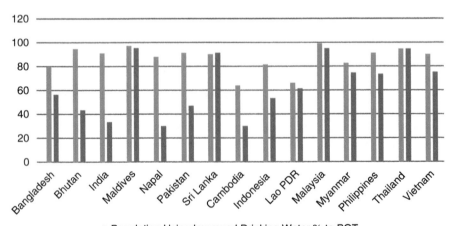

Fig. 1 Natural disaster-affected population in South and Southeast Asia 2010 (Source: Based on
data from *Statistical Year Book for Asia and the Pacific*, Environment II, UNESCAP, Bangkok
2011)

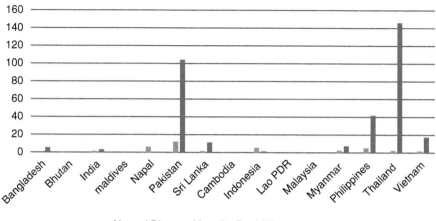

■ Natural Disaster Mortality Per Million

■ People Affected by Natural Disasters Per Million

Fig. 2 Natural disaster mortality and affected population in South and Southeast Asia 2010 (Source: Based on Data from *Statistical Year Book with for Asia and the Pacific*, Environment II, UNESCAP, Bangkok 2011)

respect to natural disasters in at least 20 years In Southeast Asia, many more people died as a result of natural disasters from 2001 to 2010 than during the previous decade, mainly because of two extreme events: the Indian Ocean earthquake and tsunami of 2004 and Cyclone Nargis in Myanmar in 2008. While the floods in Pakistan raised (UNESCAP 2011).

Figure 2 depicts natural disaster-related mortality and affected population in South and Southeast Asian countries. Pakistan, Philippines, Indonesia and Nepal had higher mortality per million in comparison with other countries. As regards the affected population due to natural disasters is concerned, Thailand and Pakistan suffered most, followed by Philippines, Vietnam and Sri Lanka in the region.

Thus it is obvious that there are inequalities resulting in the population's vulnerability to face extreme climatic events such as temperature increase and high precipitation leading to drought and flooding, respectively.

Burden of Communicable Diseases

The burden of communicable diseases is still high in the region. Dengue continues to pose a major and increasing public health problem. Chikungunya fever is reemerging and outbreaks of Nipah virus infections are being reported. Nipah virus infection (NiV) is named after a Malaysian village where it was first discovered and categorised as the emerging infectious disease of public health importance in the Southeast Asia Region. Drugresistant malaria has spread.

Although several countries have made significant progress toward increasing water supply coverage, sanitation coverage remains low – 31 % of the total population has access to improved sanitation in Cambodia and Nepal, 34 % in India and 96 % each in Malaysia and Thailand. Thus the level of vulnerability varies in the region. As a consequence of poor sanitary practices, diarrhoea causes substantial mortality. Public awareness of food hygiene related to food standards is limited, as is the food safety surveillance system. More than 70 % of workers are not covered by occupational health provisions (The World Bank 2013).

In the background of geoecology of health and disease in the region, an attempt has been made in this book by authors to assess the impact of climate change scenarios in geographically varied region in South and Southeast Asia, on human health conditions, and adaptation strategies by nations and communities in order to combat climate change impact. Part I contains chapters covering chapters on the South and Southeast Asian region. Part II consists of regional studies encompassing selected countries.

Rais Akhtar in Chap. 1 throws light on dimensions of climate change and its devastating impacts on health disasters in both developed and developing countries with focus on the South and Southeast Asian region. This chapter discusses greenhouse gas emission policies of the USA, India and China and highlights the association between climate change and human health.

The authors of Chap. 2 focus on many reasons to be concerned about the impacts of climate change on this region of South Asia, many parts of which experience apparently intractable poverty. Whilst there are a few countering, hopeful factors, it is fair to conclude that great problems lie ahead, not only for population health in South Asia but for South Asian society, unless there is an urgent and wide-scale change in policy and implementation; heatwaves are a growing problem in South Asia, and the management of vector-borne diseases (VBDs) is likely to become more problematic due to climate change. The authors assert that urgent action is therefore needed, particularly by India, an emerging great power, to transform its long-held justification for its lack of action in mitigating and preparing for climate change.

A number of groups within populations have been identified as potentially vulnerable in emergency situations by many official organisations and NGOs. Whilst climate change and effects on health tend to be longer term, similar groups tend to be more vulnerable to associated risks. Prominent among these are older persons; McCracken and Phillips in Chap. 3 look at older persons' health and climate change and explain that the health status and health vulnerability of older people in the context of climate change have physiological, psychological and social components. They emphasise that, whilst older persons on average tend to be at greater risk than others from adverse events, it is also important to recognise older persons' heterogeneity and to avoid stereotyping all as equally vulnerable or needy. Nevertheless, they note that older persons' health status will generally be more susceptible to factors such as extreme heat, as evidenced by excess mortality in the European heatwave of 2003 and elsewhere. Older persons, particularly those with pre-existing conditions such as chronic lung problems or hypertension, may also be affected by increasing indoor and external air pollution levels. Risks may be raised of conditions such as ischemic heart disease, stroke, COPD, acute lower respiratory disease and lung cancer, which affect older cohorts disproportionately. Higher temperatures may also be associated with extreme weather events such as typhoons and associated heavy rainfall, flooding and damage to property, wellbeing and even risk to life. The authors assert that older persons, especially those without families or support, may be especially at risk and unable to escape or mitigate the effects of such events. They may also be resident in places where many types of infectious diseases will be affected by temperature rises, including malaria, dengue and other parasitic or infectious conditions.

S. K. Dash in Chap. 4 identified the regions of extreme temperature events. The chapter includes an indepth analysis of daily maximum and minimum temperatures all over India spanning a number of years. Such study over India is essential not only because of scientific interest but also for getting useful information for policy formulation. India is a land of variety with large spatial and temporal variations in topography, land conditions, surface air temperature and rainfall. For scientific analysis, according to the author, the landmass of the country is usually divided into seven temperature homogeneous zones. Therefore, the degree of temperature extremes may not be the same in each homogeneous zone.

Chattopadhyay and colleagues in Chap. 5 opine that rapid change in climate through the Quaternary under the impact of global warming and its impact on the snow fields and glaciers of the inner Himalayan parts has attracted many geoscientists over the last four decades. But little work has been done so far to assess the impact of climate change and its impact on the size reduction of glaciers as well as on the human health in the surrounding settlement areas of the glaciers. Their chapter is a revelation and discussion on the present pattern of retreat of the major glaciers in the southeastfacing slopes of Kanchenjunga in the Sikkim Himalaya under the present state of global warming. Survey conducted by the authors in the three settlement areas in this western part of the Sikkim Himalaya on human health status, based on random sampling, shows that incidences of some particular diseases, like skin problems and asthma, which were rare earlier are spreading significantly among the inhabitants in this part of the Sikkim Himalaya.

R.C. Dhiman describes in Chap. 6 the situation of malaria and dengue in India. The latest trend of malaria indicates that with the introduction of rapid diagnostic kits, combination therapy for treatment and long-lasting insecticidal nets for prevention from mosquito bites from 2005 to 2009, cases have been reduced to around 50 % since 2001, and many states are qualified for malaria preelimination phase. On the other hand, the spatial and temporal distribution of dengue is increasing gradually. According to Dhiman, overall, there has been about 10 times increase in incidence and about four times death due to dengue in 2012 against the data of 2007. Further efforts should be directed to understand the dynamics of asymptomatic malaria in hard-core malarias areas, to ensure compliance to treatment and intensified intervention measures and to demonstrate malaria elimination in areas which are in preelimination phase.

The author suggests that dengue prevention warrants improvement in water supply and awareness of communities about prevention of breeding of *Aedes* mosquitoes and seeking health facility in time. Chapter 7 is the outcome of instigated thought by several studies on urbanization and prevalence of vector-borne disease in tropical areas. It is envisaged and hypothesised by Rais Akhtar and colleagues to find if any correlation exists between urbanization process, heat island generated due to urbanization (e.g. higher density of roads network, buildings and traffic) in the urban areas and the outbreak of vector-borne disease like malaria and dengue. This chapter primarily looked into the temporal data of all metropolitan cities and found that there is an increased incidence of dengue outbreak in all metropolitan cities. According to authors, there is immense variability in the rainfall in Delhi, but all other metropolitan cities have been experiencing an average rainfall over the past two decades. Review of literature led to construction of hypothesis that there exist a close association between urbanization and the increased incidence of dengue outbreak. Regression result shows that there is high possibility between the urbanization and disease outbreak.

The process of urbanization poses threats and opportunities in the urban environment. On the one hand, urbanization leads to development, but on the other, as the population increases, the land use/cover changes, pollution increases and there is alteration of urban heat balance. The combination of heat and pollution acts as a major health threat. In Chap. 8, R. B. Singh and Aakriti Grover, based on primary survey, analyse the historical growth and spread of industries in East Delhi. Further, the impact of growth of industries on human health in East Delhi is examined. The authors examine research articles on land surface temperature, and land use/cover change acts as a base to explore the existence of urban heat island in Delhi. The study found that the urban environment of Delhi has undergone rapid shift. This shift is accompanied with conversion of land use/cover and rise in air pollutants. The health hazard in these regions is high and needs urgent attention.

Chapter 9 by Bimal Raj Regmi and colleagues aims to understand the human health dimensions of climate change in Nepal. A case study approach was used to describe and quantify the association between climate factors and reported cases of typhoid and other healthrelated hazards in Nepal along with their impacts and implications. The research findings show that diseases and healthrelated hazards have increased in the country. There is an association between the incidence of climatesensitive diseases and changes in temperature and precipitation trends. The study data suggests that climate change is likely to have impacts on the health sector in Nepal. However, current adaptation policies, strategies and response measures in this sector are insufficient to address such impacts. The lack of response measures has resulted in increased risk and vulnerability among the poor and marginalised communities living in both rural and urban areas of Nepal. There is an urgent need to devise policies and strategies to fill the existing information and knowledge gaps and implement an integrated approach for better health planning and research in Nepal and to develop longterm mechanisms of addressing health issues and challenges in urban and rural areas.

S. Siva Raju and Smita Bammidi suggest in Chap. 10 that the two major global drivers of change in the twenty-first century – global warming and population ageing – will have points of both convergence and of conflict. From the experiences of the developed countries that have faced them first, we realise that these will operate within different frameworks and need to be addressed at various levels, sizes, densities, structures and functions. In South and Southeast Asian countries, capital cities like Mumbai have populations above five million and are still growing. Moreover, urbanization and globalisation are two processes that catalysed climate change and brought in further social complications for the phenomenon of population ageing. A large number of older persons now inhabit cities, and their financial and social supports are dwindling. They will face serious implications due to the rising rate of urbanization, the proximity to the coast that enhances their vulnerability to weather/climate risks and the solutions to which are energy-driven (automobiles, elevators, airconditioning). As climate change poses health, social and economic risks for older persons, discussion of population ageing vis-a-vis climate change will focus on ensuring and promotion of health and quality of life of the older persons by modifying their built or external environments and other psychosocial interventions. In order to gain an understanding on the effects of climate change for older persons in context of their rising numbers and urbanization, to identify/map the risks or costs involved, initiate possible interventions, discuss policy support and to estimate the opportunities and challenges that may arise, efforts for researchbased knowledge in this emerging field are imminent.

Excessive heat and flooding are extreme weather events associated with climate change, and they pose serious threats to global health. Uma Langkulsen and Desire Rwodzi, in the Chap. 11, reviewed literature and assess the effect of heat on human health among the Thai population. Their review showed that previous studies focused on mortality, especially among the elderly or persons with pre-existing illnesses, and overlooked the idea that physical labour adds to the heat exposure risks via the surplus heat generated due to muscular movements.

The authors also assessed the demographic profiles and existing public health resources in Bangkok and surrounding provinces as part of the wider project to develop a health and climate change adaptation resilience simulator for coastal cities. This comes following a realisation that recurrent flooding makes the Thai capital, Bangkok, and surrounding areas prone to inundation because of the low-lying topography. This had created a need for Thailand to protect its people, natural and man-made resources and productive capacities in response to the impact of climate change-induced floods.

Primary and secondary data sources were used to describe demographic characteristics of the population and to assess the adequacy of public health emergency resources. The study results showed that the study area falls short of the Southeast Asia as well as World Health Organization set standards for health worker densities and hospital beds. K. Maudood Elahi in Chap. 12 on Bangladesh asserts that the country may experience some of the more severe impacts because of its characteristic climatic and geographical conditions, coupled with high population density and poor health infrastructure. Many of the climatic events would make the climate change-induced health impacts worse as a result of newer environmental threats, such as changes in microclimate, erratic climatic behaviour and salinity intrusion in soil and water. The chapter identifies some of the possible direct and indirect impacts of climate change on the health condition of Bangladesh. In identifying such impacts secondary sources of information have been widely reviewed. It has been seen that even though climate change-induced health impacts have been gaining importance in Bangladesh, there is still a lack of research and capacity in this field and its ever-increasing level of vulnerability of the people. The linkage between climate change and the increased incidence of disease, rate of mortality and availability of safe water has not yet received the proper focus it requires.

Budi Haryanto in Chap. 13 indicates that climate change is a serious challenge currently facing Indonesia, as the impacts are already happening. Indonesia's geographic and geological characteristics are also easily affected by climate change, natural disasters (earthquakes and tsunamis) and extreme weather (long drought and floods). Its urban areas also have high pollution levels. Much evidence can be seen, ranging from an increase in global temperature, variable season changes, extremely long droughts, high incidence of forest fires and crop failures. This chapter describes some evidences of climate change in Indonesia to weather disasters such as floods, landslides and drought; burden of vector-borne diseases; air pollution from transportation and forest fires; and reemerging and newly emerging diseases. To respond to its negative impacts to human, the health adaptation strategy and efforts undertaken in Indonesia nowadays include: to increase awareness of health consequences of climate change; to strengthen the capacity of health systems to provide protection from climaterelated risks and substantially reduce the health system's greenhouse gas (GHG) emissions; and to ensure that health concerns are addressed in decisions to reduce risks from climate change in other key sectors.

The local, regional or global climatic conditions may alter atmospheric compositions and chemical processes and are implicated in increasing frequencies of extreme temperature and precipitation, which can also be intensified through varying urbanization, economic development and human activities. In Chap. 14 on Taiwan, Huey Jen Su and colleagues speculate that adverse health consequences related to such exposures are often of primary concerns for their immediate and sustaining impacts on the general welfares. The authors suggest that higher cardiovascular or respiratory mortality appears to derive more from low temperature than does from the exposure to high temperature for the general population. Rural residents are less affected than urban dwellers under extreme temperature events if cardiovascular or respiratory mortality is benchmarked. Towns and villages with a high percentage of the elderly living alone, the senior and the disabled people and the aborigines will present a higher mortality associated with extreme temperature. Extreme rainfall events may also interrupt the chain of food supply and life support and lead to reporting malnutrition of people in affected regions. In addition, contaminated water sources for drinking and recreation purposes, due mostly to flooding after extreme precipitation, are known to be associated with disease outbreak and epidemics.

Malaysia, located in Southeast Asia, has a tropical climate and abundant rainfall (2000–4000 mm annually). In Chap. 15 on Malaysia Mohammed AM Alhoot and colleagues argue that the country faces weatherrelated disasters such as floods, landslides and tropical storm attributed to the cyclical monsoon seasons, which causes heavy and regular downpours. Storms, floods and droughts lead to the rise and emergence of climatesensitive diseases due to the contamination of water, environment and creation of breeding sites for disease-carrying vector mosquitoes. The authors suggest that the effect of climate change on health is an area of substantial concern in Malaysia. This chapter examines the trends of six prominent climate-sensitive diseases in Malaysia, namely, cholera, typhoid, hepatitis A, malaria, dengue and chikungunya followed by the environmental and public health policies, programmes and plans to battle the impact on health from the climate change.

Extreme weather events, such as heavy rainfall, would be possible to damage the habitats of *Aedes* mosquitoes and also beneficial to construct them with lag time of several weeks. The role of extreme weather events in dengue transmission is studied by Tzai Hung Wen and colleagues in Chap. 16 who observe that effects of climate change on human health become increasingly a growing concern. Especially, the worldwide risk of vector-borne diseases, such as dengue fever, could be strongly influenced by meteorological conditions. Taiwan is unique in terms of its geographical location and abundant ecological environments. It makes Taiwan an unusual epidemic region for dengue disease. This chapter focuses on risk factors related to climate change and geographic characteristics which have been shown to affect dengue transmission in Taiwan. Empirical studies all concluded that the relationship between temperature and dengue incidence is highly correlated. Some studies also concluded that temperature increase has more dramatic influence to dengue incidences in southern Taiwan. However, authors suggest that the effects of other meteorological factors, including precipitation and humidity, on dengue incidence are rather inconsistent and need to be further investigated. Further studies should also put more focus on the extent of influence of meteorological conditions on human activities and what activity changes may increase or decrease the risk of dengue transmission.

New Delhi, Aligarh   Rais Akhtar

References

The World Bank (2013) Warmer world threatens livelihoods in South East Asia, Press release, June 19, Washington, DC
UNESCAP (2011) Statistical yearbook for Asia and the Pacific 2011 WHO (2008) www.who.int/globalchange/.../World ...

Acknowledgement

In the process of writing, editing and preparing this book, there have been many people who have encouraged, helped and supported me with their skill, thoughtful evaluation and constructive criticism.

First of all I am indebted to all the contributors of chapters from different countries for providing the scholarly and innovative scientific piece of research to make this book a reality. I am also thankful to the reviewers who carefully and timely reviewed the manuscripts: Prof. Sir Andy Haines, Prof. David Phillip, Dr. David Dodman, Prof. Alistair Woodward, Prof. Mei-Hui Li and the late Prof. Tony McMichael.

I first met Prof. Tony McMichael in 1995 at the London School of Hygiene and Tropical Medicine, London. Tony and I published a paper in 1996 in *The Lancet*, and my research journey in the field of medical geography and climate change and health continues unabated and with rigour. Tony McMichael has been a great source of inspiration. Tony wrote the Foreword 2 to this book, only about 2 months before his sudden and untimely demise.

I am also grateful to Professor Andy Haines who has been very kind in not only reviewing some papers but writing a Foreword 1, which adds greatly to the book with his thoughtful insights in the field of climate change and human health.

I also thank my family, wife Dr. Nilofar Izhar, daughter Dr. Shirin Rais and son-in-law Dr. Wasim Ahmad, who encouraged and sustained me in developing the structure of the book and in editing tasks, and I am deeply grateful for their support and indulgence.

Finally, and most essentially, I am deeply indebted to the Springer and the entire publishing team, without whose patience, immense competence and support this book would not have come to fruition. I would specially thank Dr. Robert K. Doe whose energising leadership ensured that this book would indeed translate to reality.

I am also thankful to Miss Sowmya Ramalingam and Ms. S. Madhuriba for their constant guidance and cooperation during the preparation and review process of the manuscript.

Sincerely,
Rais Akhtar

New Delhi, Aligarh

Contents

Contributors

Rais Akhtar International Institute of Health Management and Research (IIHMR), New Delhi, India

Mohammed A. Alhoot Medical Microbiology Unit, International Medical School, Management & Science University, Shah Alam, Selangor, Malaysia

Department of Medical Microbiology, Faculty of Medicine, University of Malaya, Kuala Lumpur, Malaysia

Smita Bammidi College of Social Work Nirmala Niketan, Mumbai, India

Colin D. Butler Faculty of Health, University of Canberra and Visiting Fellow, National Centre for Epidemiology and Population Health, Australian National University, Canberra, Australia

Guru Prasad Chattopadhyay Department of Geography, Visva-Bharati, Santiniketan, India

Mu-Jean Chen Department of Environmental and Occupational Health, College of Medicine, National Cheng Kung University, Tainan, Taiwan

National Environmental Health Research Center, National Institute of Environmental Health Sciences, National Health Research Institutes, Miaoli, Taiwan

Nai-Tzu Chen Department of Environmental and Occupational Health, College of Medicine, National Cheng Kung University, Tainan, Taiwan

National Environmental Health Research Center, National Institute of Environmental Health Sciences, National Health Research Institutes, Miaoli, Taiwan

Dilli Ram Dahal Department of Geography, Visva-Bharati, Santiniketan, India

Anasuya Das Department of Geography, Visva-Bharati, Santiniketan, India

S.K. Dash Centre for Atmospheric Sciences, Indian Institute of Technology Delhi, New Delhi, India

Ramesh C. Dhiman National Institute of Malaria Research (ICMR), New Delhi, India

K. Maudood Elahi Department of Environmental Science, Stamford University Bangladesh, Dhaka, Bangladesh

Geography and Environment, Jahangirnagar University, Dhaka, Bangladesh

Aakriti Grover Department of Geography, Swami Shraddhanand College, University of Delhi, Delhi, India

Pragya Tewari Gupta National Institute of Urban Affairs, New Delhi, India

Budi Haryanto Research Center for Climate Change, University of Indonesia, Depok, Indonesia

Uma Langkulsen School of Global Studies, Thammasat University, Pathumthani, Thailand

Mei-Hui Li Department of Geography, College of Science, National Taiwan University, Taipei, Taiwan

Min-Hau Lin Institute of Health Policy and Management, College of Public Health, National Taiwan University, Taipei, Taiwan

Wah Yun Low Dean's Office, Faculty of Medicine, University of Malaya, Kuala Lumpur, Malaysia

Kevin McCracken Department of Geography and Planning, Macquarie University, Sydney, NSW, Australia

Anil Pandit Hays Medical Center, Hays, KS, USA

David R. Phillips Department of Sociology and Social Policy, Lingnan University, Tuen Mun, Hong Kong

Bandana Pradhan Institute of Medicine, Tribhuvan University, Kirtipur, Nepal

S. Siva Raju School of Development Studies, TISS, Mumbai, India

Mala Rao Department of Primary Care and Public Health, School of Public Health, Imperial College of Science, Technology and Medicine, London, UK

Bimal Raj Regmi School of Social and Policy Studies, Flinders University, Adelaide, SA, Australia

Desire Tarwireyi Rwodzi School of Global Studies, Thammasat University, Pathumthani, Thailand

Shamala Devi Sekaran Department of Medical Microbiology, Faculty of Medicine, University of Malaya, Kuala Lumpur, Malaysia

Manpreet Singh Dalberg Global Development Advisors, Nairobi, Kenya

R.B. Singh Department of Geography, Delhi School of Economics, University of Delhi, Delhi, India

A.K. Srivastava India Meteorological Department, Pune, India

Cassandra Star School of Social and Policy Studies, Flinders University, Adelaide, SA, Australia

Huey-Jen Su Department of Environmental and Occupational Health, College of Medicine, National Cheng Kung University, Tainan, Taiwan

Wen Ting Tong Department of Primary Care Medicine, Faculty of Medicine, University of Malaya, Kuala Lumpur, Malaysia

Tzai-Hung Wen Department of Geography, College of Science, National Taiwan University, Taipei, Taiwan

Chapter 1
Climate Change and Geoecology of South and Southeast Asia: An Introduction

Rais Akhtar

Abstract Climate change is projected to impact human health in many ways including changes in water availability and quality, air quality and sanitation, availability and access to food and nutrition and transmission of vector-borne diseases. Environmental consequences of climate change, such as extreme heat waves, rising sea levels, changes in precipitation resulting in flooding and droughts, intense hurricanes (cyclones, typhoons) and degraded air quality, can affect directly and indirectly the physical, social and psychological health of humans. Climate change and human health have emerged as an important focus of research in the World Health Organization since 2008. However, the First International Conference on Health and Climate organised by the WHO in August 2014 in Geneva recognises the relevance of the impact of climate change on human health from a global-change and health perspective. This chapter also highlights policies of the United States, China and India towards GHG emission reduction and the successful climate agreement in Paris in December 2015.

Keywords Vector-borne diseases • Katrina hurricane • GHG emissions • Heat wave • USA • China • India • Paris climate agreement

Climate change represents a range of environmental hazards and will affect populations wherever the current burden of climate-sensitive disease is high – such as the urban poor in low- and middle-income countries and landless farming communities. Understanding the current impact of weather and climate variability on the health of populations is the first step towards assessing the future impact.

Nevertheless climate change has been occurring and impacting human health and welfare across the globe – both in the developed and developing regions. Its impact was experienced mainly in the last one and a half decades. The 2003 heat wave in Western and Central Europe that killed nearly 70,000 people in 2 weeks time in the month of August, Hurricane Katrina disaster that devastated southern

R. Akhtar (✉)
International Institute of Health Management and Research (IIHMR), New Delhi, India
e-mail: raisakhtar@gmail.com

© Springer International Publishing Switzerland 2016
R. Akhtar (ed.), *Climate Change and Human Health Scenario in South and Southeast Asia*, Advances in Asian Human-Environmental Research,
DOI 10.1007/978-3-319-23684-1_1

1

USA in 2005 and Hurricane Sandy in 2012; flooding and landslides in the United States and United Kingdom and bushfire in Australia and California in 2014 are examples which show that even developed countries are vulnerable to climate change impact (Akhtar 2014). It has also been predicted that the frequency of severe flooding across Europe will double by 2050 (Cannor 2014).

In the context of the United Kingdom, in a recent issue of *The Lancet Infectious Diseases* journal, scientists from the emergency response department of Public Health England suggest that mosquitoes and ticks are known to be highly responsive to changes in temperature and rainfall. Based on climate modelling, the report asserts that warmer temperatures in the United Kingdom in the future could provide ideal conditions for the Asian tiger mosquito (*Aedes albopictus*), which spreads the viruses that cause dengue and chikungunya. However, changes in vector distributions are being driven by climatic changes and changes in land use, infrastructure and environment (Medlock and Leach 2015).

We are aware about the scientific explanations that climate change occurs because excessive amount of greenhouse gases were emitted into the atmosphere due to human activity. Human influence on the climate system is clear, and recent anthropogenic emissions of greenhouse gases are the highest in history. The Earth's atmosphere has already been warmed by 0.85 °C from 1880 to 2012. Recent climate changes have had widespread impact on human and natural systems (IPCC 2014).

In order to effectively address climate change, we must significantly reduce the amount of heat-trapping emissions we are putting into the atmosphere. We should therefore delve briefly into the policies towards reducing GHG emissions in the United States, China and India before the Paris Climate Agreement in December, 2015 was signed.

1.1 US Position

Contrary to President George W. Bush, the Obama Administration has focussed on the links between climate changes and human. This follows recent, salutary experiences including a variety of severe natural disasters, e.g., Hurricanes Katrina and Sandy, flooding, landslides, forest fires (in California) and drought. George Luber, head of the Climate and Health Programme of the US Centers for Disease Control and Prevention, has been central to this rising awareness in the United States. Luber claims that the climate change has extended the pollen season in some parts of the country by as much as 26 days, increasing the risk of allergy and asthma attacks. It has also contributed to heavier rainfalls, which can jeopardise drinking water quality. "Climate change isn't just about polar bears and penguins or impacts distant in time", Luber told *The Lancet*. He went on to explain that these impacts affect our communities and people's health, so "putting the human dimension on climate change is vital not only for understanding threat we face but also for encouraging action" (Jaffe 2015).

A study by Jonathan Patz and his colleagues that analysed data based on climate models and future projections reveals that "by 2050, many US cities may experience more frequent extreme hot days. For example New York and Milwaukee may have three times their current average number days hotter than 32 °C" (Patz et al. 2014).

1.2   Perspectives from India and China

Developing countries in general and India in particular face challenges to accelerate the pace of socioeconomic development and minimise inequalities, exacerbated during the colonial period. However, even after independence, inequalities in India have risen. India is in dire need of energy (per capita electricity consumption in India is 618 KW and in the United States 14,240) to enable infrastructure development. But there are also gross global inequalities. Most developing countries, including India, have the lowest per capita CO_2 emission (1.2 t per capita in India compared to 20.6 t in the United States), as well as the low gross domestic product (GDP) per capita. Despite this, major developed nations insist that developing nations reduce their greenhouse gas emissions, fundamental to their socioeconomic development. However, under the Kyoto Protocol, the principle of "common but differentiated responsibility" was considered as fundamental for future emission negotiations. This will promote equity and human development by expanding access to energy demand of developing countries (Akhtar 2010). India's official thinking on climate change is derived from a policy advanced by Manmohan Singh, who served as the country's prime minister until 2014. In 2007, he declared at a G20 summit in Germany that India's per capita emissions will never exceed the average per capita emissions for developed countries. Right now, that gives India quite a bit of elbow room. If the United States and the European Union are able to attain the cuts they are now talking of, India's leeway shrinks; though some in the Indian government believe that even then the country could continue to increase its emissions for 15 or 20 years beyond the 2030 cap China has agreed to and still fall below the developed world's per capita average.

Emission cuts, at the moment, do not seem to be an Indian policy priority. But there is a new Prime Minister, Narendra Modi, and a new environmental minister, Prakash Javadekar. Is there a new emphasis? It would seem not, judging from a recent interview given to The New York Times after the September 2014 UN Climate Summit in New York: Prakash Javadekar said, "The moral principle of historic responsibility cannot be washed away" (Devenport 2014). This means that developed economies, chiefly the United States, which spent the last century building their economies while pumping warming emissions into the atmosphere, bear the greatest responsibility for cutting pollution. India's Minister further clarified that government agencies in New Delhi were preparing plans for India's domestic actions on climate change, but he said they would lead only to a lower rate of increase in carbon emissions. It would be at least 30 years, he said, before India would likely see a downturn.

1.3   China's Position

China's choking air pollution, including notorious smog levels in Beijing, has become a major problem in the country and poses a threat to Chinese public health. Coal combustion generates particulate matter also known as "PM". Currently Beijing is suffering from PM2.5. China now produces more pollutants than any other nation on earth, and 66 of the country's 74 largest cities still fall far short of the government's air quality standards (Larmer 2015).

Nevertheless, the Chinese government argued that the priorities of developing countries first and foremost are economic development and poverty reduction and that the international community should accommodate these growth needs. Officials have also often noted that although China is now the largest total emitter of GHG, it is not yet the largest in terms of historical cumulative emissions or per capita emissions (far from it in the latter case). After all, most developed countries undertook pollution-intensive industrialization during the nineteenth and twentieth centuries that resulted in large cumulative historical emissions (http://www.gov.cn/2011lh/content_1825838.htm).

Despite China's argument, its own domestic compulsions and international pressure are as follows: "the Chinese government proposed a carbon intensity reduction target for the first time in the 12th Five Year Plan. The attitude of Chinese policymakers toward climate change policy recently underwent a radical change, from having no explicit climate change policies to a Presidential commitment to reduce carbon intensity. need to control total energy consumption, and to gradually establish a carbon market. We posit that the motivation for this radical shift can be attributed to rising concerns about the projected impacts of climate change in China, and also the government's recognition that China's traditional development model is unsustainable, not only environmentally, but also from the standpoint of social and economic development. The Chinese central government has been searching for mechanisms to transform China's development trajectory, and climate change policy represents a new and justifiable tool that can help the government transform economic development in China, especially because climate policy can be used as a new central government instrument to guide and control the behaviour of local governments" (Xiaowei and Gallagher 2014).

Earlier in a secret negotiated climate deal, the United States and China unveiled a commitment to reduce their GHG output, with China agreeing to cap emissions for the first time and the United States committing to deep reductions by 2025. These pledges struck between US President Barack Obama and his Chinese counterpart, Xi Jingping, provide an important boost to international efforts to reach a global deal on reducing emissions beyond 2020 at the UN meeting in Paris this year. China, now the world's biggest GHG emitter, has agreed to cap its output by 2030. Previously China had only pledged to slow its rapid emission growth rate. Now it has also agreed to increase its use of energy from zero-emission sources to 20 % by 2030 (The Guardian 2014).

After having discussed the GHG emission policies of USA, China and India to combat climate change impacts, it is heartening to see the successful negotiations reached at the Paris Climate Summit on 12th of December, 2015. According to the UNFCCC, the Paris Agreement aims to combat climate change and unleash actions and investment towards a low carbon, resilient and sustainable future was agreed by 195 nations. This Agreement "underwrites adequate support to developing nations and establishes a global goal to significantly strengthen adaptation to climate change through support and international cooperation. The Paris Agreement's major objective is to keep a global temperature rise within 21st century well below 2 ° Celsius and to drive efforts to limit the temperature increase even further to 1.50 ° Celsius above pre-industrial levels." http://unfccc.int

1.4   Climate Change and Human Health

Climate change and human health have emerged as an important focus of research in the World Health Organization since 2008. However, the First International Conference on Health and Climate organised by the WHO in August 2014 in Geneva pinpoints the relevance of the impact of climate change on human health from a global-change and health perspective (Neira 2014).

Climate change may affect health through a variety of pathways. These include the effects of an increased frequency and intensity of heat waves, a reduction in cold-related deaths, more frequent and more intense floods and droughts, changes in the distribution of vector-borne diseases and risk of disasters and malnutrition (Haines, 2014, Personal communication). In another study, Andy Haines and colleagues warned that the Representative Concentration Pathways (RCP 8.5) assume that present trends of relatively unrestrained use of fossil fuels and high population growth will continue. According to this emission pathway, by 2100 the global average temperature will probably be more than 4 °C above preindustrial levels with higher average temperatures over lands (Haines et al. 2014). The IPCC Fifth Assessment Report, the Chapter on Human Health, affirms with very high confidence that the health of human population is sensitive to shifts in weather patterns and other aspects of climate change. The Chapter also projects (1) greater risks of injury, disease and death due to more intense heat waves and fires (very high confidence); (2) increased risks of food and waterborne diseases (very high confidence) and (3) increased risk of undernutrition resulting from diminished food production in poor regions (high confidence) (Smith et al. 2014).

1.5   South and Southeast Asia

The scenario for climate change in South and Southeast Asia has been a major focus in the most recent Fifth IPCC Assessment. These scenarios indicate warming trends and increasing temperature extremes consistent with those observed across most of the Asian region over the past century (high confidence) (IPCC 2014).

The Fourth IPCC Assessment Report had stated that climate change, in particular via increased floods and droughts, is expected to have severe impacts on South Asian countries, whose economies rely heavily on agriculture, natural resources and forestry and fishery sectors. About 65 % of the population of South Asia live in rural areas and 75 % of them are poor, vulnerable and at risk of harm from climate change. Risks include changes in the intensity of rainfall and a possible alteration to South Asian monsoon pattern due to El Niño effects. Higher temperatures also increase the vulnerability of humans (including labourers), crops and stock.

Recently unprecedented rainfall and hailstorms in late March–April 2015 in northern India have added to the concern that climate change will exacerbate extreme events. These storms caused extensive crop damage, in turn contributing to a number of farmers committing suicides. Sharp increases in food prices were also noted. Densely populated low-lying mega-deltas in South Asia are another area identified as a particular risk (IFAD no date). A paper by Ronald Fuchs asserts that "Asia's coastal megacities are increasingly vulnerable to flooding disasters resulting from the combined effects of climate change (manifested .as sea level rise, intensified storms, and storm surges), land subsidence, and rapid urban growth" (Fuchs 2010).

Climate change is most likely to impact one of South Asia's most precious resources: water. Per capita water availability is already under intense pressure, having fallen by 70% since 1950, due to rapid population growth and urbanization (ADB 2012–2013).

> India's water crisis has multiple causes, including poor or even no government planning, corporate privatisation, industrial and human waste (impairing quality as well as quantity) and government corruption. These pressures that are magnified by overall population increase are expected to rise to 1.7 billion by 2050. In turn high birth rates are contributed by poverty and low literacy rates, along with ongoing discrimination against women and girls. South Asia's water scarcity is mirrored globally and could lead to future international political conflict. India is not immune from these risks (Snyder 2015).

In South Asia, home to nearly 1.7 billion people, cities are already feeling the pressure of population growth and urbanization. It is estimated that 22 of 32 major Indian cities already face daily water shortages. In Nepal's capital, Kathmandu, many local residents have grown accustomed to waiting in queues for hours to obtain drinking water from the city's ancient, stone waterspouts. In Karachi, Pakistan, electricity and water shortages have led to protests and citywide unrest (Surie 2015).

In South Asia, availability of fresh water is highly seasonal, with about 75 % of the annual rainfall occurring during the monsoon months. Climate change threatens water supplies by higher temperatures, changes in river regimes and a greater incidence of coastal flooding. Water availability is expected to decrease dramatically especially in the dry season (Langton and Prasai 2012).

Focussing on Southeast Asia, Yusuf and Francisco combined data on spatial distribution of various climate-related hazards (cyclones/typhoons, droughts, landslides, sea level rise) with two human indicators (population density and adaptive capacity) to analyse 530 subnational areas in seven Southeast Asian countries

(Indonesia, Thailand, Vietnam, Lao PDR, Cambodia, Malaysia and the Philippines). Based on this mapping assessment, they identified all parts of the Philippines, the Mekong River Delta in Vietnam, almost all of Cambodia, North and East Lao PDR, Bangkok (Thailand), West and South Sumatra and West and East Java (Indonesia) as the most vulnerable regions (Yusuf and Francisco 2009). Both South and Southeast Asian countries are vulnerable due to floods, with the region experiencing the largest number of events (1171) during 1980–2008, followed by storms (930 events). The data from UNISDR show that Bangladesh suffered the most with epidemics and pandemics, with 1500,000 people affected in 1991 followed by Indonesia.

Coker and his colleagues have identified that Southeast Asia is a "hotspot" for emerging infectious diseases, including drug-resistant pathogens and some with pandemic potential. Some of these infections have already exacted a heavy public health and economic toll, including severe acute respiratory syndrome (SARS) that rapidly decimated the region's tourist industry. Influenza A (H5N1) has been particularly problematic in Southeast Asia and has had a profound effect on the poultry industry. The authors further assert that:

> The regional challenges in control of emerging infectious diseases are formidable and range from influencing the factors that drive disease emergence, to making surveillance systems fit for purpose, and ensuring that regional governance mechanisms work effectively to improve control interventions. (Coker et al. 2011)

Thus disease ecological scenario in the South and Southeast Asian region (except Singapore) is highly vulnerable. This vulnerability has acquired a serious proportion when we consider the latest IPCC Fifth Assessment Report (2014) on Asia. The Report highlights that warming trends and increasing extreme temperature are exacerbated by coastal and marine systems that are under increasing stress from both climatic and nonclimatic driver. The geography of large coastal areas here also makes South and Southeast Asia two of the many places around the world that are susceptible to several different kinds of infectious diseases. Water scarcity is also a major challenge for most of the regions because of rapid and unplanned urbanization, industrialization and select regional economic development by climate change. In addition extreme climate events will have an increasing impact on human health. Frequent and intense heat waves in Asia will increase mortality and morbidity in vulnerable groups. Increase in heavy rain and temperature will increase the risk of diarrheal diseases, dengue fever and malaria. Increases in floods (as occurred in Pakistan in 2010 and in Thailand in 2011 and drought in India in 2009) will exacerbate rural poverty in parts of the South and Southeast Asian region due to impaired crop production.

The Asian Development Report (2012–2013) has reviewed human health vulnerability in South and Southeast Asia, a region undergoing the epidemiological transition with both communicable and noncommunicable diseases on rise. The Report found that infectious and parasitic diseases comprise 13.9% of total mortality, followed by respiratory diseases with 11.5% of total deaths. To sum up, there are widespread regional inequalities in the health effects of exposures related to water, sanitation, hygiene and those caused by vectors, food and air pollution. The challenge from the climate and health perspective is to devise mitigation and adaptation policies that can lower these risks.

The Asian Development Bank, however, highlights the progress made towards achieving the Millennium Development Goals in South and Southeast Asian countries. Among 13 indicators, South Asian countries experienced slow progress for poverty reduction, reduced maternal mortality, under 5 and infant mortality, and basic sanitation. There has been no progress on the reduction of CO2 emissions nor tuberculosis incidence (ADB 2012–2013). The ADB quoted the IPCC on disaster risk and climate adaptation, concluding that there is a need for smarter development and economic policies, with a focus on disaster risk reduction and adaptation. The report highlights that "Climate change can undermine both food security and livelihoods It can depress agricultural productivity and increase food insecurity and malnutrition, particularly in children. It can also increase vector-borne diseases, multiplying the disease burden" (ADB 2012–2013).

According to the ADB, there will also be huge economic costs from unmitigated climate change. The Bank estimates that in Southeast Asia, the economic cost of climate change could be as high as 6.7 % of GDP per year by 2100. This is more than twice the world average (ADB 2012–2013)

An alternative conceptual framework is to consider the health problems of climate change in South and Southeast Asia (and elsewhere) as having three pathways – "primary", "secondary" and "tertiary". In this model, primary effects are considered the most causally direct impacts of climate change on health. They include increased mortality during heat waves as what repeatedly occurred in India and elsewhere in South Asia. In the month of May 2015 alone, more than 1200 people died due to heat wave mainly in southeastern states of Telangana and Andhra Pradesh, as well as in northern India (TOI 2015). The effects of "natural" disasters such as flooding, hailstorms, droughts and severe cyclones, all of which are arguably worsened by climate change (including sea level rise in coastal regions), are also considered "primary" in this conceptualization. "Secondary" effects include via ecological changes that alter the epidemiology of some infectious diseases and chronic diseases, for example, the 1994 malaria outbreak in the Rajasthan desert following from extremely heavy rainfall and flooding (Akhtar and McMichael 1996). The health impact of allergens and air pollution, each of which interacts with climate change, can also be considered secondary effects. Tertiary effects refer to large-scale events with complex, multidimensional causation, including migration, famine and conflict. Unfortunately, all three forms of tertiary health effects are seen in countries of the region (Singh et al. 2015). Heat stress, on a sufficient scale, could also be called a tertiary effect.

In summary, if left unchecked, there is the likelihood of severe, pervasive and irreversible impacts of climate change on people and ecosystems. The global scientific community is making an increasingly impassioned appeal to the world's policymakers to combat what is now recognised as the challenge of our times (IPCC 2014).

The present book comprises studies on South and Southeast Asian countries in general with major focus on India, Nepal, Bangladesh, Indonesia, Malaysia, Thailand and Taiwan. This book is aimed at presenting a regional analysis pertaining to climate change and human health, focusing on climate change adaptation strategies in geographically and socioeconomically varied countries of South and Southeast Asia.

References

ADB (2012–2013) Asian Development Bank, United Nations Economic and Social Commission for Asia and the Pacific, The UNDP, MDGs report 2012–2013, Asia Pacific aspiration: perspective for a post 2015, Development Agenda, Bangkok

Akhtar R (2010) CO2 emission reduction and the emerging socioeconomic development in developing countries: a case study of India. In: De Dapper M et al (eds) Developing countries facing global warming: a PostKyoto assessment. Royal Academy of Overseas Sciences, Brussels, 15–26

Akhtar R (2014) Hope the West isn't heading towards climate colonialism. Hindustan Times, New Delhi, 14 April

Akhtar R, McMichael AJ (1996) Rainfall and malaria outbreaks in Western Rajasthan. The Lancet 348:1457–1458

Cannor S (2014) Frequency of severe flooding across Europe 'to double by 2050'. The Independent, London, March 2

Coker RJ et al (2011) Emerging infectious diseases in Southeast Asia: regional challenges to control. The Lancet 377(9765):599–609, 12 February

Devenport C (2014) Emissions from India will increase, official says. The New York Times, 24 September

Fuchs RJ (2010) Cities at risk: Asia's coastal cities in an age of climate change: Asia Pacific Issue. East West Center, Honolulu

Haines A, Ebi K, Smith KR, Woodward A (2014) Health risks of climate change: act now or pay later. The Lancet 384(20):1071–1075

http://www.gov.cn/2011lh/content_1825838.htm. On China emission reduction policy

IFAD (No date) Climate change impacts: South Asia. The Global Mechanism, The International Fund for Agricultural Development, Rome

IPCC (2014) Climate change : impact, adaptation and mitigation. WG II, AR5, Chapter 24 Asia, Cambridge University Press, Cambridge, UK/New York

Jaffe S (2015) Obama steps up US campaign on climate chan. TheLancet 385:9978, 25 April, 1606–1607

Langton N, Prasai S (2012) Will conflicts over water scarcity shape South Asia's future? Issue Perspect 2(1):1–2

Larmer B (2015) How do you keep your kids healthy in smog-choked China? The New York Times Magazine, 16 April

Medlock MM, Leach SA (2015) Effect of climate change on vectorborne disease risk in the UK. The Lancet Infectious Diseases, Published on line 23 March

Neira M (2014) The 2014 WHO conference on health and climate. Bull World Health Org 92:596

Patz J, Frumkin H, Holloway T, Vimont DJ, Haines A (2014) Climate change: challenges and opportunities for global health. JAMA 312(15):1565–1580

Singh M, Rao M, Butler CD (2015) Climate change, health and future wellbeing in South Asia (Chapter 2 in this book)

Smith KR, Woodward A, Campbell-Lendrum D, Chadee DD, Honda Y, Liu Q, Olwoch JM, Revich D, Sauerbom R (2014) Human health: impacts, adaptation and cobenefits. In: Climate change 2014: impacts, adaptation and vulnerability, Part A global and sectoral aspects, WG II, Cambridge University Press, Cambridge, UK/New York, 709–754

Snyder S (2015) Water in crisis in India. The Water Project, Concord

Surie MD (2015) South Asia's crisis: a problem of scarcity amid abundance. Weekly Insight and Analysis in Asia, The Asia Foundation, San Francisco, 25 March

The Guardian (2014) US and China strike deal on carbon cuts in push for global climate change pact. London,12 November 2014

TOI (2015) Intense heatwave in many parts of India toll 1242. Times of India, New Delhi, 27 May

Xiaowei X, Gallagher KS (2014) Prospects for reducing carbon intensity in China. The Centre for International Environment and Resource Policy, Tufts University, Medford, February 2014, No. 008

Yusuf AA, Francisco H (2009) Climate change vulnerability mapping in Southeast Asia. IDRC, Ottawa

Chapter 2
Climate Change, Health and Future Well-Being in South Asia

Manpreet Singh, Mala Rao, and Colin D. Butler

Abstract About one fifth of the world's population live in South Asia. There are many reasons to be concerned about the impacts of climate change on this region, many parts of which experience apparently intractable poverty. The health problems caused by climate change in South Asia have been conceptualised here as three tiers of linked effects. In this framework, primary health effects are considered the most causally direct impacts of climate change. They include increased mortality and morbidity during heatwaves and 'natural' disasters worsened by climate change. Secondary effects include those resulting from ecological changes that alter the epidemiology of some infectious and chronic diseases. Tertiary effects refer to impacts on health of large-scale events with complex, multidimensional economic and political causation, including migration, famine and conflict.

Urgent action, both preventive and adaptive, is needed. India, an emerging great power, and the dominant nation in South Asia, must lead urgent and intense engagement with this overarching issue. Transformation of its energy system would improve population health, constitute regional leadership and be significant at the global level.

Keywords Climate change • Energy • Health inequality • South Asia • Sustainability

M. Singh (✉)
Dalberg Global Development Advisors, Nairobi, Kenya
e-mail: manpreet1@gmail.com

M. Rao
Department of Primary Care and Public Health, School of Public Health, Imperial College of Science, Technology and Medicine, London, UK
e-mail: mala.rao@imperial.ac.uk

C.D. Butler
Faculty of Health, University of Canberra and Visiting Fellow, National Centre for Epidemiology and Population Health, Australian National University, Canberra, Australia
e-mail: colin.butler@canberra.edu.au

© Springer International Publishing Switzerland 2016
R. Akhtar (ed.), *Climate Change and Human Health Scenario in South and Southeast Asia*, Advances in Asian Human-Environmental Research,
DOI 10.1007/978-3-319-23684-1_2

11

2.1 Introduction

South Asia, defined geographically as the region of Asia that includes Afghanistan, Bangladesh, Bhutan, India, Maldives, Nepal, Pakistan and Sri Lanka, holds 1.6 billion people, almost one quarter of the world's current population. Recent population projections suggest that by 2050, the population of South Asia will approximate or exceed 2.2 billion people (International Institute for Applied Systems Analysis 2012). Despite the perception of global economic growth and rising prosperity, 45 % of the world's poor people (living at or below $2 a day) reside in South Asia; over one third of the world's poor lives in India alone (Sumner 2012).

Scenarios describing the effects of climate change are detailed elsewhere in this book, but the most recent estimates, detailed in the Intergovernmental Panel on Climate Change's (IPCC) fifth assessment report, suggest that by 2100:

- Global mean surface temperatures are likely to rise by between 1.4 and 5.5 °C, compared to the average temperature between 1850 and 1900. Already the world has warmed by about 0.85 °C compared to that baseline (IPCC 2013).
- Extreme rainfall events are very likely to become more intense and more frequent in tropical regions (IPCC 2013).
- The sea level is likely to rise by between 0.26 and 0.82 m (depending on the scenario) compared to the period 1986–2005 (IPCC 2013).

Data on the predicted impacts of climate change in South Asia are limited. The IPCC models suggest that there is likely to be an increase in mean temperatures in South Asia – this represents both an increase in winter temperatures and an increase in the frequency of extremely hot days and nights during the summer. Rainfall patterns are also likely to change: models suggest an increase in the frequency of extreme precipitation events, but an overall decrease in mean annual rainfall (IPCC 2013).

As the impact of climate change deepens, the people of South Asia are vulnerable due to a combination of factors. These include aspects of the region's physical and human geography, including high population densities, large areas lying close to sea level (particularly in the Maldives and coastal Bangladesh, including the Sundarbans, the delta which extends into India) and agricultural dependence on major rivers, fed by Himalayan glaciers vulnerable to melt. This vulnerability is exacerbated by high levels of poverty and income inequality and by poorly functioning national institutions and governance structures. It is important to recognise that the human toll from the increased disasters expected from climate change can be reduced by good governance and large-scale cooperation. In South Asia, the response to some disasters has greatly improved. For example, in Bangladesh in 1970 Cyclone Bhola (a category 3 storm, a classification in which the strongest category, a measure of maximum wind speed, is 5) killed at least 300,000 people (Cash et al 2013), but category 4 Cyclone Sidr in 2007 caused fewer than 3,500 deaths (Paul 2010). However, other governance structures remain weak, as shown by the belated and inequitable response to the 2005 earthquake in Pakistan (Yasir 2009).

2.1.1 *Primary, Secondary and Tertiary Effects*

The health impacts of climate change can be conceptualised in many ways. This chapter uses a framework that groups these impacts into 'primary', 'secondary' and 'tertiary' effects (Butler and Harley 2010; Butler 2014a, b). Primary effects are the most direct and obvious; these include changes in mortality and morbidity due to increased heatwaves and also include the most direct effects (e.g. trauma or drowning) of extreme weather events worsened by climate change, such as floods or cyclones. Secondary health effects are more indirect; they include the impact of changes in the ecology and the environment such as in the epidemiology of vector-borne diseases (VBDs) or of atopic conditions. Tertiary effects refer to the health impacts of complex, large-scale events that occur at a societal level, examples of these include conflict, famine and migration. Tertiary effects also include other ways that climate change may affect overall human well-being, for example, by impacting on economic growth or by causing extinction of other species (Thomas et al. 2004; Selwood et al. 2014).

There is, of course, overlap between the primary, secondary and tertiary effects of climate change – this is particularly true during 'natural' disasters – where there are direct primary health effects (such as death by drowning during flooding), but also potential secondary and tertiary effects as a result of responses to disasters, including diarrhoeal diseases, internal displacement and migration. Furthermore, all kinds of effect are mediated by economic, social and governance factors, from the presence or absence of heat alerts and storm warning signals to the quality of primary health care and the capacity of governments to respond to refugee crises.

Although high-quality data that specifically predict the health impact of climate change in South Asia are limited, the primary and secondary health effects have been well modelled in the global context. The tertiary effects are undoubtedly harder to model and quantify, not least as their causal attribution is so challenging and debated. However, we are far from the first health authors to recognise that climate change, beyond a threshold of warming of perhaps as little as 1.5 °C, may be associated with large-scale economic collapse, mass migration and resource scarcity, with consequences including conflict (Costello et al. 2009; Watts et al. 2015; King et al. 2015).

Given the complexity of the climate and of societal responses to climate change, the likelihood and timelines of these tertiary effects will be disputed. For example, the war in Syria is increasingly recognised as in part caused by climate change (Gleick 2014; Kelley et al. 2015). If so, we could consider it a tertiary effect. The risk of these catastrophic outcomes is real and must be taken into consideration in cost-benefit analyses and when planning adaptation responses (Weitzman 2009). Avoiding a 'catastrophic' outcome is also central to the most recent Lancet commission on climate change (Watts et al. 2015).

2.1.2   The Position of the Indian Government on Climate Change

Genuine global engagement with the issue of climate change will precipitate intense attempts to lower carbon emissions and their replacement by alternative paradigms that promote genuine and equitable economic growth, at the same time reducing pressure on natural resources. However, India has long argued against specific emission reduction targets for countries in the developing world, with the justification that climate change is a result of historical emissions from industrial activity in developed countries (Ministry of External Affairs 2009). It does however need to be acknowledged that India has pledged that it will not allow its per capita GHG emissions to exceed the average emissions of the developed world, even as it pursues its socio-economic development objectives. But as a regional power and aspiring global superpower, India has the potential to divest itself from the carbon-intensive development pathways of industrialisation and urbanization pursued by the developed world and, instead, adopt an alternative path, especially involving large-scale provision of renewable energy. This will not only benefit population health in India but create a vision for other developing countries, by allowing populations to lower the rate of climate change increase and, at the same time, continue on a path towards socio-economic growth, and for societies to adapt in ways that will minimise adverse health impacts.

India is a largely vegetarian society. As the carbon footprint of meat is so high, Indians can also make an important contribution to slowing the rate of climate change through maintenance of this ancient tradition (McMichael et al. 2007). On the other hand, many people in South Asia are deficient in zinc and iron and may benefit from more animal products in their diet. But they will also benefit from improved sanitation, handwashing and the treatment and prevention of parasitic conditions, such as of hookworm.

2.2   Primary Health Effects in South Asia

As mentioned, these include health impacts as a result of changes in temperature and exposure to 'natural' disasters, such as forest fires, floods and storms.

2.2.1   Changes in Temperature

Climate change in South Asia is predicted to lead to an increase in the frequency and duration of summer heatwaves, but also an increase in mean winter temperatures (IPCC 2013). The overall direct impact on health may be mixed – warmer winter temperatures may lower mortality and morbidity in some areas (Hashizume et al. 2009), but heatwaves may also increase illness and mortality. High temperatures cause ill health through heatstroke, which can progress to death, and also through an increase in the mortality from cardiovascular and respiratory illness (Basu and Samet 2002). Some other conditions, such as multiple sclerosis, are worsened by heat. Whilst multiple sclerosis is uncommon in India, there may be other conditions that are worsened, perhaps, most obviously gastroenteritis. These risks are exacerbated by growing urbanization and the urban heat island effect, the phenomenon whereby cities show a higher mean temperature than surrounding farmland due to the building materials used and human activity within cities.

Models assessing the relationship between temperature and mortality in Delhi reveal a 3.9 % increase in cardiac and respiratory mortality and 4.3 % increase in mortality from all other causes, for every 1 °C rise in temperature (McMichael et al. 2008). The impact of heatwaves in South Asia has already been documented, such as a heatwave in the Indian state of Andhra Pradesh in 2006 that is estimated to have killed over 600 people (IPCC 2012). Heatwaves in 2015 in India and Pakistan have been widely reported in the media as causing over 3,000 deaths. Such media estimates are likely to be conservative, as they are largely based on hospital admissions. Many people are likely to have died in these heatwaves before reaching a hospital in South Asia.

A more recent study found that a heatwave in Ahmedabad, Gujarat, lasting for about a week, was associated with an increase in all-cause mortality of 43 % compared to a similar time of the year without a heatwave (Azhar et al. 2014). Heatwaves disproportionately affect the poorest and most marginalised in a society – the elderly (particularly elderly women), manual labourers and those who work outdoors (Kovats and Akhtar 2008; Kovats and Hajat 2008). High temperatures in India consistently above 30–32 °C have been linked with a significant increase in rural mortality thought to be mediated through lower rural incomes, due to lowered crop yields (Burgess et al. 2011). This could also be conceptualised as a risk factor of a tertiary health impact, potentially causing large-scale food insecurity.

2.2.2 Natural Disasters

Models suggest that the frequency of tropical cyclones is unlikely to rise, but that their intensity may increase, with more extreme precipitation (IPCC 2013). Heavy precipitation and tropical cyclones interact with rising sea level to cause primary health effects through flooding and secondary effects through an increase in diarrhoeal disease. Longer-term impacts of tropical cyclone include destruction of farm crops and salinisation of coastal soils and water supplies (IPCC 2014a), leading to the tertiary health effects of forced migration and internal displacement.

Cyclone Nargis, which killed over 130,000 people in Myanmar (not part of South Asia as we have defined it, but in close proximity) in 2008, is an example of the devastating impact that natural disasters can wreak, particularly in countries that are densely populated in coastal regions, with poor-quality building infrastructure and vulnerable populations. The impact of Nargis was exacerbated by seawater surges, which led to extensive damage to Myanmar's rice crop and salinisation of land. Over 50,000 acres of previously fertile delta land was temporarily considered unfit for planting (Webster 2008).

2.2.3 Flooding

Flooding can be caused by increased precipitation, sea level rise or glacier melt. Models suggest that climate change in South Asia is likely to result in longer monsoon seasons, with more heavy precipitation (IPCC 2013). Sea levels rise as a result of thermal expansion of the oceans, melting of ice sheets and loss of glacier mass. Models suggest that, by 2100, global sea level is likely to have risen by between 0.26 and 0.97 m (IPCC 2013). Another effect of climate change relevant to South Asia is the melting of Himalayan glaciers (IPCC 2013). In the short term, this is associated with increased water flows in the rivers supplied by Himalayan glaciers, such as the Ganges and Brahmaputra. Large coastal cities in South Asia (such as Mumbai, Dhaka and Chittagong) face risks of flooding from sea level rise, from rapid increases in river flows, and flooding caused by blocked, poorly maintained drains and sewers overwhelmed by heavy rainfall (Tanner et al. 2009).

Flooding causes ill health directly, through trauma and drowning. More significantly, it harms water and sanitation infrastructure and is associated with increased diarrhoeal disease and VBDs (Githeko et al. 2000). Urbanization, and rapid population growth without sufficient investment in sanitation infrastructure, has left South Asian megacities particularly vulnerable (Kovats and Akhtar 2008). Floods already pose a major health problem in parts of South Asia: flooding in Dhaka in 2004 inundated the whole eastern part of the city; two million people experienced drinking water that was more contaminated than normal. Flooding in Dhaka in 1998 is estimated to have resulted in 5,000 excess cases of diarrhoea (Alam and Rabbani 2007).

2.3 Secondary Health Effects in South Asia

The secondary effects of climate change on health include how changes in the environment and ecology can bring about changes in disease epidemiology. This includes increases in water-borne illnesses caused by flooding and temperature rise and shifts in the geographical distribution of VBDs. Importantly, climate change is only one of many factors which shape the incidence and impact of infectious diseases. Non-climatic factors (other than humidity, temperature and rainfall) which interact with and which complicate the epidemiology of climate change include migration, changes in land use, drug and insecticide resistance and eradication efforts (Reiter 2008). Furthermore, although climate change is predicted to increase transmission windows of malaria overall in South Asia, it will likely reduce it in some areas.

2.3.1 Diarrhoeal Disease

Climate change is likely to result in an increased incidence of diarrhoeal disease in South Asia, through two separate mechanisms. The first is that natural disasters (particularly flooding) can damage water and sanitation infrastructure resulting in contamination of drinking water supplies with infective faecal material (ten Veldhuis et al. 2010). Models suggest that in the WHO South-East Asian Region D (which comprises Bangladesh, Bhutan, North Korea, India, Maldives, Myanmar and Nepal), there may be an excess incidence of 40 million cases of diarrhoea per year, by 2030, as a result of climate change (Ebi 2008).

Another mechanism is the association of climate change with warmer sea temperatures. Some organisms (e.g. the *Vibrio cholerae* bacteria, which cause cholera) are naturally found in coastal water bodies and are concentrated in plankton. Warmer sea temperatures are associated with an increased rate of bacterial reproduction and are also associated with phytoplankton blooms (Lipp et al. 2002). As a result, the likelihood of transmission of cholera to humans is increased, either through drinking infected water or through eating shellfish which concentrates the organism.

2.3.2 Leptospirosis

Leptospirosis is a bacterial disease, carried by animal hosts, passed to humans through direct contact with infected animals or through exposure to the urine of infected animals. Several outbreaks of leptospirosis have been associated with flooding in India, through multiple hypothesised mechanisms, by bringing humans in closer contact with the bacteria and its animal hosts, through damage to sanitation infrastructure and by increasing rodent populations. In addition, rising temperatures mean that spirochaetes (the organisms which cause leptospirosis) can persist longer in the environment, also enhancing transmission risk. The incidence of leptospirosis is likely to increase in urban megacities, with their poor water and sanitation infrastructure, and where there are large numbers of animal reservoirs (such as rats) in close proximity to humans (Lau et al. 2010).

2.3.3 Vector-Borne Diseases

Changes in temperature can affect the rates of reproduction and survival of infectious organisms and insect vectors, thus affecting rates of transmission of vector-borne diseases and causing diseases to spread to new areas (Patz et al. 2005). There are six major vector-borne illnesses in South Asia (malaria, dengue, chikungunya, filariasis, Japanese encephalitis and visceral leishmaniasis). Of these, malaria has the most impact, killing over 55,000 people in South Asia in 2010 (Murray et al. 2012).

Malaria is transmitted by the *Anopheles* mosquito, an insect vector which is highly sensitive to changes in temperature. As climate change results in higher temperatures (and warmer winters), transmission windows are predicted to increase in North India, and malaria is predicted to spread to new parts of previously low-risk countries, such as Bhutan and Afghanistan. In fact, malaria has already re-emerged in parts of North Afghanistan. Climate change is one potential risk factor for this, amongst intensified rice growing and insufficient chemical vector control (Faulde et al. 2007). In addition, the development stages of *Anopheles* mosquitoes are highly temperature dependent, as is the growth and maturation of the parasite within the mosquito. The abundance of potentially infectious mosquitoes increases with warmer temperatures, up to a threshold of around 30 °C (Beck-Johnson et al. 2013). As a result of climate change, malaria transmission is predicted to extend by 2–3 months in the northern Indian states of Punjab, Haryana and Jammu and Kashmir, although its transmission windows will decrease in South India as temperatures exceed 40 °C (Dhiman et al. 2010).

The *Aedes* mosquito, a vector for the virus which causes dengue fever, is similarly sensitive to changes in temperature and humidity. A 2–4 °C rise in temperature is predicted to increase dengue transmission windows in North India (Dhiman et al. 2010); as a result of climate change and population growth, the global population at risk of dengue is predicted to rise from 1.5 billion people in 1990 to 4.1 billion people by 2055 (Hales et al. 2002).

Evidence on the effect of climate change on other VBDs is more limited. Chikungunya, a viral disease transmitted by insects including the *Aedes* mosquito, has recently spread from its origin in Africa to affect parts of Southern India, Sri Lanka and Indian Ocean islands, including Reunion, Mauritius, Madagascar and the Maldives. Like all VBDs, its aetiology is complex, also involving human mobility and even trade, such as of tyres and plants in which vectors or their eggs sometimes survive and thus accidentally migrate (Gould and Higgs 2009). Similar to dengue, as temperatures rise, transmission windows for chikungunya are likely to increase in North India (Dhiman et al. 2010). Visceral leishmaniasis, a parasitic disease spread through a climate-sensitive sandfly vector, is also re-emerging in parts of Nepal and Northern India (Dhiman et al. 2010). Whilst this is likely to be principally due to a lagged resurgence in vector populations following the reduced use of insecticides, climate change is a plausible contributing factor.

2.4   Tertiary Health Effects in South Asia

The tertiary effects of climate change occur at the intersection of climate, politics and ecology and may include large-scale societal shifts (Butler and Harley 2010). They do not fit easily into traditional quantitative models. Whilst the most speculative of the three classes of effect we have used, they may also be the most important, as realisation grows that the interaction between humans and ecosystems must be viewed through the lens of interdependent complex adaptive systems, even though outcomes can be difficult to predict (Folke et al. 2002).

The complex nature of tertiary effects may lead to a series of positive (reinforcing) feedback loops, with drastic consequences at a population level. Changes in temperature and rainfall patterns are likely to harm agriculture and food production. Sea level rise, and an increased frequency of extreme precipitation events, may worsen flooding and land inundation, contributing to forced human migration (Bowles et al 2014), such as the potential for the Sundarbans population to be compelled to migrate to West Bengal or Bangladesh (Daigle 2015). Migration and limited availability of natural resources may lead to conflict. Movement of people into internally displaced persons' (IDP) and refugee camps can be associated with localised land degradation, more food insecurity and increased scarcity of natural resources. In turn these events can be associated with further conflict. All of these indirect effects are influenced by social, economic and demographic pressures, in conjunction with climate change (McMichael 2013).

2.4.1 Food Insecurity

Energy undernutrition is directly estimated to kill 190,000 people in South Asia each year, 50,000 of them under the age of five (Global Burden of Disease 2010). The indirect effects of chronic undernutrition are much greater, including an increased susceptibility to infectious diseases (Rao and Beckingham 2013). In total, including indirect effects, undernutrition is currently estimated to result in almost a million deaths each year of children under the age of five in South Asia (Black et al. 2008).

Existing food insecurity is likely to be exacerbated by climate change, although there is a high degree of regional and crop variation. Extreme weather events are likely to dramatically reduce access to food, and temperatures above 30 °C have a large negative impact on crop yields. By contrast, elevated atmospheric CO_2 is likely to improve some crop yields in the short term, although how this interacts with other factors, including ocean acidification, is unknown (IPCC 2014b). In South Asia, higher temperatures are predicted to reduce the yield of rice production (IPCC 2014b). This impact is likely to be exacerbated by a reduction in water availability, as rainfall patterns change and as glacier melt leads to long-term reduction in river flows, particularly in the Indus and Brahmaputra River basins (IPCC 2013; Immerzeel et al. 2010). Overall, in South Asia, one estimate is that crop yields will decrease by 5–30 % by 2050, leading to an additional 5–170 million people at risk of undernutrition (Schmidhuber and Tubiello 2007). The IPCC projects that by the middle of the twenty-first century, the largest number of food-insecure people will be located in South Asia, partly caused by reductions in sorghum and wheat yields (IPCC 2014b).

As discussed previously, the eventual impact on food security is very difficult to predict as human factors, including population growth, economic growth and income inequality, are likely to have a greater overall influence on food security than climate change alone (IPCC 2014b).

2.4.2   Migration

Many effects of climate change are likely to contribute to migration in and from South Asia. Large-scale extreme weather events, including flooding and tropical cyclones, are likely to lead to sudden but temporary human movements, with the possible return of displaced people. By contrast, gradual and long-term climatic shifts, including sea level rises, increased competition for natural resources and land degradation, may result in more permanent internal and external migration of populations (Walsham 2010). As with all tertiary effects, migration is not easy to model, with multiple economic, political, social and cultural drivers influencing vulnerability and individual decisions to migrate. In South Asia, climate change-influenced and driven migration will occur in the context of ongoing population growth and rural-to-urban migration driven by a perception of greater economic opportunities.

A sea level rise of 1 m (towards the upper end of the current IPCC estimates for 2100) would impact at least five million people across South Asia, mostly in Bangladesh and Sri Lanka (Dasgupta et al. 2007). Even worse, however, these projections of sea level rise do not factor in possible feedbacks to the climate system such as from methane and CO_2 release from the melting tundra. Sea level rise may directly displace people through loss of land mass, but can indirectly displace people through saltwater intrusion and degradation of agricultural land (Walsham 2010). In Bangladesh, this may lead to internal migration to urban slums in cities such as Dhaka or Chittagong or attempted external migration to nearby India, despite the fence that currently marks the entire land boundary. Land degradation and difficulty with agricultural harvests is estimated to have already led to the emigration of 12–17 million people from Bangladesh to India, before the fence was fully constructed (Reuveny 2007).

Long-term migration is likely to be exacerbated by the short-term impact of natural disasters, including flooding from heavy precipitation and high-intensity tropical cyclones. Bangladesh is currently affected by annual inland monsoon floods, which can cause internal displacement. Severe flooding in 2007 affected around three million households (Walsham 2010). By 2050, climate change is modelled to lead to an additional 1.9 million people affected by areas newly inundated in monsoon floods (Sarraf et al. 2011).

Indian Ocean island nations are also particularly vulnerable to the health impact of climate change. The Maldives – a nation consisting of 1,200 individual islands, which lie 0–2 m above sea level – has long been considered one of the countries most vulnerable to climate change. Ongoing sea level rise as a result of climate change may result in widespread inundation of the nation (Church et al. 2006), requiring the whole population to emigrate.

The health impact of migration is varied and disproportionately affects the poorest and most vulnerable in a society. The wealthiest people are able to migrate proactively, whereas the poorest either lack the financial capital to move away from natural disasters, or are forced to move into temporary and makeshift shelter in the aftermath of disasters (The Government Office for Science 2011a). The act of migration itself is associated with impacts on health during the travel phase, which include trauma, undernutrition and increased transmission of infectious disease. It is also associated with negative health outcomes in the destination phase, including impacts on mental, reproductive and child health (Zimmerman et al. 2011). This is particularly exacerbated if populations are forced into IDP camps, refugee camps or other 'temporary' accommodation, which can be associated with transmission of infectious diseases and an increase in gender-based violence.

2.4.3 Conflict

Climate change may be associated with conflict through a variety of mechanisms, including instability following extreme weather events, competition over scarce natural resources such as water or a breakdown of international and national social and political systems (The Government Office for Science 2011b; Kelley et al. 2015). Direct empirical evidence for a causative link between climate change and conflict is limited, and no one suggests that climate change is a single driver causing conflict. The problem is significant enough that the British and American militaries (amongst others) are considering climate change as part of their long-term contingency planning. South Asia is particularly vulnerable to potential conflict, as a region with existing ethnic, religious and political tensions that have led to a number of international military engagements over the last 50 years. In addition, India and Pakistan have nuclear capabilities, and large areas of Afghanistan and Northern Pakistan are already outside state control.

At a local level, extreme weather events can drive short-term migration of people into internally displaced persons' (IDP) camps or refugee camps. These place increased strains on local environmental resources, including water supplies, and can lead to conflict between indigenous and migrant communities (Reuveny 2007).

Competition over scare natural resources can also lead to conflict; within South Asia, one particularly scarce natural resource is likely to be water, due to glacier melt and variation in monsoon rainfall. As population growth, urbanization and economic growth continue in South Asia, demands and stresses for the water supply are likely to be increased. The major Himalayan rivers traverse national boundaries (the Brahmaputra River originates in Tibet, China), and management of water across national boundaries represents another potential flashpoint for conflict, particularly as India and China are both planning large-scale river diversion projects

(UCL Hazard Research Centre and Humanitarian Futures Programme, King's College London 2010). Silt from the Brahmaputra River may be reduced this century by its proposed damming and even diversion of part of its water to eastern China (Matthew 2013). That would not only reduce soil fertility in Bangladesh but also increase the delta's vulnerability to sea level rise. Numerous other dams have been built or are planned along other rivers, especially in India; some of these are already reducing silt flow to the Sundarbans (Matthew 2013).

Also, as legitimate states fail, the opportunity for non-state actors rises, with distinct motivations. For example, there is evidence that extremist organisations provided aid in remote areas of Pakistan after the 2008 earthquake – this legitimises their role and further erodes state influence (The Government Office for Science 2011b).

2.4.4   Economic Instability

Globalisation has resulted in complex interconnected supply chains; individual countries are less self-sufficient and therefore increasingly vulnerable to external events. Disrupted supply chains can directly affect health, by limiting the supply of necessary medicines and vaccinations. Economic harm and worsening poverty also hurt health, both physical and mental.

The economic impact of extreme weather events is hard to fully assess. The disruption to banking and financial systems can be very large – floods in Mumbai in 2005 caused a temporary closure of the Stock Exchange (IPCC 2012). However, probably an even larger economic impact is felt in the informal economy – where the disruption of everyday market buying and selling can affect the livelihoods of millions of people.

2.4.5   Mental Health

Climate change is likely to have adverse mental health effects at each of the three levels of the framework that we have proposed. Any severe physical illness is likely to have an adverse mental health effect, even if just anxiety. Extreme weather events and natural disasters can lead to mental ill health, including acute traumatic stress in the short term and post-traumatic stress disorder, depression, anxiety disorders and drug and alcohol abuse in the long term (Fritze et al. 2008). Survivors of disasters can also be harmed through seeing others killed or hurt, by the loss of family members and livestock and droughts (Hanigan et al. 2012).

Some VBDs can lead to cognitive and psychiatric disturbances (e.g. cerebral malaria or post-viral fatigue and depression). Primary care services across South Asia are already stretched, and if overall demand for health services increases, mental health is likely to be further ignored. The tertiary impacts of climate change on mental health are likely to be the greatest, through a variety of levers, including resource scarcity, extreme weather events and widespread social and community disruption. Mental scars arising from the witness or experience of violent conflict can easily be transmitted for generations.

Involuntary migrants face many risks, including trafficking and other forms of exploitation. Some will languish in camps for years. Populations at the greatest risks of climate change-influenced migration are likely to be amongst the most marginalised and socially excluded. Additionally, people with existing mental illnesses (whether diagnosed or not) are amongst the most vulnerable during natural disasters and forced migrations.

Crop failure and indebtedness are already associated with thousands of farmer suicides and other forms of stress in India, particularly in Andhra Pradesh, Karnataka, Kerala and Maharashtra (Gruère et al. 2008; Sarkar and van Loon 2015). As discussed in the section on food security, climate change is likely to be associated with reduced crop yields in South Asia and, as a result, may worsen rural mental health, including suicides.

2.5 Conclusion

Climate change is already harming the health of people in South Asia – both physically and mentally and through a range of different direct and indirect mechanisms. But these effects are likely to become far worse as the environmental impact of anthropogenic activity, including from climate change, intensifies this century.

Although this chapter used a primary, secondary and tertiary framework to describe the health impact of climate change, it is stressed that these effects are interconnected, and impacts may be multiplicative. For example, high temperatures cause a primary health impact through heatstroke; they may also worsen labour productivity and also lower crop yields. These effects are also interconnected with more general societal trends – economic migrants from South Asia already face poor living conditions, including poor shelter, inadequate sanitation and possible conflict with host communities. These existing vulnerabilities will exacerbate any migration or conflict caused by climate change.

As always, the health impacts of climate change cause the greatest harm to the poorest and most vulnerable members of a society – these are people who may live in urban slums with poor access to clean water and sanitation, people who are already most susceptible to undernutrition and food insecurity and people who cannot afford to migrate proactively away from areas prone to natural disasters. Poor women and girls are at the greatest risk. These people are dependent on state responses for adaptation, as they lack the financial capacity and political power to take individual action.

India is currently publicly reluctant to support mitigating its own carbon emissions, believing that developed countries must pay the price for their historical emissions. This position, whilst understandable, allows India to avoid engagement with climate change as a political and social issue. However, India, whilst continuing to expand its coal use, is also promoting an aggressive expansion of renewable energy (Kalra and Wilkes 2015).

Following this lead, many South Asian governments are reluctant to curtail their own carbon-dependent economic growth. True global leadership on the issue would allow South Asian governments to pursue a path of 'green growth', demonstrating an alternative to the current damaging economic paradigm. A number of policies, such as decentralised solar power generation, can allow for economic growth and improvements in peoples' quality of life, whilst mitigating overall carbon emissions. There is a significant opportunity for South Asian governments to use their strengths in research on technology development to become world leaders in carbon-neutral development policies.

References

Alam M, Rabbani MG (2007) Vulnerabilities and responses to climate change for Dhaka. Environ Urban 19:81–97

Azhar GS, Mavalankar D, Nori-Sarma A, Rajiva A, Dutta P, Jaiswal A, Sheffield P, Knowlton K, Hess JJ (2014) Heat-related mortality in India: excess all-cause mortality associated with the 2010 Ahmedabad heat wave. PLoS One 9(3):e91831

Basu R, Samet JM (2002) Relation between elevated ambient temperature and mortality: a review of the epidemiologic evidence. Epidemiol Rev 24(2):190–202

Beck-Johnson LM, Nelson WA, Paaijmans KP, Read AF, Thomas MB, Bjørnstad ON (2013) The effect of temperature on anopheles mosquito population dynamics and the potential for malaria transmission. PLoS ONE 8(11):e79276

Black RE, Allen LH, Bhutta ZA, Caulfield LE, de Onis M, Ezzati M, Mathers C, Rivera J (2008) Maternal and child under nutrition: global and regional exposures and health consequences. Lancet 371:243–260

Bowles DC, Reuveny R, Butler CD (2014) Moving to a better life? Climate, migration and population health. In: Butler CD (ed) Climate change and global Health. CABI, Wallingford, pp 135–143

Burgess R, Deschenes O, Donaldson D, Greenstone M (2011) Weather and death in India. Department of Economics Manuscripts, Massachusetts Institute of Technology, Cambridge, MA

Butler CD (2014a) Famine, hunger, society and climate change. In: Butler CD (ed) Climate change and global health. CABI, Wallingford, pp 124–134

Butler CD (ed) (2014b) Climate change and global health. CABI, Wallingford

Butler CD, Harley D (2010) Primary, secondary and tertiary effects of eco-climatic change: the medical response. Postgrad Med J 86:230–234

Cash RA, Halder SR, Husain M, Islam MS, Mallick FH, May MA, Rahman M, Rahman MA (2013) Reducing the health effect of natural hazards in Bangladesh. The Lancet 382:2094–2103

Church JA, White NJ, Hunter JR (2006) Sea-level rise at tropical Pacific and Indian Ocean islands. Glob Planet Change 53:155–168

Costello A, Abbas M, Allen A, Ball S, Bell S, Bellamy R, Friel S, Groce N, Johnson A, Kett M, Lee M, Levy C, Maslin M, McCoy D, McGuire B, Montgomery H, Napier D, Pagel C, Patel J, de Oliveira JAP, Redclift N, Rees H, Rogger D, Scott J, Stephenson J, Twigg J, Wolff J, Patterson C (2009) Managing the health effects of climate change. Lancet 373:1693–1733

Dasgupta S, Laplante B, Meisner C, Wheeler D, Jianping Yan D (2007) The impact of sea level rise on developing countries: a comparative analysis. World Bank Policy Research Working Paper

Daigle K (2015) Millions at risk from rapid sea rise in swampy Sundarbans (News report). http://www.cnbc.com/2015/02/18/millions-at-risk-fromrapid-sea-rise-in-swampy-sundarbans.html

Dhiman RC, Pahwa S, Dhillon GPS, Dash AP (2010) Climate change and threat of vector-borne diseases in India: are we prepared? Parasitol Res 106:763–773

Ebi KL (2008) Adaptation costs for climate change-related cases of diarrhoeal disease, malnutrition, and malaria in 2030. Global Health 4

Faulde MK, Hoffmann R, Fazilat KM, Hoerauf A (2007) Malaria reemergence in Northern Afghanistan. Emerg Infect Dis 13:1402–1404

Folke C, Carpenter S, Elmqvist T, Gunderson L, Holling CS, Walker B (2002) Resilience and sustainable development: building adaptive capacity in a world of transformations. Ambio J Human Environ 31:437–440

Fritze JG, Blashki GA, Burke S, Wiseman J (2008) Hope, despair and transformation: climate change and the promotion of mental health and wellbeing. Int J Ment Heal Syst 2:1–10

Githeko AK, Lindsay SW, Confalonieri UE, Patz JA (2000) Climate change and vector-borne diseases: a regional analysis. Bull World Health Organ 78(9):1136–1147

Gleick P (2014) Water, drought, climate change, and conflict in Syria. Weather Clim Soc 6(3):331–340

Gould EA, Higgs S (2009) Impact of climate change and other factors on emerging arbovirus diseases. Trans R Soc Trop Med Hyg 103:109–121

Gruère GP, Mehta-Bhatt P, Sengupta D (2008) Bt cotton and farmer suicides in India. Rev Evid Int Food Policy Res Inst IFPRI Discuss Pap 808

Hales S, de Wet N, Maindonald J, Woodward A (2002) Potential effect of population and climate changes on global distribution of dengue fever: an empirical model. Lancet 360:830–834

Hanigan IC, Butler CD, Kokic PN, Hutchinson MF (2012) Suicide and drought in New South Wales, Australia, 1970–2007. Proc Natl Acad Sci U S A 109:13950–13955

Hashizume M, Wagatsuma Y, Hayashi T, Saha SK, Streatfield K, Yunus M (2009) The effect of temperature on mortality in rural Bangladesh – a population based time-series study. Int J Epidemiol 38:1689–1697

UCL Hazard Research Centre, Humanitarian Futures Programme, Kings College London (2010) The waters of the third pole: sources of threat, sources of survival. University College London/ Kings College London, London

Immerzeel WW, van Beek LPH, Bierkens MFP (2010) Climate change will affect the Asian water towers. Science 328:1382–1385

Institute for Health Metrics and Evaluation (2010) Global Burden of Disease Study 2010. IHME, Seattle

International Institute for Applied Systems Analysis (2012) Asian Demographic and Human Capital Data Sheet

IPCC (2012) Managing the risks of extreme events and disasters to advance climate change adaptation. A special report of working groups I and II of the Intergovernmental Panel on Climate Change. Cambridge University Press, Cambridge, UK/New York

IPCC (2013) Climate change 2013: the physical science basis. Cambridge University Press, Cambridge, UK/New York

IPCC (2014a) Climate change 2014: impacts, adaptation, and vulnerability. Part B: regional aspects. Contribution of working group II to the fifth assessment report of the Intergovernmental Panel on Climate Change. Cambridge University Press, Cambridge, UK/New York

IPCC (2014b) Climate change 2014: impacts, adaptation, and vulnerability. Part A: global and sectoral aspects. Contribution of working group II to the fifth assessment report of the Intergovernmental Panel on Climate Change. Cambridge University Press, Cambridge, UK/ New York

Kalra A, Wilkes T (2015) Modi says India to strike own path in climate battle. http://planetark.org/wen/73021. Accessed 8 Apr 2015

Kelley CP, Mohtadi S, Cane MA, Seager R, Kushnir Y (2015) Climate change in the fertile crescent and implications of the recent Syrian drought. Proc Natl Acad Sci U S A 112(11):3241–3246

King D, Schrag D, Dadi Z, Ye Q, Ghosh A (2015) Climate change: a risk assessment. http://www.csap.cam.ac.uk/projects/climate-change-risk-assessment/

Kovats S, Akhtar R (2008) Climate, climate change and human health in Asian cities. Environ Urban 20:165–175

Kovats RS, Hajat S (2008) Heat stress and public health: a critical review. Ann Rev Public Health 29:41–55

Lau CL, Smythe LD, Craig SB, Weinstein P (2010) Climate change, flooding, urbanisation and leptospirosis: fuelling the fire? Trans R Soc Trop Med Hyg 104:631–638

Lipp EK, Huq A, Colwell RR (2002) Effects of global climate on infectious disease: the cholera model. Clin Microbiol Rev 15:757–770

Matthew R (2013) Climate change and water security in the Himalayan region. Asia Policy 16:39–44

McMichael AJ (2013) Globalization, climate change, and human health. N Engl J Med 368:1335–1343

McMichael AJ, Powles J, Butler CD, Uauy R (2007) Food, livestock production, energy, climate change and health. Lancet 370:1253–1263

McMichael AJ, Wilkinson P, Kovats RS, Pattenden S, Hajat S, Armstrong B, Vajanapoom N, Niciu EM, Mahomed H, Kingkeow C (2008) International study of temperature, heat and urban mortality: the "ISOTHURM" project. Int J Epidemiol 37:1121–1131

Ministry of External Affairs (2009) The road to Copenhagen: India's position on climate change issues. Government of India, New Delhi

Murray CJ, Rosenfeld LC, Lim SS, Andrews KG, Foreman KJ, Haring D, Fullman N, Naghavi M, Lozano R, Lopez AD (2012) Global malaria mortality between 1980 and 2010: a systematic analysis. Lancet 379:413–431

Patz JA, Campbell-Lendrum D, Holloway T, Foley JA (2005) Impact of regional climate change on human health. Nature 438:310–317

Paul BK (2010) Human injuries caused by Bangladesh's cyclone Sidr: an empirical study. Nat Hazards 54(2):483–495

Rao M, Beckingham A (2013) Food, nutrition, and public health. In: Pielke RA (ed) Climate vulnerability. Academic, Oxford, pp 87–94

Reiter P (2008) Global warming and malaria: knowing the horse before hitching the cart. Malar J 7:S3. doi:10.1186/1475-2875-7-S1-S3

Reuveny R (2007) Climate change-induced migration and violent conflict. Polit Geogr 26:656–673

Sarkar A, van Loon G (2015) Modern agriculture and food and nutrition insecurity: paradox in India. Public Health. doi:10.1016/j.puhe.2015.04.003

Sarraf M, Dasgupta S, Adams N (2011) The cost of adapting to extreme weather events in a changing climate (No. 67845). The World Bank

Schmidhuber J, Tubiello FN (2007) Global food security under climate change. Proc Natl Acad Sci U S A 104:19703–19708

Selwood KE, McGeoch MA, MacNally R (2014) The effects of climate change and land-use change on demographic rates and population viability. Biol Rev

Sumner A (2012) Where do the world's poor live? A new update. IDS Work Pap 2012:1–27

Tanner T, Mitchell T, Polack E, Guenther B (2009) Urban governance for adaptation: assessing climate change resilience in ten Asian cities. IDS Working Paper, 01–47

Ten Veldhuis JAE, Clemens FHLR, Sterk G, Berends BR (2010) Microbial risks associated with exposure to pathogens in contaminated urban flood water. Water Res 44(9):2910–2918

The Government Office for Science (2011a) Foresight: migration and global environmental change, London

The Government Office for Science (2011b) Foresight international dimensions of climate change: Final Project Report, London

Thomas CD, Cameron A, Green RE, Bakkenes M, Beaumont LJ, Collingham YC, Erasmus BF, De Siqueira MF, Grainger A, Hannah L (2004) Extinction risk from climate change. Nature 427:145–148

Walsham M (2010) Assessing the evidence: environment, climate change and, migration in Bangladesh. International Organization for Migration, Dhaka

Watts N, Adger WN, Agnolucci P, Blackstock J, Byass P, Cai W, Chaytor S, Colbourn T, Collins M, Cooper A, Cox PM, Depledge J, Drummond P, Ekins P, Galaz V, Grace D, Graham H, Grubb M, Haines A, Hamilton I, Hunter A, Jiang X, Li M, Kelman I, Liang L, Lott M, Lowe R, Luo Y, Mace G, Maslin M, Nilsson M, Oreszczyn T, Pye S, Quinn T, Svensdotter M, Venevsky S, Warner K, Xu B, Yang J, Yin Y, Yu C, Zhang Q, Gong P, Montgomery H, Costello A (2015) Health and climate change: policy responses to protect public health. The Lancet 6736(15): 60854–60856, 23 June 2015. http://dx.doi.org/10.1016/S0140-.

Webster PJ (2008) Myanmar's deadly daffodil. Nat Geosci 1:488–490

Weitzman M (2009) On modeling and interpreting the economics of Catastrophic climate change. Rev Econ Stat 91:1–19

Yasir A (2009) The political economy of disaster vulnerability: a case study of Pakistan earthquake 2005. Munich Personal RePEc Archive

Zimmerman C, Kiss L, Hossain M (2011) Migration and health: a framework for 21st century policy-making. PLoS Med 8, e1001034

Chapter 3
Climate Change and the Health of Older People in Southeast Asia

Kevin McCracken and David R. Phillips

Abstract This chapter is essentially speculative, considering possible health-related consequences for older persons of climate change events should these occur. The Asia-Pacific region and its Southeast Asian subregion has almost a third of the world's older persons. The heterogeneity of older cohorts should be recognised and not stereotyped as uniformly vulnerable. Nevertheless, older persons will generally be more vulnerable than other age groups to events such as extreme heat and other climate stressors. Risks may be raised for ischaemic heart disease, stroke, COPD, acute lower respiratory disease and lung cancer, affecting older groups disproportionately. Higher temperatures are also associated with sudden-onset weather emergency events such as typhoons and associated heavy rainfall, flooding and damage to property and life. Many parts of the region are susceptible to such events, along with neighbouring countries in East and South Asia. Many types of infectious diseases will also be affected by temperature rises. Expansion is likely of areas at risk from malaria, dengue and other parasitic or infectious conditions. For example, small temperature increases could raise both latitude and altitude at which mosquitoes could breed and affect humans. Other risks include increasing incidence of food poisoning and threats to food security as extreme climate events and disease can disrupt food production. Finally, climate change can impact psychological and mental health. Older persons will be relatively at greater risk, especially those with dementias who may be less able to cope with adverse conditions. The severity of climate change impacts will be mediated by political and socio-economic factors and adaptive strategies, so forward planning by governments is essential.

Keywords Climate change • Health • Older persons • Extreme events • Southeast Asia

K. McCracken (✉)
Department of Geography and Planning, Macquarie University,
Sydney, NSW, Australia
e-mail: kevin.mccracken@mq.edu.au

D.R. Phillips
Department of Sociology and Social Policy, Lingnan University,
Tuen Mun, Hong Kong

© Springer International Publishing Switzerland 2016   29
R. Akhtar (ed.), *Climate Change and Human Health Scenario in South and Southeast Asia*, Advances in Asian Human-Environmental Research,
DOI 10.1007/978-3-319-23684-1_3

3.1   Introduction

This chapter is essentially speculative, centred on likely climate change in Southeast Asia, also affecting the wider Asia-Pacific, over the course of this century and the potential impacts of this change on the health of older people resident in the region. Climate is a major determinant, directly or indirectly, of human well-being in all regions around the world, and any substantial change in climate dynamics by definition will have important reverberations across all domains of human existence (United Nations Development Programme 2012). Nowhere is this truer than in the area of health (McMichael et al. 2001) and especially the health of populations that for one reason or another have particular vulnerabilities. One group generally viewed in this light is older persons (Filberto et al. 2009–2010; Gamble et al. 2013; Haq et al. 2008; IPCC 2014; WHO 2013).

3.2   Global Overview of Climate Change

It is now virtually universally accepted in the scientific community that the Earth's climate system has been undergoing significant change over the past century or so. Sound evidence assembled for the Fifth Assessment Report (AR5) of the Intergovernmental Panel on Climate Change (IPCC) clearly shows atmospheric and oceanic warming have occurred, ice sheets and glaciers have shrunk, sea levels have risen and extreme weather events have become more frequent (IPCC 2014; Hijioka et al. 2014). The present-day global mean surface temperature, for example, is around 0.8 °C warmer than that of the immediate pre-industrial period, whilst global mean sea level is estimated to have risen by 0.19 m since the beginning of the twentieth century. Whilst there is a small number of sceptics, the vast majority of climate scientists believe these global warming changes have been largely driven by increasing atmospheric greenhouse gas concentrations (e.g. carbon dioxide, methane, nitrous oxide) produced by human activities (e.g. burning fossil fuels, deforestation).

With continuing emissions of greenhouse gases, substantial further global warming and related climate system changes will occur well into and, for some components, beyond the twenty-first century. However, uncertainty regarding the exact magnitude of such emissions inevitably makes forecasting future climate change difficult. Whether nations around the world will muster the collective political resolve to bring about the necessary reduction of greenhouse gas emissions to limit global warming and other related climate change remains to be seen. In the light of this emission uncertainty, the usual modelling approach followed is to produce a range of climate change projections incorporating varying greenhouse gas concentrations and other climate drivers. Table 3.1 presents the latest such projections by the IPCC (2013) for air temperature and sea-level rise over the twenty-first century, using four different climate-forcing scenarios. Whatever the magnitude of the climate change ultimately realised, it will impact on almost all aspects of human life.

Table 3.1 Projected change in global mean surface air temperature and global mean sea-level rise for the mid- and late twenty-first century relative to 1986–2005

		Scenario[a]	Mid-twenty-first century (2046–2065)		Late twenty-first century (2000–2100)	
			Mean	*Likely* range	Mean	*Likely* range
Global mean surface temperature change (°C)	1	RCP2.6	1.0	0.4–1.6	1.0	0.3–1.7
	2	RCP4.5	1.4	0.9–2.0	1.8	1.1–2.6
	3	RCP6.0	1.3	0.8–1.8	2.2	1.4–3.1
	4	RCP8.5	2.0	1.4–2.6	3.7	2.6–4.8
Regional comment: '… increases in seasonal mean and annual mean temperatures are expected to be larger in the tropics and subtropics than in mid-latitudes'						
Global mean sea-level rise (m)	1	RCP2.6	0.24	0.17–0.32	0.40	0.26–0.55
	2	RCP4.5	0.26	0.19–0.33	0.47	0.32–0.63
	3	RCP6.0	0.25	0.18–0.32	0.48	0.33–0.63
	4	RCP8.5	0.30	0.22–0.38	0.63	0.45–0.82

Source: IPCC (2013)

[a]Scenarios (RCPs) incorporate different assumptions about future concentrations of greenhouse gases and other climate drivers. RCP = Representative Concentration Pathway, the accompanying values indicating the approximate total radiative forcing in year 2100 relative to 1750

3.3 Projected Climate Change in Southeast Asia

Beneath the overall global picture sketched above, climate change will not occur uniformly. Warming will be greater in some parts of the world than others, likewise sea-level rise. In some places precipitation will increase, whilst other regions will become drier. In short, climate change will vary around the planet. Projecting likely regional changes, however, is complicated, involving all the uncertainties of global modelling with the additional difficulty of incorporating all major relevant regional factors. In the case of Southeast Asia, making climate projections is especially challenging 'due to the complex terrain, the mix of mainlands, peninsulas, and islands, the related regional sea-land interactions, and the large number of complex climate phenomena characterizing the region' (World Bank 2013: 70).

Despite these difficulties, the broad consensus is that the region stands as one of the most vulnerable to climate change. An idea of the breadth of this vulnerability is given by Table 3.2 which outlines what are considered to be the major likely manifestations in the region of 2 °C and 4 °C global warming scenarios published by the World Bank (2013) – each quite plausible outcomes this century. Projected mean summer land surface warming for the region is somewhat below the projected overall global mean land surface temperature rise due to the important climate driving role in the region played by sea surface temperature which is increasing more slowly. The projected warming will still nonetheless be significant, both in its own

Table 3.2 Climate change risks/impacts relevant to human health, 2 °C and 4 °C global warming[a] scenarios, Southeast Asia

	Around 2 °C global warming[a] (2040s[b])	Around 4 °C global warming[a] (2080s[b])
Regional warming	Average summer warming about 1.5 °C above 1951–1980 baseline	Summer temperatures increase by about 4.5 °C above 1951–1980
Heat extremes		
Unusual heat extremes (currently virtually absent)	60–70 % of land area affected (summer months)	>90 % of land area affected (summer months)
Unprecedented heat extremes	30–40 % of land area affected (summer months)	>80 % of land area affected (summer months)
Warm days (beyond present-day 90th percentile temperature)	45–90 days per year	Around 300 days per year
Precipitation	Slight increase during dry season	No agreement across models re changes in wet season rainfall
	Increasing extreme rainfall events (contributing >10 % share of annual rainfall)	Increasing extreme rainfall events (contributing 50 % share of annual rainfall)
	Marginal increase in maximum number of consecutive dry days	About 5 % increase in maximum number of consecutive dry days
Tropical cyclones (typhoons)	Overall decrease in TC frequency, but increasing frequency of most intense (category 5) storms	Decreased number of TCs making landfall, but maximum wind velocity at the coast projected to increase
	Increase in TC-related rainfall	Increase in TC-related rainfall
Sea-level rise (above present)	30 cm–2040s	30 cm–2040s
	50 cm–2060s	50 cm–2060
	75 cm (65–85 cm) by 2080–2100	110 cm (85–130 cm) by 2080–2100
Ocean temperatures and chemistry	Increasing sea surface temperatures and acidification	Increasing sea surface temperatures and acidification

Source: Adapted from World Bank (2013). *Turn Down the Heat: Climate Extremes, Regional Impacts, and the Case for Resilience*
[a]Warming relative to the pre-industrial period. Observed warming from pre-industrial levels to the 1986–2005 benchmark used in Table 3.1 (IPCC, Fifth Assessment Report) has been about 0.7 °C
[b]Years indicate the decade during which warming levels are exceeded in a business-as-usual scenario

right and for its involvement in other elements and consequences of climate change. Within the region, the World Bank modelling suggests warming will be greatest in North Vietnam and Laos. For Asia as a whole, the 5th Assessment Report of the IPCC (Hijioka et al. 2014, Chapter 24, p. 1330) notes 'Warming trends and increasing temperature extremes have been observed across most of the Asian region over the past century (high confidence).' It also notes that water scarcity may become a major challenge and that 'Extreme climate events will have an increasing impact on human health, security, livelihoods, and poverty with the type and magnitude of impact varying across Asia (high confidence)' (Hijioka et al. 2014, Chapter 24, p. 1331). So, the directions of the numerous potential outcomes are not uniform across the region or the Southeast Asian subregion (Yusuf and Francisco 2009).

The modelling results summarised in Table 3.2 indicate that beyond straight mean temperature increase, the region will face associated problems of more frequent and widespread heat extremes, significant sea-level rise and more high-intensity tropical cyclones (typhoons), perhaps affecting wider parts of the region. Warming and acidification of the region's oceans will also be important. Precipitation futures are less certain. Extreme rainfall events, already common in many areas, are expected to increase, but there is no modelling agreement over whether the region as a whole will be overall wetter or drier. The 5th IPCC Assessment Report in its Asia chapter (Hijioka et al. 2014, Chapter 24, p. 1331) states that 'Coastal and marine systems in Asia are under increasing stress from both climatic and non-climatic drivers (high confidence)... It is likely that mean sea-level rise will contribute to upward trends in extreme coastal high water levels.'

A wide range of negative health consequences will flow from these changes for people across the entire region. Precise impacts are impossible to state due to the uncertainties over the exact climate change that will eventuate; future possible political, economic and social mediating factors; and, likewise, future adaptive strategies and capacities. Whilst all people in all countries will to one degree or another be impacted, some population groups will be particularly vulnerable and adversely affected. One major such group is older persons.

3.4   Climate Change and the Health of Older Persons

The disproportionate health vulnerability of older people to climate change has physiological, psychological and social components (Carnes et al. 2013; Confalonieri et al. 2007; Filberto et al. 2009–2010). Very frequently, the vulnerability is multiple elements of the different dimensions overlapping and interacting in both elderly individuals and specific subpopulations. At the same time, whilst the elderly in aggregate are at greater risk than the general population, it is important to recognise their heterogeneity and to avoid simply stereotyping all older persons as uniformly vulnerable. Obvious distinctions, for example, include by age ('young old', 'older old' and 'oldest old', say 60–70, 70–80 and 80+), gender, location (rural/urban; coastal/inland), economic resources (rich/poor), household structure and support (living alone/living with others), health status (fit/frail), etc.

Physiological vulnerability will be particularly important in relation to the 'heat extreme' changes (i.e. projected for the region) as older people are more sensitive to high temperatures and prone to heat stress (hyperthermia) than younger people. The human body copes with very hot weather through sweating or evaporation of the sweat off the skin cooling the body. With ageing this self-regulating temperature mechanism declines in efficiency, and under exposure to sustained extreme temperatures, the body can lose the ability to sweat and accordingly overheats and experiences heat exhaustion or life-threatening heat stroke. Many older people also have pre-existing chronic medical conditions and often co-morbidities such as heart disease, hypertension, diabetes and kidney disease that can increase the body's vulnerability to heat stress. Additionally, some medications (e.g. beta blockers, diuretics) to treat these diseases can impair the body's ability to regulate its temperature. In turn, heat stress can aggravate those existing conditions and damage other organ systems.

Physical mobility impairments are a further factor heightening vulnerability. These impairments increase with advancing age and compromise the ability to respond to severe rapid-onset climate dangers such as tropical cyclones and flooding by evacuating to safer locations. Ageing-related diminishing sensory awareness, particularly of hearing and vision, also adds to vulnerability to extreme weather events.

Psychological vulnerability stems from the decline in mental functioning that occurs with ageing, the progressive deterioration in mental acuity creating the risk that many elderly will not appreciate the nature and magnitude of climate change threats nor how to best cope with them. Alzheimer's disease and other dementias are particular concerns in this regard due to the increasing proportion of 'old-old' persons in the elderly population and the prevalence of dementia approximately doubling every five years after age 65. Major increases are predicted for the middle- and lower-income countries of the world, which include several in this subregion.

Vulnerability to climate change is also influenced by the social and economic circumstances people live within. Many older people only have a low level of education, are poor, live by themselves and in poor housing and are socially isolated, affecting both exposure to climate change and coping capacity. Whilst members of all population age groups may also experience these circumstances, they tend to be more common amongst the elderly and amplify age-based physiological and psychological sensitivities. This is also seen *within* the older population in male-female differences in vulnerability. Older women, for instance, tend to be more physiologically sensitive than men to high temperatures due to having fewer sweat glands, but their vulnerability to such climatic conditions is commonly amplified by being more socially and economically disadvantaged than their male counterparts. The predominance of females, especially in the older and oldest-old groups, is well recognised (Mujahid 2006), and older females typically have both lowest education levels and the fewest economic resources. They are often very dependent on family and local community support. If their living environment is affected by climate change, they may be especially vulnerable. This has been termed 'feminisation', but whilst the predominance of females in the older-age groups is very clear, there is a projected slight weakening of this trend over the course of this century. This will likely slightly narrow the gap in numbers and percentages of older men and women.

Health will be affected both directly and indirectly by climate change. The more frequent, intense and longer-lasting extreme weather events such as heatwaves and cyclonic storms projected for the region will have the most direct and immediately visible health effects, with some almost certainly claiming large injury, illness and death tolls. The WHO (2008) notes the extra vulnerability of older groups to 'emergencies', many of which could potentially be linked to such extreme weather events. The greater health burden of climate change, however, is likely to come from indirect effects, for example, from adverse effects on food production and nutrition due to the loss of productive low-lying coastal agricultural areas through inundation by rising seas and associated coastal land and aquifer salinisation, through forced displacement of populations from those coastal areas to life in unhealthy urban slums and exposure to new epidemiological locales, from warmer temperatures increasing health-threatening ground-level ozone and particulate matter air pollution problems in urban areas and from warmer conditions resulting in some infectious disease vectors and pathogens developing faster and accordingly raising the risk of infection (McCracken and Phillips 2012).

The bulk of attention in health-oriented climate change research has been on physical health consequences – deaths, injuries and illnesses. However, a serious mental health burden can also be expected (Page and Howard 2010; Doherty and Clayton 2011). Older persons with pre-existing mental disorders will likely be most affected, but the psychological well-being of many with no prior record of such illness will also be adversely impacted. Some caught up, for example, in very severe cyclones, flooding or heatwaves will almost certainly suffer post-traumatic stress disorder, plus acute anxiety about possible future such calamitous events. Those who experience the personal loss of family or friends in such climatic disasters will be prone to long-lasting depression. Climate change-forced population displacement will similarly likely generate psychological distress and anxiety in affected populations.

3.5 Elderly Populations of Southeast Asia

The Asia-Pacific region is already home to almost a third of the world's older persons, and the Southeast Asia subregion has a range of nations with various proportions of older people (Phillips 2000, 2011; Mujahid 2006; Phillips et al. 2010). Table 3.3 indicates the situation at the moment, based on 2012 UN data, and the major feature looking forward is the very considerable growth in the 60+ populations, the UN cut-off for old age, in almost every country by 2050. The data do however show the picture is by no means uniform across the subregion. For example, countries such as Singapore and Thailand currently have 12–14 % of their populations aged 60+ now, and they, plus Vietnam, are likely to have over 30 % in this age group by 2050. By contrast, some countries, such as the Philippines (with its substantial population base), Laos and small Timor-Leste, will only have between 5 and 15 % aged 60+ by mid-century. Therefore, the percentages of older persons exposed to risk of climate change do and will continue to vary considerably amongst the nations in this subregion, but large absolute numbers will still be exposed in almost all countries. Looking forward to the turn of next century, which is a

Table 3.3 Expanding older populations (persons aged 60 or over), Southeast Asia, 2010–2100

Region/country	Persons aged 60+ (thousands)			Percentage of total population aged 60+		
	2010	2050[a]	2100[a]	2010	2050[a]	2100[a]
Southeast Asia	48,542	176,120	240,796	8.1	22.4	31.8
Brunei Darussalam	25	155	194	6.2	28.3	38.6
Cambodia	1038	4775	8306	7.2	21.2	35.2
Indonesia	18,212	67,738	101,468	7.6	21.1	32.2
Laos	361	1662	3817	5.6	15.7	34.2
Malaysia	2193	9747	14,760	7.8	23.1	34.8
Myanmar	3998	13,049	13,008	7.7	22.3	27.4
Philippines	5474	21,535	49,691	5.9	13.7	26.5
Singapore	716	2506	2776	14.1	35.5	46.0
Thailand	8580	23,148	15,875	12.9	37.5	39.2
Timor-Leste	54	107	812	5.0	5.1	24.9
Vietnam	7891	31,699	30,090	8.9	30.6	37.6

Source: United Nations Department of Social and Economic Affairs, Population Division (UNDESA-PD) (2013), *World Population Prospects: The 2012 Revision*. New York: UNDESA-PD. http://esa.un.org/wpp/
[a]Projections

timescale engendered in climate change, the picture in all countries shows from 25 to 46 % of populations likely to be in the 60+ age group.

Within these aggregate data for countries, we must remember the considerable scope for subnational variation in age profiles. Many rural areas, for example, may have concentrations of older persons, whereas some cities, the magnets for migration for those of working age, may have fewer. The rural areas are very important, as they may well be the most exposed to extreme weather events by nature of agricultural and water-based occupations and the difficulty of providing services and infrastructure across wide areas. In some large countries (in terms of geography and population), internal variations in the percentages of older persons can be very large, rendering national average figures misleading. For example, in Indonesia, of 30 provinces, the 'top 10' in terms of ageing had 6–12.9 % of population aged 60+, whilst the 'lowest 10' provinces had only 2–4.8 % of population aged 60+. This considerable inter-province range reflected differences in fertility, migration and mortality. There were also interprovincial variations in projections of older population increase by 2025, though virtually all provinces are likely to expect a 5–10 percentage point increase in this period (Abikusno 2007). There are similar interstate variations in the percentages of older persons in Malaysia, ranging from 3.1 % in F.T. Labuan and 3.8 % in Sabah to 9.2 % and 9.3 % in Perlis and Perak states aged 60+. In Malaysia, there are also considerable variations in the ethnic composition of older groups, Malay, Chinese and Indian, in different states (Ong et al. 2009). These data alert us to consider variations in potential needs for climate change-related services and possible cultural social support practices within nations.

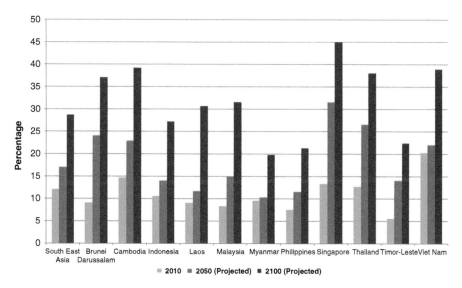

Fig. 3.1 The 'oldest old': percentage of older population (persons aged 60+) aged 80+, Southeast Asia, 2010–2100 (Source: United Nations Department of Economic and Social Affairs, Population Division (UNDESA-PD). 2013), *World Population Prospects: The 2012 Revision*. New York: UNDESA-PD. http://esa.un.org/wpp/)

Two particularly important features of the demographic ageing of the populations in health and health-care service terms are the growth of the 'oldest-old' cohorts, and the continued dominance of females in these cohorts referred to earlier, even if the future gender imbalance slightly reduces. Figure 3.1 shows the potentially huge increase in the proportions of various nations' populations aged 60+ who will be in the 80+ age cohorts (often called the 'oldest old') by mid-century and especially by 2100. The figures commonly of around 10 % in the oldest-old cohorts in 2012 may well reach 20–30 % by 2050 in several countries. This has great significance given the potential vulnerabilities of such very old people, and the steps that need to be taken to provide services and support for their needs, in both 'routine' warm-climate environments and especially when affected by extreme climate-linked events. Whilst many in the very oldest cohorts may indeed be relatively healthy, members of these cohorts are statistically most likely to be affected by disabilities, psychological and physical, and many may suffer from dementias. The existence of co-morbidities in this age group (the existence of several causes of illness at the same time) is widely recognised. Co-morbidities often mean that the persons in question need more health and welfare services; they may be disadvantaged in what they can do and co-morbidities may render such persons less well able to cope with, say, extreme heat and less able to recognise and avoid dangers from, say, extreme weather events and flooding.

The other notable feature is the dominance of females in old age cohorts. Figure 3.2 shows regional and country-specific information on sex imbalances in the older populations. Whilst the excess of older females over males is projected to reduce

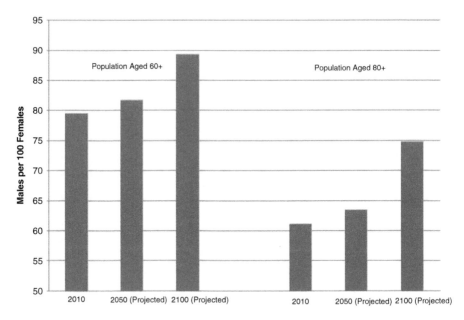

Fig. 3.2 Sex ratio (males per 100 females), population ages 60+ and 80+, Southeast Asia Region, 2010–2100 (Source: United Nations Department of Economic and Social Affairs, Population Division (UNDESA-PD). 2013), *World Population Prospects: The 2012 Revision*. New York: UNDESA-PD. http://esa.un.org/wpp/)

somewhat in the future (mainly due to improving male life expectancy), it seems that there will always be a considerable excess of older females, especially amongst the oldest old. The actual extent of this will of course vary from country to country in reflection of their specific demographic histories. In Myanmar and the Philippines, for instance, the UN (2012) projections point to their perhaps only being around 60 or so men aged 80 and over per 100 similarly aged women at the end of the century. In other countries the imbalance will be less marked, but substantial. Many of the oldest-old women will also be single, mainly being widowed, and may be with or without surviving children, though some may never have married and often therefore be in a particularly vulnerable or dependent economic situation. Older females in the region are also far more likely to be illiterate than older males (Mujahid 2006), with the attendant economic and social risks this can entail.

Looking at World Population Prospects projections (UNDESA-PD 2013), life expectancy increases across many countries in Southeast Asia over the next two to three decades appear likely to mirror those in East Asia, where many people now reach older age. Perhaps most significant is life expectancy at age 60, as this suggests the length of time people will be in potentially weaker physical, psychological and economic situations and in varying degrees of dependency.

3.6 Current Health Status of Older Populations in the Region

Health status is always a fairly difficult item for which to have comprehensive and reliable data, especially internationally. Nevertheless, based on the indicator of life expectancy at age 60 (Fig. 3.3), substantial variations are evident in Southeast Asia. Currently, these range from about 15 to 16 years on average in several countries such as Cambodia, Laos, Myanmar and Timor-Leste to around 20 years or more in Brunei, Thailand, Vietnam and Singapore. As ever, females have a greater life expectancy than males. These years of life expectancy have considerable implications for the numbers of people who will have to handle rising temperatures with climate change and the need to provide adequate living environments for them.

Looking at people's health status, in terms of morbidity (illness), we also see variations (World Health Organization 2004). Within individual countries, Indonesian data showed about 30 % of people aged 60+ had reported an illness (ranging from minor to more serious) in the previous month. The incidence was very slightly higher in rural than in urban areas and again slightly higher amongst women than men (Abikusno 2007). Comparing Thailand and Myanmar, Knodel (2013) noted that 47 % of older Thai respondents compared with only 33 % of older Burmese respondents reported their health as good or very good. Using 2012 Myanmar survey data, Knodel's analysis found, perhaps unsurprisingly, that persons in their 60s reported better health and memory than those aged 70+. Again, such data indicate the importance of the demographic ageing of the older groups, as future impacts of climate change are very likely to be felt more severely amongst older and oldest-old cohorts.

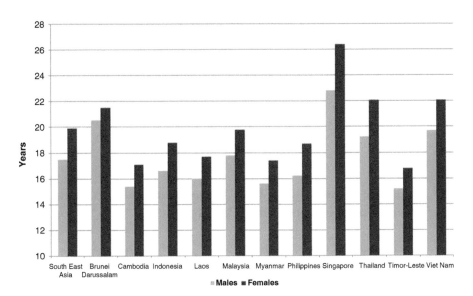

Fig. 3.3 Life expectancy (years) at age 60, Southeast Asia, 2011 (Sources: World Health Organization, Global Health Observatory (GHO), life tables. www.who.int/gho/mortality_burden_disease/life_tables/life_tables/en/)

Table 3.4 Marital, living and labour force situation of older population (persons aged 60+), Southeast Asia, early 2000s

Region/country	Persons aged 60+					
	% currently married		% living independently		% in labour force	
	Males	Females	Males	Females	Males	Females
Southeast Asia	82	44	22	21	56	33
Brunei Darussalam	86	52	–	–	28	7
Cambodia	87	50	14	12	76	55
Indonesia	84	38	24	24	63	34
Laos	83	46	–	–	54	32
Malaysia	86	50	22	20	51	14
Myanmar	73	39	12	9	53	36
Philippines	80	49	18	17	58	36
Singapore	84	51	–	–	43	20
Thailand	81	52	21	17	50	28
Timor-Leste	75	46	–	–	64	26
Vietnam	85	44	30	27	45	34

Source: United Nations Department of Social and Economic Affairs, Population Division (UNDESA-PD). (2012b), *Population Ageing and Development 2012 (Wallchart)*. New York: UNDESA-PD. http://www. un.org/esa/population/publications/2012PopAgeingDev_Chart/2012AgeingWallchart.html

Table 3.4 indicates the living and working arrangements of people over 60 in the region and also shows variability. It can be seen that substantial proportions are 'living independently', though with, for example, a range from 12 % of males in Myanmar to 30 % in Vietnam. The proportion living alone however is highly likely to rise with age, especially amongst females whose spouses die before them. Currently, married percentages amongst females ranged from 38 % in Indonesia to over half (52 %) in Thailand and Brunei Darussalam. This has considerable implications for the vulnerability and coping abilities of older single persons living without families. Indeed, the percentage of older people living alone is either stable or gradually increasing in many countries in the region. As we discuss below, in the case of heatwaves and extreme weather events such as typhoons and flooding, older and alone groups almost always come off worst.

3.7   Climate Change and Likely Impacts on Older Persons' Health in Southeast Asia: The Prospects?

3.7.1   Temperature Rises and Extreme Heat Events

Direct evidence for climate change effects in the region and especially their links with health are limited at the moment. However, based on extrapolation of experience in the wider Asia-Pacific and elsewhere, we can assume changes such as extreme heat will have adverse effects on older persons. The WHO (2013) (after Robine et al. 2003) note that extreme heat in Europe in August 2003 caused more than 70,000 excess deaths, mainly amongst older persons, with cardiovascular and respiratory disease being particular underlying causes. Vandentorren et al. (2006) looked at the deaths in France of people aged 65+ living alone during this extreme heatwave using case-control methods. They found lack of mobility and pre-existing medical conditions to be major risk factors, along with environmental characteristics such as lack of thermal insulation in housing, sleeping on the top floor and building temperature. These risk factors in the French case are confirmed by Poumadère et al. (2005), who also note that people aged 75+ made up almost 83 % of the excess mortality in this heatwave. The unfavourable living conditions and other environmental risks could be ameliorated, for example, by dressing lightly and using cooling devices. However, many older people lived alone and in poor economic conditions, so might have been unable to recognise the risks and, even if they did, unable to take evasive actions. Vandentorren et al. suggest information and advice on how to react during heatwaves would be important. Perhaps equally important, the French event did cause a shift in perception of risk to acknowledge dangerous climate and that such excess heat events pose an 'unambiguous danger' (Poumadère et al. 2005, p. 1492).

Learning from such experiences, there will be a need to focus on public attitudes or awareness concerning likely risks and dangers of climate change and related events and ways to prepare for or minimise the consequences. Therefore, authorities in Southeast Asia should recognise and provide understandable and useful information and education for older persons and their families (as well as for all citizens). In the context of extreme heat, for example, the rapid growth of urbanization and ever-larger megacities in this region will almost certainly be associated with increasing urban heat island effects (Campbell-Lendrum and Corvalan 2007) that are likely to be magnified by climate warming (particularly in heatwaves), from the use of air conditioners, power generation, vehicles etc., which can raise urban temperatures. Moreover, as noted, almost all countries and cities will have greater numbers of older persons and probably much higher elderly percentages, so the populations at risk will almost certainly be significantly greater in coming decades.

As we do not currently have much direct evidence on the possible effects of temperature rises on older persons in the region, we can look to research and experiences in other parts of the Asia-Pacific for possible guidance. In South Korea, for example, Son et al. (2012) have examined excess mortality for heatwaves in seven major cities between 2000 and 2007. Their study does not focus on older persons specifically but they do identify older groups as being at greater risk. Total mortality increased by just over 4 % on heatwave days compared with non-heatwaves. Timing was important, as mortality seemed to increase more when heatwaves occurred in early summer, or they lasted longer or were more intense. The capital city, Seoul, was particularly affected and experienced a larger rise in mortality than other cities. Similar to studies elsewhere, they estimated a larger impact of heatwaves on respiratory than cardiovascular mortality. Risks were higher for males than females and for older people rather than younger people.

Son et al. (2012) also found Seoul residents with lower education were more at risk of dying in heatwaves, and those outside hospitals were also at greater risk than those in hospitals, which could reflect both health status and access to insurance. They also examined the effects of air pollution levels which tend to be higher in very hot weather and could impact more on those without air-conditioning and/or air filtering systems. Indeed, urban air pollution is estimated by WHO (2013) to be responsible for 1.2 million deaths annually. WHO (2014) further estimated that seven million deaths globally in 2012 were from the joint effects of household and ambient air pollution, although there are considerable difficulties in estimating risks and exposures for these factors. They also estimated the proportions of deaths from diseases potentially attributable to these two types of pollution globally were ischaemic heart disease (36 %), stroke (33 %), COPD (17 %), acute lower respiratory disease (8 %) and lung cancer (6 %). So, the effects are important though varied and difficult to pin down, as many such conditions may also be influenced by factors such as occupational exposure and smoking. Climate change is nevertheless very likely to exacerbate especially ambient air pollution. The University of Hong Kong publishes an online environmental index that details sources of air pollution, districts exceeding WHO limits and the cumulative health impacts, doctor visits, hospital bed days and deaths and economic losses due to air pollution (School of Public Health, University of Hong Kong 2014). Such detailed data on location of pollution and its consequences could be extremely useful for cities in Southeast Asia that are increasingly suffering from this problem.

Air pollution is important as a risk in itself and also as it can especially affect older people with pre-existing circulatory and breathing problems, particularly COPD (chronic obstructive pulmonary disease). With extreme heat, the pollution could trigger initial attacks. More research is needed as to whether outdoor air pollution increases the incidence of problems such as COPD, but Chau et al. (2011) note that it certainly exacerbates existing cases, impacting on quality of life and increasing medical costs. WHO (2011) earlier estimated that about one million people globally die from COPD associated with indoor pollution, the majority of whom will be older persons, although the regional breakdown needs to be investigated.

COPD incidence increases with age, and Chau et al.'s study in Hong Kong, likely to be relevant for Southeast Asian populations, indicated annual incidence rates of over 27 per 1,000 amongst males aged 75+ compared with 13.6 per 1,000 for those aged 65–74. All in all, such data mean that older people are likely to be affected more than others in the region if air quality worsens, as seems to be occurring in many cities in Southeast Asia and the Asia-Pacific generally. The WHO (2014) found that the WHO Southeast Asian region was the second worst affected (after the Western Pacific Region) by the combined effects of household and ambient air pollution, with 2.3 million of the estimated 7 million deaths from these effects globally in 2012, as noted above. The majority of these deaths (91 %) occurred in the over-25 age group, and it must be assumed that a high proportion of these would be amongst older persons in the region.

There has been further interest in air pollution and links between COPD, age and socio-economic development in Hong Kong. For example, Chen et al. (2011) found COPD declined with improving economic development across the generations, though gains could be reduced by increases in air pollution or smoking. COPD increased when air pollution increased and declined when air quality improved. Like others, they also found mortality from COPD increased with age and especially amongst postmenopausal women. The association between mortality and various types of air pollution is further confirmed by Wong et al. (2008). They found health outcomes (nonaccidental, cardiovascular and respiratory mortality) were more strongly associated with SO_2 and NO_2 amongst people living in areas where a social deprivation index (SDI) was high when compared with people in areas of middle or low SDI. The authors point out that, given the relationship between socio-economic status, air pollution and mortality, policymakers need to consider the impacts of socio-economic disadvantage and industrial, economic development and environmental management. This can include climate change insofar as local policies can assist in amelioration of heat increases and other aspects of change. Wong et al. (2015) note that, in Asia, air quality is poor and often deteriorating, and most recently, their methodologically novel research followed a cohort of over 66,000 people aged 65+ for over a decade to 2011. Participants were contacted via all 18 public Elderly Health Centres in Hong Kong. Their home addresses were geographical coded so satellite remote sensing data on particulate matter (PM) could be used to examine their long-term exposures to and effects on health of aerodynamic diameter PM2.5 which could be an important cause of morbidity and mortality. Respondents' social and lifestyle characteristics and morbidity were gained from interviews; mortality was examined through death registry records. The study found that 'exposure to PM2.5 was significantly associated with mortality from natural and cardiovascular causes, and with mortality due to IHD and cerebrovascular disease specifically, in people aged ≥65 years' (Wong et al. 2015, p. 11). Clearly, these Hong Kong studies underscore a worrying issue given the ageing populations in most of this region and the potential for climate change to be associated in some circumstances with deteriorations in air quality.

In addition, over recent decades, extensive areas of Southeast Asia, both rural and urban, have frequently been blanketed by air pollution from smoke from forest and peat fires (generically locally termed 'haze'). 'Haze' has affected vast tracts of Indonesia (especially Sumatra), Peninsular and East Malaysia, Singapore, Thailand and other parts of the region. Haze almost certainly has health impacts. In a study in Klang Valley, west coast of peninsular Malaysia, haze was found to increase the risk of respiratory mortality, probably from exposure to toxic particles (Asia Research News 2015). Daily mortality rates became higher than normal by the sixth day of haze. This ongoing research emphasises the need to give appropriate public health advice and assistance to vulnerable groups including older persons and those with existing respiratory and cardiac disease. Serious haze originating from Indonesia from July to late October 2015 affected southern Thailand, Malaysia, Singapore, Brunei, Vietnam and Indonesia. Singapore's National Environment Agency (2015) issued a September health advisory that "The elderly, pregnant women and children should minimise outdoor activity, while those with chronic lung or heart disease should avoid outdoor activity". Things may be getting worse, linked to climatic conditions. For example, fires and haze in Sumatra in March 2014, at an unusual time and in spite of government attempts to regulate burning, have been linked by World Resources Institute (2014) to extreme drought, making burning easier and fires very difficult to control. If temperatures become more extreme and cultivation practices alter, it can be anticipated that air pollution from such events will also increase and impact on everyone's health and particularly that of older persons.

In the Korean heatwaves, Son et al. note they were working with relatively small numbers (of both heatwaves and deaths), but their findings do seem to have implications in terms of public health policy and targeting of health promotion at some more vulnerable groups and devising of health alerts for extreme heat events. Their evidence on impacts of heatwaves from an Asian country is important, as are findings from places such as Hong Kong, which has both hot weather and cold weather warning systems in place aimed especially at the older and vulnerable populations.

Under the influence of climate change, we may well expect extreme weather events to increase, and in most of Southeast Asia, the effects are likely to be towards increasing heatwave frequency and intensity. Higher temperatures are also associated with sudden-onset weather emergency events such as typhoons and associated heavy rainfall, flooding and damage to property and life. In this respect, the Philippines, Vietnam and its neighbouring countries, plus Southern China (including Hong Kong and Macau) and Taiwan through their location, are currently at the greatest risk. In November 2013, Typhoon Haiyan (Yolanda), one of the deadliest to date, struck the Philippines late in the year causing an estimated more than 6,000 deaths and almost 30,000 injuries, affecting millions. Its effects displaced four million from their homes, devastating large areas of Luzon and almost destroying the city of Tacloban. Whilst all ages were affected, HelpAge International (2013) reported that people aged 60+ made up two-fifths of deaths from Haiyan even though they only comprise 8 % of the population. Many apparently decided not to evacuate as they felt their homes could withstand the storm as in the past. Older people without families, and living alone, were at most risk, often being both poor and with no one to help them, whereas those with families were less vulnerable. As is common in such typhoons, the storm also went on to cause flooding and deaths in Southern China.

Typhoons and heavy rain are often associated with flooding. Such high-intensity events are expected to increase with temperature increase and also general sea-level rises (see Table 3.2). Looking at Southeast Asian major cities' population exposed to coastal flooding alone, currently a total of almost five million people are currently exposed to coastal flooding in the top five cities ranked according to this risk (Table 3.5). Estimates for the year 2000 of the total urban dwellers in each country of the region at risk of coastal and inland flooding are in turn shown in Table 3.6. Clearly many people already face serious physical and psychological risks from flooding episodes. Future urbanization and population growth, especially when settlements occur in particularly risk-prone sites, are only likely to exacerbate the problems.

Table 3.5 Cities in Southeast Asia ranked in terms of population exposed[a] to coastal flooding in the 2070s

Country	Urban agglomeration	Future exposed population 2005	Current exposed population 2070[b]
Vietnam	Ho Chi Minh City	1,931,000	9,216,000
Thailand	Bangkok	907,000	5,138,000
Myanmar	Rangoon	510,000	4,965,000
Vietnam	Hai Phong	794,000	4,711,000
Indonesia	Jakarta	513,000	2,248,000

Source: Nicholls, R.J., Hanson, S., Herweijer, C., Patmore, N., Hallegatte, S., Corfee-Morlot, J., Chateau, J. and Muir-Wood, R. (2007). *Ranking of the World's Cities Most Exposed to Coastal Flooding Today and in the Future: Executive Summary*, Paris: OECD. http://www.oecd.org/environment/cc/39721444.pdf
[a]Exposure refers to a 1 in 100-year surge-induced flood event (assuming no flood defences)
[b]Projections

Table 3.6 Urban population at risk of coastal and inland flooding, Southeast Asia, 2000

Region/country	Coastal flood risk[a] Urban population at flood risk	% Urban population at flood risk	Inland flood risk Urban population at flood risk	% Urban population at flood
Southeast Asia	63,919,387	36.1	25,484,820	14.7
Brunei Darussalam	24,965	11.2	1634	0.7
Cambodia	281,944	15.0	1,428,121	76.0
Indonesia	22,720,666	27.9	4,394,972	5.4
Laos	0	0	302,825	34.0
Malaysia	3,687,052	26.5	495,254	3.6
Myanmar	4,512,823	36.2	2,361,353	19.0
Philippines	6,807,578	27.4	3,713,398	14.9
Singapore	550,057	14.0	0	0
Thailand	12,471,874	60.0	6,070,291	29.2
Timor-Leste	1369	4.2	869	2.7
Vietnam	12,862,429	73.9	6,716,973	38.6

Source: Asian Development Bank (ADB). (2012). *Key Indicators for Asia and the Pacific 2012*, Metro Manila: Asian Development Bank. http://www.adb.org/publications/key-indicators-asia-and-pacific-2012
[a]Coastal flooding risk is based on population within low-elevation coastal zone (<10 m rise elevation)

3.7.2 Climate Change and Infectious Diseases in Southeast Asia

Infectious diseases, especially vector and pathogen effects, are likely to be affected by temperature rises, so we anticipate climate change in terms of warming will increase the potential spread for some vectored infectious diseases such as malaria and dengue. In particular, WHO and others have noted that global warming is likely to enlarge areas such as the global malarial mosquito transmission zones north and southwards. This is unlikely to affect Southeast Asia to any great extent, as at lower altitudes, the subregion's territories are already almost wholly within potential malarial zones. It is well recognised that malaria does not spread even where conditions are suitable in, for example, the dry season, at high altitudes or where antimalarial programmes have been successful (CDC 2014). However, increased precipitation could mean malaria control programmes become less effective. With global warming, temperature increases of a few degrees could have the potential to raise the altitude at which mosquitoes could breed and affect humans. Recent evidence from Africa and South America suggests that the altitude at which malaria can occur increases in warmer years and that a temperature rise of 1–3° could, if sustained, result in many more cases (Siraj et al. 2014). This could adversely impact older persons' health insofar that they may live in greater numbers in such affected rural areas. Within Southeast Asia, parts of Indonesia and Thailand and elsewhere in Indo-China where malarial transmission does not currently often occur or has been halted could become affected or reaffected again. Moreover, for example, some of the most northerly parts of the Philippines could be brought more definitively into being year-round malarial zones.

Another recognised risk from temperature rises is the potential for increasing the incidence of food poisoning, which again could affect people who do not have the resources to refrigerate their food or cook it thoroughly. Poorer older persons are thereby a potential group at greater risk from food poisoning. However, more general and worrisome in this regard in the region could be threats to food security. Many countries of the region, including Thailand, Vietnam, the Philippines and Indonesia, are major growers of staple foods such as rice. There is evidence that rice and other food crop production has already been affected by heavy rain, typhoons and flooding particularly in several areas of the Philippines and Vietnam and Myanmar. If extreme weather events such as temperature and precipitation changes, prolonged flooding or droughts appear, for longer and/or more frequently, this could impact food supply and increase the price of food. This again is likely to have the major impacts on the poorest groups in all these countries, who often include older persons who do not have the economic resources to buy more expensive food. Their short- and long-term nutrition and health could therefore be affected. Similarly, climate change can have ozone effects on plants and productivity. Climate change is also likely to affect food productivity of land (e.g. through flooding or seawater salinisation of former arable areas), and ocean catches may also be impaired. All these may well decrease food availability and increase costs, impacting on poorer older people, for whom any extra costs could be impossible to meet. Moreover, agricultural jobs, which have often employed older people, may be reducing (Mujahid 2006), and therefore also their sources of income are concomitantly falling.

3.7.3 Mental Health Effects

As mentioned earlier, evidence from other areas suggests a probable impact on mental health of climate change (Doherty and Clayton 2011). Swim et al. (2011) introduce a conceptual model of how climate systems and human systems can relate to cognition, affect and motivation and require mitigation or adaptation. One can speculate that older persons will be relatively unable to contribute to mitigation and possibly be the least well able to adapt to the human consequences of climate change, for physiological, psychological or socio-economic reasons. So, impacts on mental health could potentially operate through a number of ways, such as increasing stress and leading to outcomes such as depression. Extreme weather events will have an almost inevitable impact on health-care services and may reduce the availability of and attention to mental health services which are already 'Cinderella services', poorly provided in many parts of Southeast Asia. Older people tend to be more vulnerable than younger groups to stress, unipolar depression and paranoia. These can be related sometimes to their precarious social and economic situations, and sometimes to marginalisation socially and within the family, as well as from physical illness and bereavement. It is acknowledged that attempts at this time to link increases in mental health problems directly to climate change in this subregion would be rather speculative. Nevertheless, it is probably safe to predict that the growing numbers of older persons will increase both incidence of new cases and the prevalence of mental health problems in the community, which are likely to be exacerbated by issues related to climate change.

In terms of psychological and mental health interactions with climate change, as mentioned earlier, a particular challenge is likely to come from the growing numbers of people with dementias. Whilst dementia is essentially a label for a very mixed syndrome (group of symptoms) with a variety of causes, its general course involving memory loss, confused thinking, personality change and decreasing ability to self care is well known (see, e.g. Alzheimer's Disease International 2014; McCracken and Phillips 2012). Persons with dementia are often unable to take decisions about (say) keeping cool in hot weather, avoiding dangerous situations and taking medications. Therefore, they need huge amounts of personal care and support, and often specialised accommodation, all of which can be impacted strongly by extreme weather events and even gradual environmental warming.

Almost a decade ago, Access Economics (2006) was predicting that enormous increases in numbers will be affected by dementia in the Asia-Pacific, reaching almost 65 million by 2050. This phenomenon is likely to be especially important in Southeast and East Asia, given the rapid ageing of many countries in the region noted earlier and the rapidly changing family, socio-economic and occupational circumstances occurring in almost all member countries. If the changing nature of families and societies potential to care is considered, then climate change is one major factor that could adversely affect the welfare of people with dementias and the ability of their carers to look after them.

3.8 The Future: Possible Changing Vulnerabilities of Older People to Climate Change?

In concluding this chapter, we provide some summary predictions extrapolated from research elsewhere in terms of climate events, probable climate change and their impacts on the health and welfare of older persons. We have grouped these into broadly potentially 'negative' and potentially 'positive' forces, although of course the exact nature and impacts of many such forces are difficult to predict accurately. Whilst there is a tendency to be gloomy and focus mainly on the negative, at both individual and aggregate levels, some forces may even help to improve coping capacities.

(a) *Negative forces?*

- Gradual ageing of the aged populations will see increased numbers at the more advanced end of physiological and psychological deterioration. There will also very likely be increasing multi-morbidity prevalence associated with this population ageing, even if there may be certain age-specific health improvements.
- In particular, with generally growing proportions of 'oldest-old' cohorts in the elderly populations, dementia numbers will almost certainly increase. This will likely have severe implications for many countries in terms of size of population at risk and their coping capacities. Persons with dementias and those with several concurrent health conditions are much more likely to be adversely affected by trends such as temperature warming and less well suited to cope with extreme weather events (typhoons, flooding etc.).
- As in many populations globally and also in Southeast Asia, increasing overweight and obesity in the older cohorts will lead to increasing non-communicable disease impacts. Evidence elsewhere suggests that NCDs such as diabetes and cardiovascular and mobility problems associated with overweight may appear at ever-younger ages in future cohorts. These sorts of chronic health problems will almost certainly impact older population's ability to handle warmer temperatures and extreme events. For example, people with leaner body mass generally handle heat better than obese ones, who tend to retain more body heat. They may also be more mobile and better able to handle challenging environmental living conditions than overweight or obese persons.

- Increasing urbanization in many countries of the region (UNDESA-PD 2012a) will expose more older people to urban heat island conditions (Campbell-Lendrum and Corvalan 2007).
- Unless concerted efforts are undertaken by governments to reduce it, urban air pollution, localised air pollution and trans-boundary air pollution will increase in incidence, duration and severity in coming years. This will have the harmful effects on health of all age groups and especially older persons in the region. It is likely to affect those with conditions such as COPD and may well trigger cardiovascular events amongst those at risk.

(b) *Positive forces?*

- Hopefully, declining poverty and better living standards amongst older cohorts and their families will lead to improved coping capacity, for example, better accommodation standards, ventilated living, access to air conditioning, better environmental conditions and drainage, etc.
- Widening and improved levels of social protection may develop which will add to the above increases in living standards.
- Future elderly cohorts in Southeast Asia will have better educational levels than many current older groups, especially the older females who make up the majority especially amongst the oldest old. Improved general and female education will have major impacts on knowledge of health and well-being in future decades, although this will take some whilst to filter through to the 60+ cohorts.
- Improved planning standards and investment in environmental conditions to reduce warming and heat island effects (sources of electricity, control of traffic, etc.), better drainage and other infrastructural improvements may enhance regional countries' capacity to handle climate change and extreme weather events. These should ultimately benefit all social and demographic groups.
- Improved use of and affordability of technology such as air conditioning will make living with temperature rises more bearable for those with access to such amenities.
- Finally, we predict that health planning will be extended and improved to include education about, and risk avoidance connected to, climate change.

The ultimate solution, of course, will be stopping warming and hopefully to halt and even reverse climate change; and some argue that climate change might already be slowing. However, the growing populations in the Southeast Asia subregion countries and cities, almost all of whom have or desire greater access to 'modern' facilities, vehicles and utilities, are more likely to push climate change along ever more rapidly. Here, we can look at current evidence from countries in the Asia-Pacific such as China, and some in Southeast Asia such as Thailand, Vietnam and Indonesia, which have already had substantial economic development that has involved enormous environmental impacts (almost all being adverse at the macro-environmental level). These impacts have included almost unrestrained and poorly planned urbanization, with accompanying pollution, air quality deterioration and flooding risk. Therefore, in most of Southeast Asia, taking practical steps to address climate change is a long way down the path. For today's elderly cohorts, and all future older persons alive today, a warming world appears the most likely prospect, with considerable implications for their health and welfare.

References

Abikusno N (2007) Older population in Indonesia: trends, issues and policy responses, Papers in population ageing no.3. UNFPA, Bangkok

Access Economics (2006) Dementia in the Asia Pacific Region: the epidemic is here. http://www.fightdementia.org.au/common/files/NAT/20060921_Nat_AE_FullDemAsiaPacReg.pdf

Alzheimer's Disease International (2014) About dementia. Alzheimer's Disease International, London. http://www.alz.co.uk/about-dementia

Asia Research News (2015) South-east Asian haze increases risk of respiratory mortality. Asian Res News, 19 March, p 12. http://www.researchsea.com/html/article.php/aid/8655/cid/3/research/medicine/researchsea/south-east_asian_haze_increases_risk_of_respiratory_mortality_.html

Asian Development Bank (ADB) (2012) Addressing climate change and migration in Asia and the Pacific: final report. ADB, Mandaluyong City

Campbell-Lendrum D, Corvalan C (2007) Climate change and developing-country cities: implications for environmental health and equity. J Urban Health 84(1):i109–i117

Carnes BA, Staats D, Willcox BJ (2013) Impact of climate change on elder health. J Gerontol A Biol Sci Med Sci. doi:10.1093/gerona/glt159

Centers for Disease Control and Prevention (CDCs) (2014) Where malaria occurs. CDCs, Atlanta. http://www.cdc.gov/malaria/about/distribution.html. Accessed 6 Mar 2014

Chau PH, Chen J, Woo J, Cheung WL, Chan KC, Cheung SH, Lee CH, McGhee SM (2011) Trends of disease burden consequent to chronic lung disease in older persons in Hong Kong: implications of population ageing. The Hong Kong Jockey Club, Hong Kong

Chen J, Schooling CM, Johnston JM, Hedley AJ, McGhee SM (2011) How does socioeconomic development affect COPD mortality? An age-period-cohort analysis from a recently transitioned population in China. PLoS One 6(9):e24348. doi:10.1371/journal.pone.0024348

Confalonieri U, Menne B, Akhtar R, Ebi KL, Hauengue M, Kovats RS, Revich B, Woodward A (2007) Human health. In: Parry ML, Canziani OF, Palutikof JP, van der Linden PJ, Hanson CE (eds) Climate change 2007: impacts, adaptation and vulnerability. Contribution of Working Group II to the fourth assessment report of the Intergovernmental Panel on Climate Change. Cambridge University Press, Cambridge, U.K./New York pp 391–431

Doherty TJ, Clayton S (2011) The psychological impacts of global climate change. Am Psychol 66(4):265–276

Filberto D, Wethington E, Pillemer K, Wells NM, Wysocki M, Parise JT (2009–2010). Older people and climate change: vulnerability and health effects. Generations 33(4):19–25

Gamble JL, Hurley BJ, Schultz PA, Jaglom WS, Krishnan N, Harris M (2013) Climate change and older Americans: state of the science. Environ Health Perspect 121(1):15–22

Haq G, Whitelegg J, Kohler M (2008) Growing old in a changing climate. Stockholm Environment Institute, Stockholm

HelpAge International (2013) Older persons disproportionately affected by Typhoon Haiyan. HAI Newsroom, 12 December. http://www.helpage.org/newsroom/latest-news/older-people-disproportionately-affected-by-typhoon-haiyan/

Hijioka Y, Lin E, Pereira JJ, Corlett RT, Cui X, Insarov GE, Lasco RD, Lindgren E, Surjan A (2014) Asia. In: Barros VR, Field CB, Dokken DJ, Mastrandrea MD, Mach KJ, Bilir TE, Chatterjee M, Ebi KL, Estrada YO, Genova RC, Girma B, Kissel ES, Levy AN, MacCracken

S, Mastrandrea PR, White LL (eds) Climate change 2014: impacts, adaptation, and vulnerability. Part B: regional aspects. Contribution of Working Group II to the fifth assessment report of the Intergovernmental Panel on Climate Change. Cambridge University Press, Cambridge/New York, pp 1327–1370

IPCC (2013) Summary for policymakers. In: Stocker TF, Qin D, Plattner GK, Tignor M, Allen SK, Boschung J, Nauels A, Xia Y, Bex V, Midgley PM (eds) Climate change 2013: the physical science basis. Contribution of Working Group 1 to the fifth assessment report of the International Panel on Climate Change. Cambridge University Press, Cambridge,U.K./New York. http://www.ipcc.ch/report/ar5/wg1/

IPCC (2014) In: Field CB, Barros VR, Dokken DJ, Mach KJ, Mastrandrea MD, Bilir TE, Chatterjee M, Ebi KL, Estrada YO, Genova RC, Girma B, Kissel ES, Levy AN, MacCracken S, Mastrandrea PR, White LL (eds) Climate change 2014: impacts, adaptation, and vulnerability. Part A: global and sectoral aspects. Contribution of Working Group II to the fifth assessment report of the Intergovernmental Panel on Climate Change. Cambridge University Press, Cambridge, U.K./New York. https://ipcc-wg2.gov/AR5/images/uploads/WGIIAR5-PartA_FINAL.pdf

Knodel J (2013) The situation of older persons in Myanmar. Results from the 2012 survey of older persons. HelpAge International, Chiang Mai/Yangon

McCracken K, Phillips DR (2012) Global health: an introduction to current and future trends. Routledge, London/New York

McMichael A, Githeko A, Akhtar R, Carcavallo R, Gubler D, Haines A, Kovats RS, Martens P, Patz J, Sasaki A (2001) Human health. In: McCarthy JJ, Canziani OF, Leary NA, Dokken DJ, White KS (eds) Climate change 2001: impacts, adaptation and vulnerability. Contribution of Working Group II to the third assessment report of the Intergovernmental Panel on Climate Change. Cambridge University Press, Cambridge, U.K./New York pp 453–485

Mujahid G (2006) Population ageing in East and South-East Asia: current situation and emerging challenges, Papers in population ageing no.1. UNFPA, Bangkok

Ong F-S, Phillips DR, Hamid TA (2009) Ageing in Malaysia: progress and prospects. In: Fu TH, Hughes R (eds) Ageing in East Asia: challenges and policies for the twenty-first century. Routledge, London, pp 138–160

Page LA, Howard LM (2010) The impact of climate change on mental health (but will mental health be discussed at Copenhagen?). Psychol Med 40(2):177–180

Phillips DR (2000) Ageing in the Asia-Pacific region. Routledge, London

Phillips DR, Chan ACM, Cheng ST (2010) Ageing in a global context: the Asia-Pacific region. In: Phillipson C, Dannefer D (eds) The Sage handbook of gerontology. Sage, London/New York, pp 430–446

Phillips DR (2011) Overview of health and ageing issues in the Asia-Pacific region. In: Chan W-C (ed) Singapore's ageing population: managing healthcare and end of life decisions. Routledge, London, pp 13–39

Poumadère M, Mays C, Sophie Le Mer S, Blong R (2005) The 2003 heatwave in France: dangerous climate change here and now. Risk Anal 25(6):1483–1494

Robine JM, Cheung SLK, Le Roy S, Van Oyen H, Griffiths C, Michel J-P, Herrmann FR (2003) Death toll exceeded 70,000 in Europe during the summer of 2003. Les Comptes Rendus/Série Biologies 331(2):171–78

School of Public Health, University of Hong Kong (2014) Hedley environmental index. http://hedleyindex.sph.hku.hk/html/en/

Singapore National Environment Agency (2015) Haze Situation Update (September 14, 2015).

Siraj AS, Santos-Vega M, Bouma MJ, Yadeta D, Ruiz Carrascal D, Pascual M (2014) Altitudinal changes in malaria incidence in highlands of Ethiopia and Colombia. Science 343(6175):1154–1158

Son J-Y, Lee J-T, Anderson GB, Bell ML (2012) The impact of heat waves on mortality in seven major cities in Korea. Environ Health Perspect 120(4):566–571

Swim JK, Stern PC, Doherty TJ et al (2011) Psychology's contributions to understanding and addressing global climate change. Am Psychol 66(4):241–250

United Nations, Department of Economic and Social Affairs, Population Division (UNDESA-PD) (2012a) World urbanisation prospects: the 2011 revision. United Nations, New York. http://esa.un.org/unpd/wup/index.htm

United Nations, Department of Economic and Social Affairs, Population Division (UNDESA-PD) (2012b) Population ageing and development 2012 wallchart. United Nations, New York. http://www.un.org/esa/population/publications/2012PopAgeingDev_Chart/2012PopAgeingandDev_WallChart.pdf

United Nations, Department of Economic and Social Affairs, Population Division (UNDESA-PD) (2013) World population prospects: the 2012 revision. United Nations, New York. http://esa.un.org/unpd/wpp/index.htm

United Nations Development Programme (2012) Asia-Pacific human development report. One planet to share: sustaining human progress in a changing climate. Routledge, London. http://asiapacific-hdr.aprc.undp.org/climate-change

Vandentorren S, Bretin P, Zeghnoun A, Mandereau-Bruno L, Croisier A, Cochet C, Ribéron J, Siberan I, Declercq B, Ledrans M (2006) August 2003 heat wave in France: risk factors for death of elderly people living at home. Eur J Public Health 16(6):583–91

Wong CM, Ou CQ, Chan KP, Chau YK, Thach TQ, Yang L, Chung RYN, Thomas GN, Peiris JSM, Wong TW, Hedley AJ, Lam TH (2008) The effects of air pollution on mortality in socially deprived urban areas in Hong Kong, China. Environ Health Perspect 116(9):1189–1194. doi:10.1289/ehp.10850

Wong CM, Lai HK, Tsang H, Thach TQ, Thomas GN, Lam KBH, Chan KP, Yang L, Lau AKH, Ayres JG, Lee SY, Chan WM, Hedley AJ, Lam TH (2015) Satellite-based estimates of long-term exposure to fine particles and association with mortality in elderly Hong Kong residents. Environ Health Perspect, advance pub. doi:10.1289/ehp1408264

World Bank (2013) Turn down the heat: climate extremes, regional impacts, and the case for resilience. A report for the World Bank by the Potsdam Institute for Climate Impact Research and Climate Analytics.World Bank, Washington, DC. http://www.worldbank.org/content/dam/Worldbank/document/Full_Report_Vol_2_Turn_Down_The_Heat_%20Climate_Extremes_Regional_Impacts_Case_for_Resilience_Print%20version_FINAL.pdf

World Health Organization (WHO) (2004) Health of the elderly in South-East Asia: a profile. WHO, New Delhi

World Health Organization (WHO) (2008) Older people in emergencies: an active ageing perspective. WHO, Geneva

World Health Organization (WHO) (2011) Indoor air pollution. Fact sheet. WHO, Manila. http://www.wpro.who.int/mediacentre/factsheets/fs_201109_air_pollution/en/

World Health Organization (WHO) (2013) Climate change and health. Fact sheet no. 266. WHO, Geneva. http://www.who.int/mediacentre/factsheets/fs266/en/

World Health Organization (WHO) (2014) Burden of disease from the joint effects of household and ambient air pollution for 2012. Summary. WHO, Geneva. http://www.who.int/phe/health_topics/outdoorair/databases/AP_jointeffect_BoD_results_March2014.pdf?ua=1

World Resources Institute (2014) Fires in Indonesia spike to highest levels since June 2013 haze emergency. World Resources Institute, Washington, DC. http://www.wri.org/blog/fires-indonesia-spike-highest-levels-june-2013-haze-emergency

Yusuf AA, Francisco HA (2009) Climate change vulnerability mapping for Southeast Asia. International Development Research Centre (IDRC); Swedish International Development Cooperation Agency (Sida); Canadian International Development Agency (CIDA), Singapore. http://www.preventionweb.net/files/7865_12324196651MappingReport1.pdf

Part I
Regional Studies

Chapter 4
Extreme Temperature Regions in India

S.K. Dash

Abstract Identifying the regions of extreme temperature events involves an in-depth analysis of daily maximum and minimum temperatures all over India spanning a number of years. Such study over India is essential not only because of scientific interest but also for getting useful information for policy formulation. India is a land of variety with large spatial and temporal variations in topography, land conditions, surface air temperature, and rainfall. For scientific analysis, the landmass of the country is usually divided into seven temperature homogeneous zones. Therefore, the degree of temperature extremes may not be the same in each homogeneous zone. In this study, temperature extremes of different categories are examined in detail. Here we study the extreme temperature events in India based on $1° \times 1°$ resolution gridded datasets of the India Meteorological Department over the period 1969–2005. This study analyzes trends in daily maximum (TX) and daily minimum (TN) temperatures based on a set of six temperature indices for warm days and nights (TX90p, TX95p, TX99p, TX01p, TX05p, TX10p) and another set of six indices for cold days and nights (TN90p, TN95p, TN99p, TN01p, TN05p, TN10p). In total, 357 grids with $1° \times 1°$ resolution each have been selected, and the above 12 temperature indices have been mapped over the whole of India. The entire country is found to have scattered grids with signatures of warming in terms of increase in the occurrence of warm days and warm nights and decrease in cold nights and cold days. In addition, there are two clusters of grids covering large areas in the west and peninsular India which may be considered as two hotspots. There is also a similar small patch in the north India.

Keywords Temperature extremes • Warm and cold days • Warm and cold nights • Gridded data • Hotspots

S.K. Dash (✉)
Centre for Atmospheric Sciences, Indian Institute of Technology Delhi,
Hauz Khas, New Delhi 110016, India
e-mail: skdash@cas.iitd.ernet.in

© Springer International Publishing Switzerland 2016   55
R. Akhtar (ed.), *Climate Change and Human Health Scenario in South
and Southeast Asia*, Advances in Asian Human-Environmental Research,
DOI 10.1007/978-3-319-23684-1_4

4.1   Introduction

There is a growing belief that global warming is happening. Several sectors of the society are being affected by such warming of the atmosphere. Human health is one of the important sectors which show signs of being affected by the changes in climatic parameters. Here health includes physical, social, and psychological well-being. Extreme cases of temperature, precipitation, wind, and moisture which may occur as a result of changing climate patterns may affect health both directly and indirectly. Of all the climatic parameters projected by the climate models, rising temperature is the most robust. Since warming atmosphere can hold more moisture, the chances of having extreme rainfall events are also more. It is observed that heat and cold waves are killing large number of people and severely affecting the lives of several others. The 2003 European heat wave killed about 21,000–35,000 people, and in the same year, about 3000 people were dead due to heat wave conditions in Andhra Pradesh (Bhadram et al. 2005). The super-cyclone of Odisha in 1999 caused death of about 10,000 and affected the lives of more than 10–15 million people. There are many examples of such cases of extreme weather events which directly cause death to living beings and bring about large-scale destruction. More intense and frequent extreme events such as extreme temperature and precipitation can directly give rise to several weather-related diseases. For example, it is noted that excess temperature brings in cases of asthma and cardiovascular diseases. If the nighttime temperature does not decrease in addition to increase in the daytime temperatures, the situation becomes extremely unbearable and can become a killer event. Extreme temperature and humidity conditions may make breathing more difficult. Warm air, sunshine, and pollutants especially from power plants and cars combine to produce ground-level ozone and smog. Changes in water, air, and food (both quality and quantity), ecosystems, agriculture, livelihoods, and infrastructure indirectly affect the hygiene and hence the health of the human being. Rising temperatures further bring a host of conditions that can adversely affect the health of the people. For example, ozone levels may rise in urban areas. Increase in desertification may create more dust and there may be more pollen in the atmosphere. In addition, disease-carrying insects may be more prevalent. Today, we find more cases of infections and vector-borne diseases in different areas of the world that were never seen before. As an example, it is found that diseases that have commonly been found within the Mediterranean are now being seen as far as the Scandinavian countries (Pinkerton et al. 2012 reported by DeCapua 2012). Insects previously prevented by cold winters have already started moving to higher latitudes, maybe toward the poles.

The effects of global warming on health are currently small but are projected by the Intergovernmental Panel on Climate Change (IPCC 2007) to progressively increase in all countries and regions. According to IPCC (2007), climate change has (1) altered the distribution of some infectious disease vectors (medium confidence), (2) altered the seasonal distribution of some allergenic pollen species (high confidence), and (3) increased heat wave-related deaths (medium confidence). Further, IPCC has projected that there will be (a) an increase in the number of people suffering from death, disease, and injury from heat waves, floods, storms, fires, and droughts (high confidence) and (b) a continuation in the change in the range of some infectious disease vectors and associated cardiorespiratory morbidity and mortality (high confidence). One can think of some benefits to health too. This includes fewer deaths from cold, although it is expected that these will be outweighed by the negative effects of rising temperatures worldwide, especially in developing countries (high confidence) (Costello et al. 2009).

Brunkard et al. (2008) examined the relationship between dengue prevalence in Mexico and the temperature and precipitation increases. They found that dengue has increased by 2.6 % 1 week after every 1 °C increase in weekly maximum temperature and by 1.9 % 2 weeks after every 1 cm increase in weekly precipitation. It is seen that floods are frequently followed by diseases. There is evidence that diseases transmitted by rodents sometimes increase during heavy rainfall and flooding due to various facts. First of all, rodents are driven out of their burrows. Mosquito breeding sites increase. Fungus growth increases in houses. Overall, the environment becomes conducive for the rodents to multiply.

Sikka and Kulshrestha (2001) carried out a comprehensive study of climate-related health issues in India and stated that there have been health and climate linkages in India in the earlier centuries also. This study has reported connection between seasonally locked disease patterns and disease dynamics such as flu, respiratory diseases in winter, dengue, malaria, and waterborne diseases especially in summer. Dash and Kjellstrom (2011) have highlighted that in India there is a need to pay more attention to public health related to heat stress. Earlier, some studies have identified heat stress as an important health risk in agricultural (Nag et al. 1980a, b) as well as in the industrial sectors in India. In our country several agricultural laborers work for a long time exposed to the strong sun amidst high temperature and humidity conditions. Groups working in rural and semi-urban-based industries such as ceramics and pottery, iron and stone quarry, automotive and automotive parts manufacturing industries, and coal mines are potentially at risk being exposed during peak summer months. As per the present classification by the India Meteorological Department (IMD), some heat events do not come under the category of heat waves and hence go unnoticed, but they can become dangerous for the exposed workers. In connection with the workplace heat stress in India, so far the major industrial sectors looked into include automotive (Ayyappan et al. 2009), coal mines (Mukherjee et al. 2004), ceramics and pottery (Parikh et al. 1978; Srivastava et al. 2000), iron works, stone quarry (Nag and Nag 2009), and textiles (Sankar et al. 2002). Several other labor-oriented sectors can also be explored where heat stress is one of the important issues concerning the workers who are completely or partially exposed.

4.2 Important Studies Showing Evidence of Regional Warming

It is well known that the connection between the global and regional climate changes is complex. Comparison of the global and regional temperature variations shows that in some regions (Orr et al. 2004) the trends may be significantly greater than the corresponding global trends. In some other regions, local trends may also be less than global trends or may even exhibit opposite behavior, especially in the case of precipitation. When one considers the cities, it should be understood that the observed warming is not only due to the increase of greenhouse gases (GHG) but also due to urbanization (Nasrallah and Balling 1993). A study by Jones et al. (1990) on urbanization and related temperature variation indicates that the impact of urbanization on the mean surface temperature is up to 05 °C per 100 years. On the other hand, a study by Fujibe (1995) reports a rising trend of 2–5 °C per 100 years in minimum temperature at several large cities of Japan. For uniform scientific analysis across the world, the Expert Team on Climate Change Detection and Indices (ETCCDI) has defined some important temperature and precipitation indices. ETCCDI modified some guidelines that are used to examine the changes in climatic extremes over a period of time and is explained by Klein Tank et al. (2009). Several studies (e.g., Alexander et al. 2006; Yan et al. 2002; New et al. 2006; Sterl et al. 2008) have used ETCCDI criteria in their analysis of extreme events. Alexander et al. (2006) have reported about 70 % of total land areas sampled across the world having significant increase and decrease in the annual occurrence of warm nights and cold nights, respectively, during the period 1951–2003. Yan et al. (2002) examined the temperature exceedences in Europe and China and found increase in warm extremes since 1960 particularly during the summer months. Gradual decrease in cold extremes in winter has persisted during the twentieth century. Based on IPCC model simulations under SRES A1B emission scenarios, it is inferred that along with the enhancement in atmospheric warming, the ratio of record highs to record lows will continue to increase. Sterl et al. (2008) examined ECHAM5/MPI-OM coupled climate model simulations and projected global mean surface warming of 3.5 K by the end of twenty-first century and more severe extreme annual maximum temperatures in densely populated areas like India and the Middle East. Meehl et al. (2009) have reported more increase in the record high maximum temperatures than record low minimum temperatures since late 1970s. Further, they have found 30 % more warming in the model simulations than in the observations by early twenty-first century.

4.3 Spatial and Temporal Changes in Indian Temperature

Due to the large geographical spread of India cutting across several latitude circles and meridians, the country experiences large spatial and temporal variations in temperature and rainfall. During the 4 months from June to September, the southwest summer monsoon prevails over the country, and hence the dominant weather systems include cloudy sky, humid atmosphere, rainy days, strong winds, cyclonic systems, and several such systems usually associated with monsoon climate. Often there are incessant rainfall and sometimes heavy downpours leading to waterlogging and urban flooding. During October to December, the northeast winter monsoon and the weather systems called the western disturbances affect the weather of the country in general. In the winter months of January and February, the northern part of the country sometimes experiences cold wave conditions. Chilled wind accompanied by rainfall makes the life difficult for the people living in the northern part of India. Some people having limited or no access to a comfortable shelter also lose their lives. We experience different types of extreme weather events occurring in different parts of India. Cold waves, prolonged fog, heavy snowfall, and flash floods are mostly confined to the northern part of India; cyclonic storms, storm surges, heavy rainfall, and floods are usually observed in the coastal regions whereas heat waves, dry spells, and droughts largely occur in the central India. In the last decade or so, heat waves are becoming more prevalent in the states like Odisha and Andhra Pradesh in the east coast (Dash et al. 2007), and there are incidents of very heavy rains in Mumbai and Balmer in the west.

In order to conduct regional climate change studies in India, it is convenient to divide the entire country into six rainfall homogeneous zones (Parthasarthy et al. 1995) and seven temperature homogeneous zones (www.tropmet.res.in). So far as rainfall extremes are concerned, Dash et al. (2002) examined the seasonal mean monsoon rainfall over India and five of its rainfall homogeneous zones and showed large variability in the spatial distribution of rainfall over the country. Further, Dash et al. (2009) used gridded rainfall datasets prepared by the IMD for the period 1951–2004 to study the possible changes in the frequency of rain events in India with respect to their duration and intensity on a daily basis. They classified the duration of rainfall as short, long, dry, or prolonged dry spells. Likewise, they categorized the intensity of rainfall as low, moderate, or heavy. They concluded that the frequency of heavy rain days increased significantly in the North East India. Dash et al. (2011) also studied the changes in the long and short spells of different rain intensities over entire India by utilizing IMD daily gridded rainfall dataset for the period 1951–2008. In addition, they have also examined regional changes in rainfall in different homogeneous zones as well as in nine agrometeorological divisions. Their study shows that over the entire country, short spells of heavy-intensity rain have risen and long spells of moderate- and low-intensity rain have decreased during the summer monsoon season in the past 50 years.

According to IPCC (2007), globally extreme temperature events will increase in frequency and duration owing to global warming. In the context of India, it may be noted that all homogeneous zones do not exhibit similar changes in the characteristics. There are regional disparities depending upon the climate types of the region of concern. Based on local climatic conditions, IMD defined heat and cold wave conditions which are given in Table 4.1, and they operationally use these criteria to declare heat and cold wave conditions where necessary. Several scientists have used these criteria to study the changes in the extreme temperature events in India. As

Table 4.1 Definitions of heat/cold-wave conditions used by the IMD for their operational purposes

Heat-wave conditions		Cold-wave conditions	
Heat wave condition is considered when max temperature (T_{max}) of a station in the plains reaches at least 40 °C; in other regions it is 30 °C		The actual min temperature (T_{min}) of a station should be reduced to windchill effective min temperature (WCTn); cold-wave condition is considered when WCTn $\leq$ 10 °C	
Normal T_{max}	Departure of T_{max} from normal	Normal T_{min}	Departure of T_{min} from normal
$\leq$40 °C	5°–6 °C: heat wave	$\geq$ 10 °C	From –5° to –6 °C:cold wave
	$\geq$7 °C: severe heat wave		–7 °C or less: severe cold wave
>40 °C	4°–5 °C: heat wave	<10 °C	From –4° to –5 °C:cold wave
	$\geq$6 °C: severe heat wave		–6 °C or less: severe cold wave
When actual T_{max} remains $\geq$ 45 °C, heat-wave condition is declared irrespective of normal T_{max}		When WCTn is $\leq$0 °C, cold-wave condition is declared irrespective of normal T_{min} of the station	

From http://www.imd.gov.in/

compared by De and Mukhopadhyay (1998), the heat wave conditions were comparatively higher in the decade ending 1979 than those in the previous decade 1979–1988. In contrast to the heat wave conditions in the central India plains and coastal regions, the cold wave conditions are usually observed in the hilly regions in north India and adjoining plains. Pai et al. (2004) have reported significant increase in the frequency, persistence, and spatial coverage of extreme temperature events in the decade 1991–2000 compared to the two earlier decades 1971–1990. Pai et al. (2004) have also inferred that cold wave conditions were most experienced in the west Madhya Pradesh, Jammu and Kashmir, and Punjab in the decades 1971–1980, 1981–1990, and 1991–2000, respectively. De et al. (2005) based on observations have inferred that the occurrence of cold wave conditions in the last century was maximum in the Jammu and Kashmir region followed by Rajasthan and Uttar Pradesh. Dash et al. (2007) and Dash and Hunt (2007) based on observational data have examined the changes in the characteristics of surface air temperatures during the last century over seven homogeneous regions in India. Dash et al. (2007) identified maximum increase in the daily maximum temperatures in the West Coast in comparison to other homogeneous regions. They have discussed the heat wave conditions that occurred in the East Coast of India during the period 19 May to 10 June in the year 2003 and have identified four stations where the maximum temperatures crossed their respective 100-year maximum values by about 1 °C or so. Similar unusual severe heat wave conditions were reported to have occurred in a large part of India in the second half of May 1998. Maximum number of heat wave conditions were reported between the years 1980 and 1998.

Kothawale et al. (2010) have reported significant increase in the numbers of hot days in the East Coast, West Coast, and Interior Peninsula and decrease in the numbers of cold days over the Western Himalaya and West Coast during March to May. In their study, hot nights showed significant increasing trends in East Coast, West Coast, and North West. For India as a whole, their study shows that hot days and nights have increasing trends, and cold days and nights have decreasing trends. In several parts of India, warm weather conditions are also noticed in the months of June and July. In the recent past, IMD has developed a high-resolution daily (Rajeevan et al. 2006) gridded temperature dataset at 1×1 degree resolution over Indian land area. Srivastava et al. (2009) in their study have used measurements at 395 quality-controlled stations and interpolated the station data into grids with the modified version of Shepard's angular distance-weighting algorithm (Shepard 1968). Further, while preparing IMD gridded temperature data, the time series of the actual station data and the interpolated grid data for the period 1969–2005 were cross-validated by calculating the root-mean-square errors (RMSE). These errors associated with the interpolation scheme used in preparing gridded temperature values over the plains were found to be up to 0.5 °C. However, the errors were relatively large in the hilly regions such as Jammu and Kashmir and Uttarakhand where there were sparse sources of observational data. Dash and Mamgain (2011) used these gridded data in their study and have defined summer and winter seasons based on threshold approach over entire India and its seven temperature homogeneous regions and categorized temperature exceedences as moderate and severe. Their study is based on actual warm/cold periods across all the months. It may be noted that durations of weather extremes are as important as their intensities. Mild-intensity heat or cold waves persisting for longer duration can create equal or sometimes more harm than high-intensity events of short duration. In their study, Dash and Mamgain (2011) have emphasized on the changes in the characteristics of long and short spells of temperature extremes. They have followed the guidelines suggested by ETCCDI to characterize the extreme categories of surface air temperatures over the period 1969–2005. They have examined the changes in the numbers of warm days and nights, cold days and nights, Warm Spell Duration Index (WSDI), and Cold Spell Duration Index (CSDI) for the seven different temperature homogeneous regions of India. The method of identifying extreme temperature events had been cross-validated by Dash and Mamgain (2011) with the actual heat and cold wave events declared by IMD at some selected stations as per the actual observations. Nevertheless, they have found differences in the characteristics of extreme temperature events occurring in the different regions of India.

The report prepared by INCCA (2010) presents the sectoral and regional analyses of the climate of the country in the current years of 1970–2005 and also during the coming years of 2030. This report infers that the frequency of cold nights has decreased at a rate of 0.9 nights per decade during the period 1970–2005. Their results indicate that the two homogeneous zones Western Himalayas and North East have significantly decreasing trends in the frequency of cold nights. Also, the frequency of hot days has increased over all the homogeneous regions except in North East India.

It is well known that although climate change is a global phenomenon, its impacts are local. There have been costly impacts on city's basic services, infrastructure, housing, human livelihoods, and health. Hence, it is very essential to examine in detail the current climate of the big cities and also to critically estimate their future projections. These scientific results may help the local bodies at the city level as guidelines to formulate their future growth plan. So far as warming of Indian cities is concerned, several studies have looked into the impact of urbanization and industrialization. Rupa Kumar and Hingane (1988) had reported the results of the analysis of long-term trends of surface air temperatures of six industrial cities in India. They concluded that Calcutta, Bombay, and Bangalore showed significant warming trends, whereas Madras and Pune did not show any significant trend. In their study, Delhi showed a significant cooling trend. De and Prakash Rao (2004) examined the trends in the climatic parameters in 14 urban cities in India, which have population exceeding 10 lakhs each. Nine stations show increasing significant trends in annual and monsoon rainfall during the period 1951–2000 when rapid development took place over these urban locations. Prakasa Rao et al. (2004) examined the effect of urbanization on the meteorological parameters at 15 Indian cities and found that radiation values, bright sunshine hours, wind speeds, and total cloud amounts have a decreasing tendency during the last 40–50 years whereas relative humidity and rainfall amounts show increasing tendency in some cities. It may be noted that increasing amounts of aerosols in the atmosphere can cut the amount of sunlight reaching the ground and hence lead to decrease in bright sunshine hours. Prakasa Rao et al. (2004) have examined the effect of urbanization in 15 cities in India during the second half of the last century and concluded that the frequency of occurrence of summertime maximum temperature more than 35 °C has decreasing trend in north India and increasing trend in south India. They also inferred that occurrence of wintertime minimum temperatures with values less than 10 °C has shown increasing trend in northern Indian cities. However, they did not find statistical significance in case of all the cities considered. Rao et al. (2005) analyzed the extreme weather events such as high and low temperatures and heavy rainfall in connection with the climate change over India and concluded that during summer, 60–70 % of the coastal stations show increasing trends in the extreme values of maximum day- and nighttime temperatures.

In order to examine the changes in the temperature at city level in India, Dash et al. (2011) selected four cities, Howrah, Visakhapatnam, Madurai, and Kochi, and used temperature values from various sources such as gridded data prepared by IMD and daily observed values obtained from the respective IMD (www.imd.gov.in) meteorological observatories and also from the simulations of various IPCC models. The study has been carried out for the current climate, the near future, and at the end of the century. The temperature patterns are also analyzed for the meteorological subdivisions and homogeneous zones where the four cities are located. Meteorological subdivisions include Gangetic West Bengal, Coastal Andhra Pradesh, Tamil Nadu and Pondicherry, and Kerala where the cities Howrah, Visakhapatnam, Madurai, and Kochi are located, respectively. Howrah and Kochi are situated in North East and West Coast temperature homogeneous zones, respectively. The two cities Madurai and Visakhapatnam are situated in East Coast temperature homogeneous zone. It is well known that climate change has large uncertainties especially when one examines climate change at a particular location, for example, in a city, and the uncertainties may become larger. While identifying the changes in the climatic factors at a place, it is also necessary to examine the corresponding changes in the same factors in the surrounding areas. The robustness of any change at a place depends on similar changes in the larger area surrounding that place. Hence, in their study, Dash et al. (2011) have considered the changes occurred in the respective meteorological subdivisions and homogeneous zones in addition to the cities. For example, increase in temperature is considered to be a robust indication in a city if similar increase is found in its meteorological subdivision and temperature homogeneous zone. If the nature of increase in a city matches with that either in the meteorological subdivision or homogeneous zone, it is termed as somewhat robust. The changes in the extreme temperature types have mixed results at the four cities. Results show that at Howrah, warm days of the types TX90p, TX95p, and TX99p and warm nights of types TN90p, TN95p, and TN99p do not show any significant trend. At Visakhapatnam and Kochi, warm days TX90p, TX95p, and TX99p and warm nights TN90p, TN95p, and TN99p show increasing trends without any statistical significance. Cold days (TX10p, TX05, and TX01p) and cold nights (TN10p, TN05p, and TN01p) at Visakhapatnam show insignificant decreasing trends. At Madurai, warm nights (TN90p, TN95p, and TN99p) show increasing trends without any statistical significance. The statistically significant results are as follows. At Howrah, cold days of all the three types TX10p, TX05, and TX01p and cold nights of the types TN10p and TN05p show increasing and decreasing trends significant at 95 % and 90 % confidence levels, respectively. Cold nights show decreasing trends significant at 10 % level. At Madurai, warm days of the types TX90p and TX95p show increasing trends significant at 95 % confidence level and cold days of the types TX10p and TX05p show decreasing trends significant at 95 % confidence level. Cold nights TN10p and TN05p show decreasing trends significant at 90 % and 95 % confidence levels, respectively. At Kochi, cold days of the types TX10p and TX05p and cold nights TN10p show decreasing trends significant at 95 % and 90 % confidence levels, respectively.

4.4 Identification of Extreme Temperature Regions in India

As mentioned above, for the convenience of scientific analysis, India has been divided (www.tropmet.res.in) into seven homogeneous temperature regions (Fig. 4.1) on the basis of surface air temperature characteristics across the country. These regions are named as the East Coast (EC), Interior Peninsula (IP), North Central (NC), North East (NE), North West (NW), Western Himalaya (WH), and West Coast (WC). TCCDI has defined a core set of 27 extreme indices for temperature and precipitation, out of which 16 are temperature related. The list of the temperature indices calculated in this study is given in Table 4.2. As noted earlier, in our study, IMD gridded maximum and minimum temperature datasets for the years 1969–2005 are used. Maximum temperatures are used to estimate the number of warm and cold days and minimum temperatures are used to find out the warm and cold nights. We calculate these indices over each grid point in the country based on $1° \times 1°$ resolution IMD gridded datasets. This exercise is conducted for a total of 357 grid points over the Indian mainland. It is to be noted that, along the northern part of India in the state of Jammu and Kashmir, few grid points have no or undefined

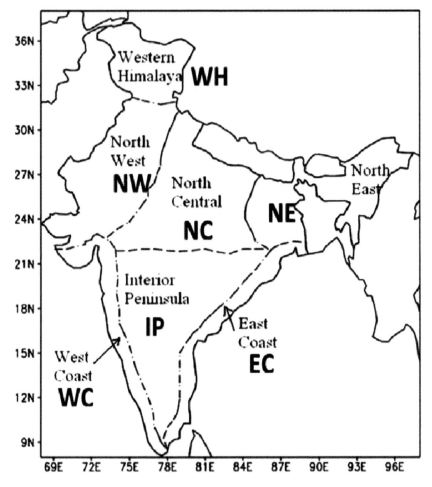

Fig. 4.1 Seven temperature homogeneous regions of India (Source: www.tropmet.res.in)

Table 4.2 Definitions of extreme-temperature indices used in this study on the basis of gridded data.

Temperature Indices	Names	Definition
TX90p, TX95p, TX99p	Warm days	Count of days on which max temperature TX > 90th, 95th, and 99th percentile, respectively
TN90p, TN95p, TN99p	Warm nights	Count of days on which min temperature TN > 90th, 95th, and 99th percentile, respectively
TX10p, TX05p, TX01p	Cold days	Count of days on which max temperature TX < 10th, 5th, and 1th percentile, respectively
TN10p, TN05p, TN01p	Cold nights	Count of days on which min temperature TN < 10th, 5th, and 1th percentile, respectively

data due to difficult terrain and lack of data availability at higher-altitude locations.

The exact lengths of summer and winter seasons differ from one region to the other in our country. Hence, in this study, first the annual data series for each region are used to separate those above the 75th percentile and those below the 25th percentile. Then the extents of summer and winter periods for that region are identified by analyzing threshold values of temperatures above the 75th percentile and below the 25th percentile, respectively. Thus identified summer period is then used to calculate 90th, 95th, and 99th percentile. Similarly, winter period is used to calculate 10th, 5th, and 1st percentiles. IMD recorded temperatures at synoptic hours 0530 h (0000 UTC) and 1430 h (0900 UTC) Indian standard times (IST) are accepted as daily minimum and maximum temperatures, respectively. For detection of trends, the Mann–Kendall rank-based test (Kendall 1975) has been done.

In Fig. 4.2, warm days of the type TX90p show increase at 27 and 31 grid locations significant at 90 % and 95 % confidence levels, respectively, over entire India. However, TX90p also decreases significantly at seven grid locations. The southern part of India shows more increase in warm days significant at 95 % confidence level than northern parts of India. Maharashtra and Andhra Pradesh are identified as the hotspots for the warm days of the type TX90p. A small area covering some parts of Punjab, Himachal Pradesh, and Uttarakhand also shows increase in warm days of the type TX90p, but at lesser significance. There are also isolated grids in Arunachal Pradesh, Tripura, Mizoram, and Manipur indicating decrease in warm days of the type TX90p significant at 90 % and 95 % confidence levels. In addition there are some grid locations in Uttar Pradesh, Bihar, and Rajasthan showing significant increase in warm days. Warm days of the type TX95p indicate increasing trends (not shown in figure) at a few locations in Maharashtra and Rajasthan, the same decrease at a few locations in Meghalaya and Manipur.

Figure 4.3 shows the occurrence of cold days of the type TX10p decreasing at 35 and 37 grid locations significant at 90 % and 95 % confidence levels, respectively, scattered across the entire country. However, TX10p also increases significantly at 15 grid locations. The southern part of India shows more decrease in cold days significant at 95 % and 90 % confidence levels compared to the northern part. The most prominent hotspot for decrease in cold days of the type TX10p in peninsular India covers major part of Andhra Pradesh, North Karnataka, southern Maharashtra, and small part of Tamil Nadu. In the north, a smaller area covering Punjab, Uttarakhand, and Himachal Pradesh also indicates significant decrease in cold days. In the North East region, some scattered grids in Arunachal Pradesh, Tripura, Mizoram, and Manipur also show significant decrease in cold days. However, some grid locations near Assam and Meghalaya show significant increase in cold days significant at 95 % confidence level. Similarly, grid locations in Madhya Pradesh show significant increase in cold days significant at 95 % confidence level. Cold days of the type TX05p indicate decreasing trend at few locations in Andhra Pradesh. Over some grid locations in Maharashtra, Andhra Pradesh, Punjab, Himachal Pradesh, and Bihar, both warm days of the type TX90p and cold days of the type TX10p increase and decrease, respectively. These results indicate significant increase in the daytime temperatures over these regions.

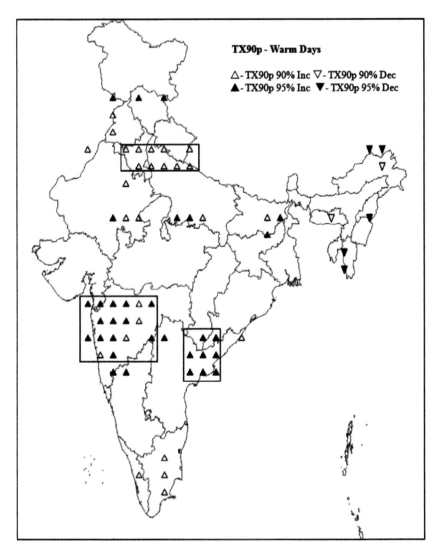

Fig. 4.2 Warm days of the type TX90p based on IMD gridded maximum temperature during
1969–2005

Number of warm nights of the type TN90p indicates increasing (Fig. 4.4) trend
at 15, 9, and 4 grid locations significant at 90 %, 95 %, and 99 % confidence levels,
respectively, over entire India. However, TN90p also decreases significantly at eight
grid locations. A small area in the south consisting of regions from Tamil Nadu,
Kerala, and Karnataka shows more increase in warm nights significant at 95 % and
90 % confidence levels than the northern and other parts of India. Some locations in
Maharashtra, Gujarat, and Tamil Nadu in the south and Uttarakhand and Himachal
Pradesh in the north show more increase in warm nights. In the North East region,
some isolated grid locations in Arunachal Pradesh, Meghalaya, and Manipur also
indicate significant increase in warm nights at 90 % confidence level. Some grid

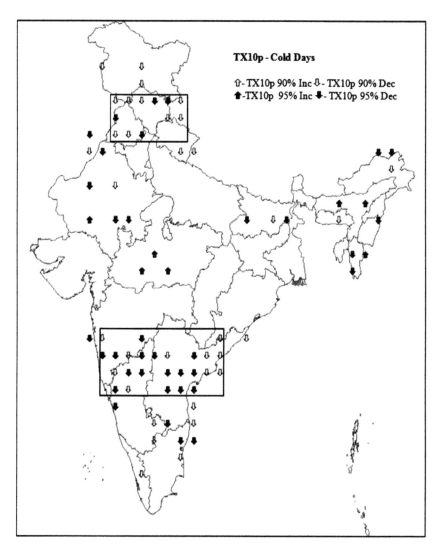

Fig. 4.3 Cold days of the type TX10p based on IMD gridded maximum temperature during 1969–2005

locations in Madhya Pradesh and Chhattisgarh show significant decrease in warm nights at 95 % confidence level.

Figure 4.5 shows the changes in the number of cold nights of the type TN10p. As depicted in this figure, TN10p decreases at 9, 9, and 8 grid locations significant at 90 %, 95 %, and 99 % confidence levels, respectively. The prominent region of decrease in cold nights significant at 95 % and 99 % confidence levels lies in the southwest India which includes Gujarat and Maharashtra. In addition, some locations in Tamil Nadu and Madhya Pradesh indicate decrease in cold nights. Also cold nights show decrease in isolated grids in Himachal Pradesh, Uttarakhand, Bihar, and Arunachal Pradesh. These results indicate an increase in nighttime temperature over these regions.

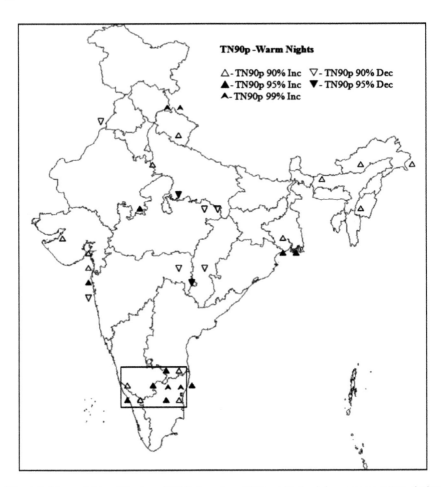

Fig. 4.4 Warm nights of the type TN90p based on IMD gridded minimum temperature during 1969–2005

Figure 4.6 depicts extreme cases of warm nights TN95p and cold nights TN05p. As seen in this figure, increase in the number of warm nights of the type TN95p is confined to the area in the northwest consisting of Gujarat and Maharashtra. Similarly there is another area in the south consisting of Tamil Nadu, Kerala, and Karnataka where warm nights show significant increasing trend. Some locations in Madhya Pradesh show decrease in cold nights significant at 95 % confidence level. At few locations in Maharashtra, Gujarat, and Madhya Pradesh, cold nights of the type TN05p decrease in number. Over some grid locations in Gujarat and Maharashtra, both warm nights of type TN90p and cold nights of type TN10p show increase and decrease, respectively. These signals point to significant increase in the nighttime temperatures over these regions.

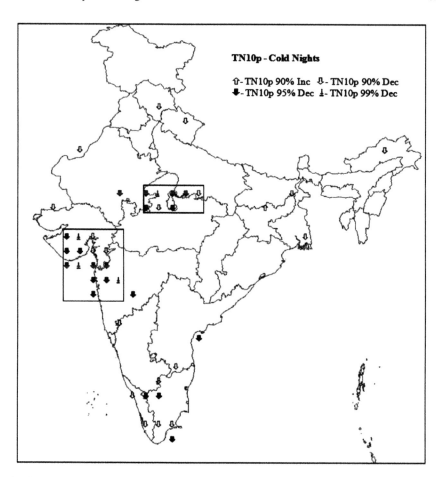

Fig. 4.5 Cold nights of the type TN10p based on IMD gridded minimum temperature during 1969–2005

4.5   Conclusions

This study has been undertaken to identify extreme temperature regions in India based on IMD gridded dataset during the period 1969–2005 at $1° \times 1°$ resolution. Statistical analysis yields that isolated grid points with increasing occurrence of warm days and nights and decreasing trends in cold days and nights are scattered all over the country. In addition, there are some large areas (with clusters of grid points) showing signatures of temperature extremes which can be identified as hotspots. For example, Maharashtra and Andhra Pradesh are identified as the hotspots for the warm days of the type TX90p. A small area covering some parts of Punjab, Himachal Pradesh, and Uttarakhand also shows increase in warm days TX90p, but at lesser

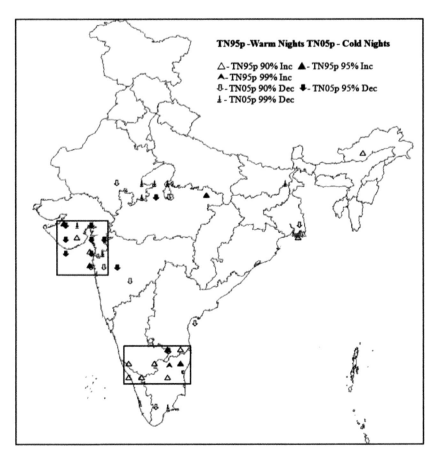

Fig. 4.6 Warm nights of the type TN95p and cold nights of the type TN05p based on IMD gridded minimum temperature during 1969–2005

statistical significance. The most prominent hotspot for decrease in cold days of the type TX10p in peninsular India covers major part of Andhra Pradesh, North Karnataka, southern Maharashtra, and small part of Tamil Nadu. In the north, a smaller area covering Punjab, Uttarakhand, and Himachal Pradesh also indicates significant decrease in cold days. Increase in the number of cold nights is confined to a small area in the south consisting of regions from Tamil Nadu, Kerala, and Karnataka. The prominent region of decrease in cold nights significant at 95 % and 99 % confidence levels lies in the southwest India which includes Gujarat and Maharashtra. Increase in the number of warm nights of the type TN95p is confined to an area in the northwest consisting of Gujarat and Maharashtra. Similarly there is another area in the south consisting of Tamil Nadu, Kerala, and Karnataka where warm nights show significant increasing trend. Some locations in Madhya Pradesh show decrease in cold nights significant at 95 % confidence level.

Acknowledgments The gridded rainfall and temperature datasets are obtained from the India Meteorological Department (IMD). Part of this research work has been undertaken with support from a research project sponsored by the Department of Science and Technology (DST), Government of India. The author wishes to thank all his coauthors in the base papers referred to in this article. Special thanks are due to Ms. Vaishali Saraswat and Ms. Neha Sharma for their efforts in preparing some of the figures.

References

Alexander LV et al (2006) Global observed changes in daily climate extremes of temperature and precipitation. J Geophys Res 111:D05109. doi:10.1029/2005JD006290

Ayyappan R, Sankar S, Rajkumar P, Balakrishnana K (2009) Work-related heat stress concerns in automotive industries: a case study from Chennai, India. Global Health Action 2:58–64

Bhadram CVV, Amatya BVS, Pant GB, Krishnakumar K (2005) Heat waves over Andhra Pradesh: a case study of summer 2003. Mausam 56(2):385–394

Brunkard JM, Cifuentes E, Rothenberg SJ (2008) Assessing the roles of temperature, precipitation, and ENSO in dengue re-emergence on the Texas-Mexico border region. Salud Publica Mex 50(3):227–234

Costello A, Abbas M, Allen A, Ball S, Bell S, Bellamy R, Freil S et al (2009) Lancet-University College London Institute for Global Health Commission. Managing the health effects of climate change. Lancet 373:1693–1733

Dash SK, Hunt JCR (2007) Variability of climate change in India. Curr Sci 93(6):782–788

Dash SK, Kjellstrom T (2011) Workplace heat stress in context of rising temperature in India. Curr Sci 101(4):496–503

Dash SK, Mamgain A (2011) Changes in the frequency of different categories of temperature extremes in India. J Appl Meteor Climatol 50:1842–1858. doi:http://dx.doi.org/10.1175/2011J AMC2687.1

Dash SK, Shekhar MS, Singh GP, Vernekar AD (2002) Relationship between surface fields over Indian ocean and monsoon rainfall over homogeneous zones of India. Mausam 53:133–144

Dash SK, Jenamani RK, Kalsi SR, Panda SK (2007) Some evidence of climate change in twentieth-century India. Clim Chang 85:299–321. doi:10.1007/s10584-007-5079305-9

Dash SK, Kulkarni MA, Mohanty UC, Prasad K (2009) Changes in the characteristics of rain events. J Geophys Res 114:D10109. doi:10.1029/2008JD010572

Dash SK, Nair AA, Kulkarni MA, Mohanty UC (2011) Characteristic changes in the long and short spells of different rain intensities in India. Theor Appl Climatol 105(3-4):563–570

De US, Mukhopadhyay RK (1998) Severe heat wave over the Indian subcontinent in 1998, in perspective of global climate. Curr Sci 75:1308–1315

De US, Prakash Rao GS (2004) Urban climate trends-Indian scenario. J Indian Geophys Union 8(3):199–203

De US, Dube RK, Prakasa Rao GS (2005) Extreme weather events over India in last 100 years. J Indian Geophys Union 9(3):173–187

DeCapua J (2012) Rising temperatures, rising health problems. Voice of America, 3 April 2012

Fujibe F (1995) Temperature rising trends at Japanese cities during the last hundred years and their relationships with population, population increasing rates and daily temperature ranges. Pap Meteorol Geophys 46:35–55

Indian Network for Climate Change Assessment (INCCA) (2010) Climate change and India: a 4 x 4 assessment (A sectorial and regional analysis for 2030s), Ministry of Environment and Forests, Government of India, pp 164

IPCC (2007) Fourth assessment report. Intergovernmental Panel on Climate Change, Geneva. Cambridge University Press, Cambridge/New York

Jones PD, Groisman PY, Coughlan M, Plummer N, Wang W-C, Karl TR (1990) Assessment of urbanization effects in time series of surface air temperature over land. Nature 347:169–172

Kendall MG (1975) Rank correlation methods, 4th edn. Charles Griffin, London

Klein Tank AMG, Zwiers FW, Zhang X (2009) Guidelines on analysis of extremes in a changing climate in support of informed decisions for adaptation. Climate Data and Monitoring WCDMP-No.72,WMO-TDNo.1500,56 pp. http://www.clivar.org/organization/etccdi/etccdi.php

Kothawale DR, Revadekar JV, Rupa Kumar K (2010) Recent trends in pre–monsoon daily temperature extremes over India. J Earth Syst Sci 119:51–65

Meehl GA, Coauthors (2009) Decadal prediction. Bull Am Meteor Soc 90:1467–1485

Mukherjee A, Bhattacharya S, Saiyed HN (2004) Assessment of respirable dust and its free silica contents in different Indian coal mines. Ind Health 43:277–284

Nag PK, Nag A (2009) Vulnerability to heat stress scenario in western India. WHO, APW N. SO 08 AMS 6157206, National Institute of Occupational Health, Ahmedabad, India

Nag PK, Sebastian NC, Malvankar MG (1980a) Effective heat load on agricultural workers during summer season. Indian J Med Res 72:408–415

Nag PK, Sebastian NC, Malvankar MG (1980b) Occupational workload in Indian agricultural workers. Ergonomics 23:91–102

Nasrallah HA, Balling RC (1993) Spatial and temporal analysis of middle eastern temperature changes. Clim Chang 25:153–161

New M et al (2006) Evidence of trends in daily climate extremes over southern and west Africa. J Geophys Res 111:D14102. doi:10.1029/2005JD006289

Orr A, Cresswell D, Marshall GJ, Hunt JCR, Sommeria J, Wang CG, Light M (2004) A 'low-level' explanation for the recent large warming trend over the western Antarctic Peninsula involving blocked winds and changes in zonal circulation. Geophys Res Lett 31(6):L06204. doi:10.1029/2003GL019160

Pai DS, Thapliyal V, Kokate PD (2004) Decadal variation in the heat and cold waves over India during 1971–2000. Mausam 55(2):281–292

Parikh DJ, Ghodasara NB, Ramanathan NL (1978) A special heat stress problem in ceramic industry. Eur J Appl Physiol Occup Physiol 40:63–72

Parthasarthy B, Munot AA, Kothawale DR (1995) Monthly and seasonal rainfall series for all India homogeneous regions and meteorological subdivisions: 1871–1994, Research report RR-065. Indian Institute of Tropical Meteorology, Pune, p 113

Pinkerton KE et al (2012) An official American Thoracic Society workshop report: climate change and human health. Proc Am Thorac Soc 9(1):3–8

Prakasa Rao GS, Jaswal AK, Kumar MS (2004) Effects of urbanization on meteorological parameters. Mausam 55(3):429–440

Rajeevan M, Bhate J, Kale JD, Lal B (2006) High resolution daily gridded rainfall data for the Indian region: analysis of break and active monsoon spells. Curr Sci 91(3):296–306

Rao GSP, Murty MK, Joshi UR (2005) Climate change over India as revealed by critical extreme temperature analysis. Mausam 56:601–608

Rupa Kumar K, Hingane LS (1988) Long-term variations of surface air temperature at major industrial cities of India. Clim Chang 13:287–307

Sankar S, Padmavathi R, Rajan P, Ayyappan R, Arnold J, Balakrishnan K (2002) Job-exposure-health profile for workers exposed to respirable dusts in textile units in Tamil Nadu, India: results of preliminary investigations. Epidemiology 13:846

Shepard D (1968) A two-dimensional interpolation function for irregularly spaced data. In: Proceedings of the 23rd national conference, Association for Computing Machinery, New York, pp 517–524

Sikka DR, Kulshrestha SM (2001) Climate and health studies: status and scope in India. Joint COLA/CARE Tech Rep 5:P74

Srivastava A, Kumar R, Joseph E, Kumar A (2000) Heat exposure study in a glass manufacturing unit in India. Ann Occup Hyg 44:449–453

Srivastava AK, Rajeevan M, Kshirsagar SR (2009) Development of a high resolution daily gridded temperature dataset (1969–2005) for the Indian region. Atmos Sci Lett 10(4):249–254

Sterl A, Coauthors (2008) When can we expect extremely high surface temperatures? Geophys Res Lett 35:L14703. doi:10.1029/2008GL034071

Yan Z, Coauthors (2002) Trends of extreme temperatures in Europe and China based on daily observations. Clim Change 53:355–392

Chapter 5
Rapid Deglaciation on the Southeast-Facing Slopes of Kanchenjunga Under the Present State of Global Climate Change and Its Impact on the Human Health in This Part of the Sikkim Himalaya

Guru Prasad Chattopadhyay, Dilli Ram Dahal, and Anasuya Das

Abstract Rapid change in climate through the Quaternary under the impact of global warming and its impact on the snow fields and glaciers of the Inner Himalayan parts have attracted many geoscientists over the last four decades. But little work has been done so far to assess the impact of climate change and its impact on the size reduction of glaciers as well as on the human health in the surrounding settlement areas of the glaciers. This paper is a revelation and discussion on the present pattern of retreat of the major glaciers in the southeast-facing slopes of Kanchenjunga in the Sikkim Himalaya under the present state of global warming. Studies on the calculation of trim line of the fresh moraines of the two main glaciers, namely, Rathong and Onglaktang, have revealed that these glaciers have retreated remarkably over the last 50 years experiencing a volume reduction of more than 70 % during this limited time period. These glaciers are considered to be among the fastest retreating glaciers in any part of the Himalaya. This feature can be attributed to the rapid climatic amelioration in the present day. According to meteorological data made available, a rise of about 1 °C in the atmospheric temperature in this part of the Himalaya has taken place. The discharged pattern of the rivers, fed by the meltwater from these glaciers, has been very erratic and temperamental, thus causing flash floods particularly over the last two decades. Snowfall on the higher parts of the mountain has diminished in recent years, and the winter season of the year has shortened to a marked extent. Survey in the three settlement areas in this western part of the Sikkim Himalaya on human health status, based on random sampling, shows that incidences of some particular diseases, like skin problems and asthma, which were rare earlier, are spreading significantly among the inhabitants in this part of the Sikkim Himalaya.

G.P. Chattopadhyay (✉)
Professor, Department of Geography, Visva-Bharati, Santiniketan 731235, India
e-mail: cguruprasad@rediffmail.com

D.R. Dahal, Ph.D. • A. Das, Ph.D.
Research Scholar, Department of Geography, Visva-Bharati, Santiniketan 731235, India

© Springer International Publishing Switzerland 2016   75
R. Akhtar (ed.), *Climate Change and Human Health Scenario in South and Southeast Asia*, Advances in Asian Human-Environmental Research,
DOI 10.1007/978-3-319-23684-1_5

Keywords Sikkim Himalaya • Quaternary • Deglaciation • Glacier retreat • Firn line • Trim line • Glaciofluvial landforms • Global warming • *Jökulhlaup* • Human health

5.1 Introduction

Over the last 100 years, the earth's average surface temperature has increased by about 0.8 °C (ICIMOD 2009) with about two thirds of the increase occurring over just the last three decades. A number of previous workers have reported conditions of retreat of the Himalayan glaciers and the resultant topographic features, viz. from the Western Himalaya by de Terra and Hutchinson (1936), from the Central (Nepal) Himalaya by Fuji and Higuchi (1976) and from the Eastern Himalaya by Kar (1969). Recent works by this author (Chattopadhyay 1985, 1998a, b, 2008, 2009, 2011a, b, 2012a, b, c) have thrown some light on the characteristic features of the Quaternary environmental changes in this part of the Eastern Himalaya. An attempt has been made here to describe the present state of deglaciation in the Kanchenjunga summit area of the Sikkim Himalaya and its impact on the surrounding mountainous terrain of the Himalaya.

5.1.1 *Physical Environment of the Study Area*

The study was concentrated upon the area around the southeast-facing slopes of the Kanchenjunga summit complex through the valleys of Onglaktang and Rathong glaciers around Dzongri Pass in the Sikkim Himalaya (Fig. 5.1). The area is a rugged mountainous tract formed by the main ridges of Singalila that ascends gradually to Kanchenjunga summit (8598 m). The two main streams in this part, fed by the glaciers, are Churang Chhu and Prek Chhu, the headwaters of the Rangit River. The timberline rests at about 4000 m and the perpetual snow line begins from about 5500 m. Geologically and structurally this area falls within the region of 'Darjeeling Gneiss' (Gansser 1964).

The ground surface of the first 200 m above the timber line resembles an Alpine meadow. On the bare mountain slopes, frost-shattered (periglacial) detritus, comprising mostly massive gneissic boulders, occur extensively (Chattopadhyay 1985). Numerous morainic ridges of pre-existing glaciers markedly traverse the main river valleys and the ridge walls ascending up to the snow line. The two main valley glaciers that exist today are the Onglaktang glacier, the source of Prek Chhu stream, and the Rathong glacier, the source of Churang Chhu stream. Both the glaciers are fed by the ice of the main neveé field of Kanchenjunga summit complex at its south face.

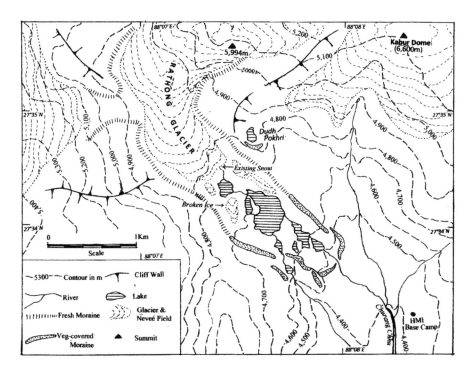

Fig. 5.1 Map showing the environment around the snout of the Rathong glacier with the existence of meltwater lakes

In order to ascertain the maximum previous extent of the Onglaktang and Rathong glaciers, several glaciofluvial landforms like successive loops of lateral and terminal moraines and other drift limits were studied in detail. These features have been used for reconstructing the glacial history of this region.

5.1.2 Methodology of Study Followed

The extent and limits of the main valley glaciers, other miniature valley-wall glaciers and the morainic ridges around them were marked on the base map prepared from the topographic map of Sikkim on 1:150,000, made available from the University of Bern, Switzerland. For the purpose of detailed cartographic representation of glacial features, the original map was enlarged to 1:75,000. This map was corrected in the field by through field investigations. Landsat and IRS data were used to detect the temporal change of the conditions of the glaciers. The two glaciers, selected for detailed study, were Rathong and Onglaktang glaciers. Former maximum volume of these two glaciers was calculated from reconstruction through the projection of height and extent of the trim lines of the existing fresh lateral and terminal morainic drifts.

5.2   Pattern of Retreat of the Rathong and Onglaktang
         Glaciers

5.2.1   Retreat of the Rathong Glacier

Rathong glacier, having its source from the neveé field around the Kabru Dome
(6600 m) summit, descends over the west-facing slope of the mountain. The source
of this glacier lies at an elevation of 6200 m (27°36′ N, 88°06′ E). With a total
length and average width of about 4 km and 800 m, respectively, it extends south-
eastward through the main trough between Koktang (6147 m) and Kabru Dome
(6600 m) down to about 4800 m from where the headwater of the Churang Chhu
stream rises.

    In a glacier-occupied valley, maximum height of the fresh lateral moraine (trim
line) usually corresponds to the height of the surface of the glacier mass at its maximum volume increase in the past. This parameter has been used here to estimate the
former thickness of the valley glaciers. The fresh morainic ridge, showing the evidence of its recent retreat following the Stage III advance, occurs at an approximate
distance of 1.2 km from the snout. A reconnaissance survey of the present condition
of the Rathong glacier was conducted on foot from the Himalayan Mountaineering
Institute Base Camp (4500 m) upward over the fresh (Stage III) morainic ridges and
the glacier body itself, up to the corrie lake of Dudh Pokhari. With close observation
and measurements with reference to the base map, the existing average length,
width and depth of this glacier have been recorded as 4000 m, 800 m and 50 m,
respectively. This gives the existing volume of the glacier as (4000 m × 800 m × 50 m =)
160 million m³ (Plates 5.1 and 5.2).

    The extent of fresh lateral morainic loop of the Rathong has been used as the basis
for calculating the former maximum volume. Here, the average height of the fresh
lateral moraine, corresponding to Stage III, from the glacier surface is about 70 m,
and the average distance between the highest parts of the lateral morainic ridge on
both sides of the glacier is about 1500 m. On the basis of this observation, the reconstructed former depth, width and length of the glacier would be (50 m + 70 m =)
120 m, 1500 m and 5200 m, respectively. Therefore, the calculated former volume of
the Rathong would be (5200 m × 1500 m × 120 m =) 936 million m³. This suggests
that, with the recent retreat (following the Stage III advance), the volume of this
glacier has reduced to about 17.09 % (less than one fifth) of its original size.

    The most remarkable morphological feature around the Rathong glacier is the
accumulation of fresh hummocky moraines mingled with detached parts of the glacier ice, which cover extensive part of the recently deglaciated valley for a distance
of about 1 km. This gives a certain indication of the existing rapid rate of glacier
retreat. Field observation of its upper part also revealed that after the complete dis-
appearance of the original neveé field around the Rathong pass, this glacier now
owes its supply only from the neveé field on the southerly slope of the Kabru
summit.

Plate 5.1 Rapidly melting Rathong glacier

5.2.2 Retreat of the Onglaktang Glacier

Onglaktang glacier, situated in the northeastern part of the study area, is nourished by the same ice field as the Rathong glacier, around the Kabru Dome (6600 m). It extends to the south-southeast through the valley for a distance of about 5 km. The source of this glacier lies at an elevation of 6000 m (27°36′ N, 88°08′ E). Its average width is about one kilometre and its snout rests at about 4500 m from where the headwater of the Prek Chhu stream originates. Its two original tributary glaciers, one extending from the Forked Peak (6108 m) to the west and the other from the Pandim (6691 m) summit to the east, are now completely detached as a result of their apparent retreat.

The existence of fresh terminal morainic ridge, at a distance of about 1 km downstream from the present snout, clearly suggests that this glacier has retreated about 1 km following the Stage III advance. For a thorough investigation of the present condition and the stretch of this glacier, the adjacent valley area beyond the Samiti Lake up to Jemathang for a distance of about 6 km was traversed on foot. With close observation and measurements in the field using the base map, average width and depth of the Onglaktang glacier have been recorded as 800 m and 60 m,

Plate 5.2 Recently deglaciated valley floor of the Onglaktang glacier

respectively. Hence, considering its present length (5,000 m), width and depth, the approximate volume of the Onglaktang glacier would be (5000 m × 800 m × 60 m =) 240 million m³.

The average height of the fresh lateral moraine (corresponding to Stage III) from the existing surface of the Onglaktang glacier is about 70 m, whilst the average distance between the highest part of the lateral moraines on both sides across the glacier is about 1,000 m. This indicates that the former average depth and width of the glacier were (60 m + 0 m =) 130 m and 1,000 m, respectively. Considering the recent retreat of 1,000 m, the former length of the glacier would be (5,000 m + 1,000 m =) 6,000 m. Hence, the reconstructed volume of the Onglaktang glacier, at its maximum during the Stage III advance, would be (6,000 m × 1,000 m × 130 m =) 780 million m³. This suggests that with the recent retreat (after the Stage III advance), the volume of the glacier has reduced to about 30.77 % (i.e., less than one third) of its original size. Other fresh morainic evidences on the adjacent areas show that the valley-wall glaciers extending from the neveé field of the Pandim summit to the east used to feed the Onglaktang during their maximum extent in the past. With marked retreats, these glaciers have now been completely detached from the Onglaktang glacier.

5.2.3 Interpretation of Field Data Collected from the Areas Around the Two Glaciers

The quantitative data and associated information collected through field investigations reveal the alarming rate of retreats of the major glaciers and reduction and shifting of the neveé fields around the southwest-facing slopes of the Kanchenjunga summit complex. The environmental degradation, as a consequence of this rapid decay of glaciers and neveé fields, is conspicuous around the Rathong glacier. The establishment of Himalayan Mountaineering Institute Training Camp (in early 1960s) in the immediate vicinity of the Rathong and rapid increase in the number of mountaineers and trekkers since then caused marked environmental degradation in recent years. Regular visits made by thousands of trainees taking part in the mountaineering courses have caused eco-degradation in this part in various ways. From 1960s through 1980s, camp activities have completely destroyed the trees around the Base Camp and the Dzongri and dumped garbage all around the glacier snout. This caused local atmospheric warming resulting in the gradual decrease in snowfall. Rapid decay of glacier body often gives rise to debris-dammed lakes adjacent to the snout. At certain intervals, catastrophic dam bursts can cause flooding, disturbing the equilibrium of the valley profile in the upper catchment areas of the glacier-fed rivers. Such disaster has already struck the upper Onglaktang valley on many of occasions, as in 1988 and 1994. If such rapid rate of retreat of the glaciers continues around this part of the Himalaya, it is quite reasonable to suspect that most of the major rivets in North Bengal will increasingly suffer from a state of hydrological misbalance with marked decrease in discharge during the drier months and excessively heavy flooding during the monsoons.

5.3 Hazards Occurred as a Result of the Formation of Meltwater Lakes

We could gather that the receding glacier has caused lot of ecological disorder in the region. It is to be noted that glaciers are lifeline of high-altitude lakes. Lakes are so beautiful and crystal clear that had been existing in the bosom of rocky Himalayan region since time immemorial having ecological niche for many life forms. Receding glacier caused disconnection of water supply to lakes and lakes are dying its untimely death. As a result of elimination of so many lakes at the high Himalaya, many lives have been badly affected causing ecological imbalance in the region.

In the north of Sikkim Himalayan, a most fascinating lake, 'Green Lake', was in existence, and it was very much popular among the mountaineers, trekkers of the western countries now totally eliminated due to the fact that the lake virtually ceased to get enough water from the melting glacier. The glacier which was feeding this lake now receded to such an extent that there is no linkage at all between the lake and the glacier.

A beautiful lake in the west Sikkim Himalaya, a pilgrimage for Buddhist follow-ers mostly Sikkimies and Tibetans called 'Dudh Pokhari' (Milky lake), is now at a vulnerable stage as its southern wall has become very thin due to falling down of rocks and sands that are regularly under the influence of frost shattering, freezing cold and thawing. It is feared that in the near future, this lake might blast off sud-denly causing considerable damages to forest, wildlife and human lives and proper-ties. Natural lakes like Lam Pokhari (long lake), Mayur Pokhari, Laxmi Pokhari, Kala Pokhari, Dallay Pokhari, Nir Pokhari, Memenchu and other so many smaller lakes are at their dying stage due to reasons explained already.

5.3.1 Potential Dangers of Glacial Lake Outburst Flood (GLOF) in the Alpine Regions Especially in the Himalaya

On high mountain slopes, behind fragile dams of rock and ice created by gla-ciers, water collects to form lakes that loom precariously, threatening to empty at any time and deluge the valleys below and all who live there. New studies by glaciologists and geographers show the menace of these glacial lakes to be an increasingly critical factor, too often ignored, in plans for the economic develop-ment of the world's mountainous regions, particularly the Alps, Andes and Himalaya. A recent study done by ICIMOD and UNEP (2002) reported 27 poten-tially dangerous lakes in Nepal Himalaya which are supposed to cause huge and destructive flash flood in the near future. Glacial geomorphologists have identi-fied this as glacial lake outburst flood (GLOF) or *jökulhlaup*, an Icelandic term to describe any abrupt and voluminous release of subglacial water (Post and Mayo 1971; Rana et al. 2000).

Glaciologists fear that the outburst of a glacial lake in 1985 is a foretaste of things to come in the Himalaya. The 1985 disaster occurred due to an ice avalanche. A wall of ice and rock crashed down the high slope and plunged into a glacial lake, sending a wave of water over the natural dam formed by the glacier. The dam col-lapsed, and the flood rushing down the Dudh River destroyed a power plant, wiped out 14 bridges and killed at least five people. Jack D. Ives, a professor of geography at the University of Colorado at Boulder, and June Hammond, a graduate student at the university, have identified a large lake on the Imja glacier, six miles south of Everest, that poses a direct threat to several Nepalese villages, the Buddhist monas-tery at Thyangboche and two favourite climbers' trails, to the base camp for Everest expeditions and to Island Peak. Ives said that in the Himalaya, the potential hazard of glacial lakes has become critical only in recent years, with the increased con-struction of roads, bridges and power plants and other economic development. In the recent past, flash floods have occurred in the Thimphu, Paro and Punakha-Wangdue valleys. Of the 2,674 glacial lakes in Bhutan, 24 have been identified by a recent study as candidates for GLOFs in the near future. In October 1994, a GLOF, 90 km upstream from Punakha Dzong, caused massive flooding on the Pho Chhu River, damaging the dzong and causing casualties. In 1994 a glacier lake outburst in the Lunana region of Bhutan flooded a number of villages below endangering the lives of thousands of people. The burst of Dudh Koshi Lake in Nepal in 1997 made similar hazards. Subglacial lakes with about 450,000 cubic metres of accumulated water (up till now) developed in the recently deglaciated Rathong glacial valley have threatened the HMI Base Camp to wash away at any time.

5.3.2 Evidences of Glacial Lake Outburst Flood (GLOF) in the Himalaya

In the glaciated parts of the high-altitude Himalayan region, behind fragile dams of terminal moraine and segregated ice produced by glaciers, water get accumulated to form lakes that loom precariously, threatening to empty at any time and deluge the valleys below and all who live there. New studies by glaciologists and geographers have showed the menace of these glacial lakes to be an increasingly critical factor, too often ignored, in plans for the economic development of the Himalaya as well as the other glaciated alpine regions of the world (Chattopadhyay 2007). A recent study done by the International Centre for Integrated Mountain Development (ICIMOD and UNEP 2002; ICIMOD 2009) and the United Nations Environment Programme (UNEP) turned attention towards the problems of rapid glacier melting in the Himalaya and reported 27 potentially dangerous lakes in Nepal Himalaya which are supposed to cause huge and destructive flash flood in the near future.

Nepal, covering the entire part of the Central Himalaya, has experienced several glacial lake outburst floods originating from numerous glacial lakes, some of which are even located outside its territory. Although other natural disasters such as rainfall floods, earthquakes, landslides or wildfires have claimed the lives of thousands of Nepalese in recent decades, glacial lake outbursts are feared for the potential devastation from a single large event (Rana et al. 2000; Kattelmann 2003). At least six times in the last 40 years, glacial lakes in the Khumbu region of the Nepal Himalaya, near Everest, have drained without warning, causing devastating floods of water and rock (Fushimi et al. 1985). The GLOF event of 4 August 1985 from *Dig Tsho* glacial lake in Eastern Nepal, in a valley next to Mount Everest, had especially brought about awareness of potentially dangerous glacial lakes in the high Himalaya, nationally and internationally. It was observed that an ice avalanche slumped down into the lake and generated a wave about 5 m high which overtopped the moraine dam. The lake, roughly measuring an area of 1,500 m × 300 m at a depth of 18 m, drained almost completely within 4–6 h. The flood destroyed bridges, large and small buildings, agricultural land and the nearly completed Namche Hydropower Plant, 2 weeks before its inauguration, which resulted in a loss of billions of rupees.

Remarkably, only 4–5 people lost their lives in this flood itself because a Sherpa festival was in progress and few people were walking the trails at the time (ICIMOD/ UNEP 2002; Kattelmann 2003).

Glaciologists fear that the outburst of a glacial lake in 1985 is a foretaste of things to come in the Himalaya. Jack D. Ives (Ives 1986), a Professor of Geography at the University of Colorado at Boulder, and June Hammond, a graduate student at the university, have identified a large lake on the Imja glacier, six miles south of Everest, that posed a direct threat to several Nepalese villages, the Buddhist monastery at Thyangboche and two favourite climbers' trails, to the base camp for Everest expeditions and to Island Peak. Ives observed that, in the Himalaya the potential hazard of glacial lakes has become critical only in recent years, with the increased construction of roads, bridges and power plants and other economic development. In the recent past, flash floods have occurred in the Thimphu, Paro and Punakha-Wangdue valleys. Of the 2,674 glacial lakes in Bhutan, 24 have been identified by a recent study as candidates for GLOFs in the near future. In October 1994, a GLOF, 90 km upstream from Punakha Dzong, caused massive flooding on the Pho Chhu River, damaging the dzong and causing casualties. In 1994, a glacier lake outburst in the Lunana region of Bhutan flooded a number of villages below endangering the lives of thousands of people. The burst of Dudh Koshi Lake in Nepal in 1997 made similar hazards. Subglacial lakes with about 450,000 cubic metres of accumulated water (up till now) developed in the recently deglaciated Rathong glacial valley on the southeast-facing slope of Kanchenjunga in Sikkim Himalaya, have threatened the HMI Base Camp to wash away at any time. This paper records the observation on the changing forms of retreating glaciers in this part of the Himalaya and development moraine-dammed lakes with potential dangers.

5.3.3 Recent Retreat of Rathong Glacier and Dangers of Moraine-Dammed Lake Outburst

This study is based on regular field observations made by these authors on the glaciers extending over the southeast-facing slopes of Kanchenjunga in the Sikkim Himalaya during 1984–2009 (a span of 25 years). The two glaciers concentrated upon were Rathong glacier, the source of Rathong Chhu River above the HMI Base Camp, and Onglaktang glacier, the source of Prek Chhu River, beyond Thangsing Camping ground. Both Rathong Chhu and Prek Chhu are the headwaters of Rangit River, the merged water of which meets River Tista further down in South Sikkim, to the north of Tista Bazar. It has been explored through repeated observations over the years that both Rathong and Onglaktang glaciers have been reduced to 17.09 % and 30.77 %, respectively, as a result of recent retreats presumably under climatic amelioration, i.e., global warming. Major part of reduction in glacier size has taken place during the last three to four decades (Plates 5.3 and 5.4).

Plate 5.3 A recently developed melt-water lake in the deglaciated valley of Rathong glacier

5.4 Climate Change and Its Impact on the Human Health

The less developed poorer nations with very rapid population growth, poverty, meagre health care, economic dependency and isolation will be quite vulnerable to the human health effects of climate change. It can directly affect health because high temperature places an added stress on human physiology. Changes in temperature and precipitation followed by extreme weather events in the hilly terrain can result in deaths directly or, by altering the environment, result in an increase of infectious diseases. J. Patz of Johns Hopkins School of Hygiene and Public Health argued (*pers. comm.*) that 'the spread of infectious diseases would be the most important public health problem related to climate change'. It is perceived that most of the vector-borne diseases will probably spread even further in response to global climate change. The diseases whose hosts are temperature sensitive and currently restricted to the tropical belt, viz. malaria, schistosomiasis, yellow fever and dengue fever, will spread poleward into more temperate regions.

There is a growing evidence of global climate change and its impact on health and well-being of citizens in countries throughout the world.[1] Recent evidence suggests that the associated changes in temperature and precipitation are already

---

[1] *cf. McMichael, A.J. et. al.(1996) Climate change and human health. Geneva, Switzerlan.*

Plate 5.4 Deposition of fresh debris in the lower part of the Rathong glacier resulted from discharge of a melt-water lake

adversely affecting population health (IPCC 2007). In the field of human health, climate change poses a major and largely unfamiliar challenge (WHO 2005). The perceived picture of the climate change is being visualised by the climatologists, meteorologist, glaciologist, geologist, palaeontologist, etc. The global scale of climate change varies fundamentally from the many other familiar environmental concerns facing human kind Thus, climate change is tremendously influencing the Himalayan ecosystem and its upstream and downstream communities. The public health too depends on the efficient functioning of such ecosystem.

The health of the communities is closely determined by safe drinking water, sufficient food, secured shelter and stable social conditions. The adverse effects of climate change in Sikkim Himalaya include glacier retreat; burst of the debris-dammed lake; upward shift firn line altitude; incident of snow avalanche and forest fire; disappearing and drying of rural springs/permanent springs *turning* seasonal; and outbreak and spread of new diseases like *bird flu*, *swine flu*, *kala-azar*, *malaria*, *etc.* As per the report of the World Health Organization (2003), climate change was already estimated to be responsible for approximately 2.4 % of the worldwide diarrhoea and 6 % of malaria in some middle-income countries in 2000. It is a certain fact that the impact of climate change on human health has not been studied, analysed and determined in detail in Sikkim Himalaya, India at the present age.

5.4.1 Demographic Pattern of Sikkim

Sikkim is a small state in the Eastern Himalaya. It has a population of 607,688 persons comprising of 321,661 males and 286,027 females (Census of India 2011 – Provisional population totals). The important feature is the increase in the population from 30,458 in 1981 to 607,688 persons in 2011 with a variation in decadal growth from 93.76 % (1901) to 12.36 % in 2011.

5.4.2 Morbidity Profile in Sikkim

Information on the morbidity aspects is generally available from the health as well as from the nutrition survey reports of the government of Sikkim. Deodhar et al. (1982) reported the ten major causes of morbidity for the general population for the period of July 1978 to June 1979. The major morbidity factors described were intestinal worms, goitre, skin diseases (mostly scabies), measles, venereal diseases, pulmonary tuberculosis, respiratory illness, cough and fever, malnutrition and asthma including bronchitis. However, Chuttani and Gyasto (1993) argued that the major causes of morbidity in Sikkim were respiratory infections, gastrointestinal infection, worm infection, skin infestation, eye and ear infection, hepatitis and iodine deficiency diseases.

5.4.3 People's Perception on Human Health and the Quality of Drinking Water

An attempt has been made under the present study to explore the perceptions of the people about their presently growing health problems. This study has been done in the second half of the year 2012 to assess the impact of climate change on the health status, and the well-being of stakeholders of the three selected settlement areas of West Sikkim namely Yoksum, Tashiding, and Pelling.

This study has been based upon structured questionnaires and plan to include the stakeholders directly and indirectly affected by climate change and to assess the impact of climate change on the health of downstream communities vis-a-vis depleting quality of drinking water owing to climate change and allied activities.

It is to be noted in this context that in 2008 four cases and one death occurred owing to *kala-azar*. Whilst, only two cases were reported till March 2009 (Sikkim State Report), a number of fresh cases of *kala-azar* have been reported afterwards (in 2012) from Pendam and Tokal Bermoik in East and South Sikkim by the DMO of Singtam hospital (*pers com*). In remote areas occurrence and outbreak of such diseases often get unreported.

Table 5.1 Peoples' perception on occurrence of diseases in Yoksom, Tashiding and Pelling villages (*Sample size 50 in each settlement*)

Major health problems as perceived	Yuksom	Tashiding	Pelling	Total	Mean	SD	Std error
Respiratory problem	8	4	8	20	6.67	2.31	0.88
Skin disease	32	13	8	53	17.67	12.66	2.05
Stomach upset	16	16	16	48	16	0	0.00
Respiratory problem and skin diseases	7	2	3	12	4	2.65	0.94
Skin disease and stomach problem	10	4	3	17	5.67	3.79	1.12
Respiratory, skin and stomach problem	4	1	2	7	2.33	1.53	0.71

Source: Data generated through field survey, 2012

They further argued that outbreak of *swine flu fever among pigs* in Rinchenpong, West Sikkim, is a new type of infection among the pigs in Sikkim. They added that Upper Hathi Dhunga, Zeel and Lower Hathi Dhunga/12 Mile have been categorised as the 'Infected Zone', whilst Lower and Upper Sangadorjee and Lower Zeel and Tapel have been identified as 'Hazardous Zone' in West Sikkim.

Report of the questionnaire survey (through random sampling) on human health has been displayed in the Table 5.1.

From the recent report of study, as has been displayed in the table above, it has been revealed that out of the total of 157 people interviewed from the three villages formidable proportions suffer from skin disease and stomach upset problems, being 33.76 % and 30.57 % respectively. In addition marked proportions of the people suffer from respiratory problem and skin as well as stomach problems, being 12.74 and 10.83 %. Another feature of health hazards, as explored from this study, is that in Yoksom village the proportion of people suffering from skin disease is considerably high (41.56 %) among the total suffering from various health hazards. It is highly likely that the people of Yoksom, living nearest to the glaciated parts (thus traditionally being used to colder weather condition), have become more vulnerable to skin disease under the impact of climate change. However, the water-borne diseases like stomach problems have fairly uniform impact on the people of all the three settlements and this appears to be the second most important health problem of the residents.

5.5   Discussion

Glaciers are considered as the natural thermometer which readily responds to any atmospheric temperature change (John 2009). Hence, glacier retreat is the direct implication of global temperature rise. People living in the villages within the Kanchenjunga National Park in West Sikkim close to the presently retreating Rathong, Onglaktang and other glaciers in this part have been found to be affected by marked physical ailments like skin disease, stomach upset and others. Outbreak of these health problems among the people particularly those who live in Yuksom village (nearest to the snow and glaciers of Kanchenjunga) can be attributed to the unusual and marked rise in atmospheric temperature in recent years which in turn have contributed to rapid deglaciation (glacier retreats) in the southeast-facing slopes of Kanchenjunga. It has also been found that the perennial springs which were the regular source of drinking water for the village people continues to dry up or become feeble under drying environmental condition causing scarcity of drinking water.

It is believed that further studies with more detailed information in this regard will reveal much clearer picture of the changing health status of the people in these high-altitude areas of the Sikkim Himalaya.

Acknowledgements The entire work was conducted under the financial support of the University Grants Commission, New Delhi, for which the authors remain indebted. Authors also wish to record their gratitude to Mr. Animesh Bose, Chief Coordinator, Himalayan Nature and Adventure Foundation (HNAF), and Siliguri and his teammates for extending help during the field investigations in 1995, 2009 and 2010. The additional work on health status of the selected villages was done in November–December 2012. The villagers of Yuksom, Tashiding and Pelling extended sincere cooperation which has been gratefully acknowledged.

References

Census of India (2011) Provisional population totals. Paper 2, Vol 2 of 2011, rural-urban distribution of population series 12, p 6

Chattopadhyay GP (1985) Nature and extent of periglacial phenomena in the Kanchenjunga area, Sikkim Himalaya. Göttinger Geographische Abh (Göttingen, Germany) 81:145–152

Chattopadhyay GP (1998a) Geomorphological processes and landforms in the glacial and periglacial environments. In: Ghosh AR (ed) Environmental issues in geography. Univ. of Calcutta, Kolkata

Chattopadhyay GP (1998b) Nomenclature and mapping of glacial and periglacial phenomena in Arctic and Alpine environments. Indian Cartograph 18:55–56

Chattopadhyay GP (2007) Deglaciation and hazard of glacial lake outburst in the Alpine regions: some observations around the glaciers on the southeast-facing slopes of Kanchenjunga, Sikkim Himalaya. In: Basu SR, Dey SK (eds) Issues in geomorphology and environment. ACB Publications, Kolkata, pp 79–90

Chattopadhyay GP (2008) Recent retreats of glaciers on the southeast-facing slopes of the Kanchenjunga summit complex in the Sikkim Himalaya. Himal Geol 29(2):171–176

Chattopadhyay GP (2009) Rapid deglaciation and hazard of glacial lake outburst in the Alpine regions: some observations around the glaciers on the southeast-facing slopes of Kanchenjunga, Sikkim Himalaya. In: Hydro-geomorphic hazards. DST Publication, New Delhi

Chattopadhyay GP (2011a) Observations on the quaternary landform features around the Dzongri-HMI base camp area on the southeast-facing slopes of Kanchenjunga in Sikkim Himalaya. Adv J Geogr World I:1–9

Chattopadhyay GP (2011b) Quaternary environmental changes around the southeast-facing slopes of Kanchenjunga: an assessment with geomorphological mapping. In: Bandyopadhyay S et al (eds) Landforms processes & environmental management, vol 18, Indian Cartographer. INCA, NATMO, Kolkata, pp 55–56. ISBN 81-87500-58-1

Chattopadhyay GP (2012a) The Malanese community in Himachal Pradesh and the present trend of disastrous impact of Globalisation on their life and culture. In: Desai M, Halder S (eds) Coping with disasters. Abhijit Publications, New Delhi, pp 277–288

Chattopadhyay GP (2012b) Glacier retreats in the Himalaya and dimension of environmental hazards. In: Bhaduri S (ed) Emerging issues in geography – proceedings of the eleventh refresher course in geography. Academic Staff College and Department of Geography, University of Calcutta, Kolkata, pp 63–68

Chattopadhyay GP (2012c) Snow avalanche hazard and their mitigation in Sikkim Himalaya. Himalayan J Soc Sci 2(3):63–68. Himalaya College of Education, Gangtok

Chuttani CS, Gyasto TR (1993) Health status of women and children in Sikkim. UNICEF, New Delhi, p 112

de Terra H, Hutchinson GE (1936) Data of post glacial climatic change in northwestern India. Curr Sci 1:5–10

Deodhar NS, Srivastava SS, Phadke S, Sen AK (1982) Health conditions of mothers and children in Sikkim. Report of All India Institute of Hygiene and Public Health, Kolkata, p 147

Fuji Y, Higuchi K (1976) Ground temperature and its relation to permafrost occurrences in the Khumbu Himal and Hidden Valley, Nepal Himalayas. Seppyo 38(Special Issue):125–128

Fushimi H, Ikegami K, Higuchi K, Shankar K (1985) Nepal case study: catastrophic floods, vol 149. AHS Publication, New Delhi, pp 125–130

Gansser A (1964) Geology of the Himalaya. Wiley, London

ICIMOD (2009) Ongoing debate on the rate of the melting of the Himalayan glaciers. Comments recorded

ICIMOD, UNEP (2002) Poster presentation made at the tenth meeting of the conference of parties to the convention on biological diversity, Nagoya, Japan, 18–19 Oct 2002

IPCC (2007) Summary for policy makers. In: Adger N et al (eds) Working group II contribution to the Intergovernmental Panel on Climate Change Fourth Assessment Report, 23 pp. (unpublished). Available from http://www.ipcc.org. Accessed 13 May 2007

Ives J (1986) Glacial lake outburst floods and risk engineering in the Himalaya, Occasional paper no. 5. ICIMOD, Kathmandu

John HT (2009) Climate change – causes, effects and solutions. John Wiley and Sons, London, pp 171–184

Kar NR (1969) Some observations on periglacial features in the Sikkim and Darjeeling Himalaya. Biuletyn Peryglacjalny 21(18):43–67

Kattelmann R (2003) Glacial lake outburst floods in the Nepal Himalaya: a manageable hazard. Nat Hazards 28:145–154

Post A, Mayo LR (1971) Glacier dammed lakes and outburst flood in Alaska. Report of the hydrological investigations. US Geological Survey Publications, Washington, DC, 20242

Rana B, Shrestha AB, Reynolds JM, Aryal R, Pokhrel AP, Budhathoki KP (2000) Hazard assessment of the Tsho Rolpa Glacier Lake and ongoing remediation measures. J Nepal Geol Soc 22:563–570

WHO (2005) Using climate to predict infectious disease epidemics. Report authors: Kuhn K, Campbell-Lendrum D, Haines A, Cox J. World Health Organization, Geneva. P.7 of 54 pp

Chapter 6
Scenario of Malaria and Dengue in India: Way Forward

Ramesh C. Dhiman

Abstract The paper describes the situation of malaria and dengue in India. The latest trend of malaria indicates that with introduction of rapid diagnostic kits, combination therapy for treatment, and long-lasting insecticidal nets for prevention from mosquito bites since 2005 to 2009, cases have been reduced to around 50 % against the year 2001, and many states are qualifying for pre-malaria elimination. On the other hand, the spatial and temporal distribution of dengue is increasing gradually. Overall, there has been about ten times increase in incidence and about four times in death due to dengue in 2012 against the data of 2007. Further efforts should be directed to understand the dynamics of asymptomatic malaria in hard core malarious areas, to ensure compliance to treatment and intensified intervention measures, and to demonstrate malaria elimination in areas which are in pre-elimination phase. Dengue prevention warrants improvement in water supply, awareness of communities about prevention of breeding of *Aedes* mosquitoes, and seeking health facility in time.

Keywords Malaria • Dengue • India • Distribution • Public health control

Vector-borne diseases are a considerable public health problem globally. Malaria which constitutes about 60 % of the reported cases of Southeast Asia ranks number one in India. From the viewpoint of rapid spatial distribution, dengue takes the lead. In recent years the potential threat of climate change has been well emphasized all over the globe (WHO 2003). In countries like India, almost the whole country is endemic for malaria except a few hilly areas in Himalayan region. Recent study has projected vulnerability of Himalayan states like Jammu and Kashmir and Uttarakhand in opening the window of transmission of malaria in a few foci and intense transmission in northeastern states by the year 2030 (Dhiman et al. 2011).

R.C. Dhiman (✉)
National Institute of Malaria Research (ICMR), Sector-8, Dwarka, Delhi 110077, India
e-mail: dhimanrc@icmr.org.in

© Springer International Publishing Switzerland 2016
R. Akhtar (ed.), *Climate Change and Human Health Scenario in South and Southeast Asia*, Advances in Asian Human-Environmental Research,
DOI 10.1007/978-3-319-23684-1_6

In order to prepare ourselves for addressing the threat of climate change on malaria or other vector-borne diseases, there is need to undergo situation analysis of the existing problem so as to get better preparedness. In this context, the situation of malaria and dengue is analyzed from the viewpoint of latest introduction of new tools of intervention in India.

6.1 Situation Analysis of Malaria

Over the last decade, there has been gradual reduction in incidence of malaria. In 2001, there were around 2 million cases (2.08 with 50 % *P . falciparum*), while in 2012 for the first time since 1970, total malaria cases came down to 1.01 million with 476 deaths. This journey of success, in spite of increase in population, has been witnessed since 2006 and more particularly after 2010.

In 2006, there were 1.78 million cases with 1707 deaths (Fig. 6.1). The proportion of *P . falciparum* was 47 %. The highest API was reported from Andhra Pradesh (37.1), Meghalaya (12.98), and Mizoram (11.7) (Fig. 6.2a). As per records of National Vector Borne Disease Control Programme, in Meghalaya and Assam there were 74251 and 66557 cases of *P . falciparum* in 2009, while in 2012, only 19585 and 30945 cases were reported, respectively. Deaths due to malaria have also reduced as evident by a maximum of 95 deaths reported from Maharashtra state in 2012 as compared to 247 in 2009. However, in 2009 and 2010, there is slight increase in cases (Fig. 6.1) which may be explained due to better diagnostic tools and surveillance through ASHA workers. Overall there has been around 50 % reduction in deaths.

In 2012, there were 1.06 million cases of which 49.9 % were *P . falciparum*. Almost all the states in India recorded API not exceeding ten except Dadra and Nagar Haveli (Fig. 6.2b). There was spectacular reduction in northeastern states. Only 519 deaths were recorded.

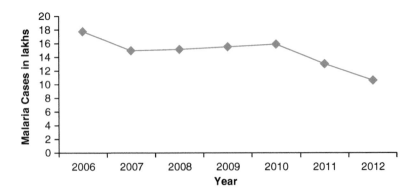

Fig. 6.1 Situation of malaria in India (2006–2012) (Source: www.nvbdcp.gov.in)

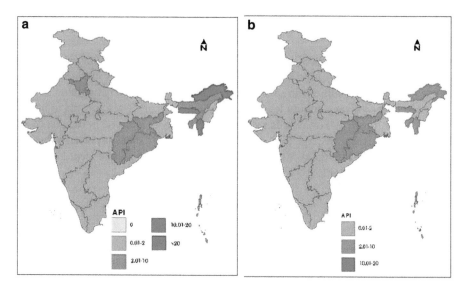

Fig. 6.2 Incidence of malaria in 2006 (**a**) and 2012 (**b**) (Source: NVBDCP)

It is evident that there has been gradual success in reduction of incidence of malaria since 2006. Hard core areas in Arunachal Pradesh, Assam, Mizoram, Meghalaya, and Manipur have witnessed sharp decline. The reason of this spectacular success is due to the augmentation of Intensified Malaria Control Programme-II launched in northeastern states with introduction of RDT in the program in 2005, ACT in areas showing chloroquine resistance in falciparum malaria in 2006 and further extended to high *P. falciparum* predominant districts covering about 95 % *P. falciparum* cases, and introduction of LLNs in 2009. The IMCP-II covered overall 94 districts (65 districts in 7 NE states, 7 districts in Jharkhand, 6 districts in West Bengal, and 16 districts in Orissa) with focus on early case detection and prompt treatment in remote and inaccessible areas through rapid diagnostic kits and involvement of Accredited Social Health Activist (ASHA) and enhancement of awareness about malaria control through community mobilization and private sector participation.

In spite of remarkable success of malaria control in northeastern states, reduction in Odisha, Jharkhand, and West Bengal was not commensurate with the efforts. It may be due to less coverage of districts, difference in usage of LLNs, educational status of communities and insurgency, etc. The reason of persistent malaria in eastern foci is basically due to climate suitability, forest cover providing suitable Relative Humidity (RH) thus stability of transmission in almost all the months. Inaccessibility to hilly areas, jhum cultivation in northeastern states, scattered human dwellings, language barrier in tribal areas, and lack of awareness about the usefulness of intervention instituted by the health authorities are some of the reasons of persistence. Technically, the existence of asymptomatic infections, noncompliance to treatment, and nonacceptance of IRS or plastering of houses with whitewash/mud are the main impediments in control efforts.

6.2   Situation Analysis of Dengue

On the other hand, the incidence of dengue fever (DF) has been increasing, and
more and more states are reporting dengue fever (DF) with inroads in villages. With
better economical conditions in some geographic areas particularly in plains, urbanization with piped water supply leading to water storage practices appears to be the
major reason. From 2007 onwards there has been gradual increase in dengue cases
in the country except in 2011 (Fig. 6.3). In 2007, only 18 states reported a total of
5534 cases of DF (Fig. 6.4a), while in 2012 all the states of India except Nagaland
reported 49602 cases (Fig. 6.4b).

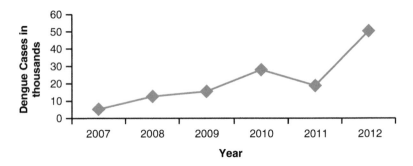

Fig. 6.3 Current situation of dengue in India (2007–2012) (Source: NVBDCP)

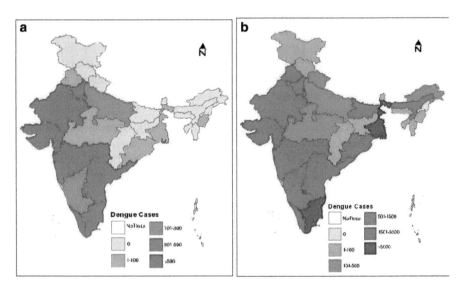

Fig. 6.4 Incidence of dengue in 2007 (**a**) and 2012 (**b**) (Source: NVBDCP)

Almost the whole country is climatically suitable for occurrence of indigenous dengue (unpublished work of the first author undertaken under the aegis of NATCOM II, Ministry of Environment and Forests, Government of India). With urbanization the spatiotemporal distribution is further expanding. Of the 34 states in India, only the state of Nagaland is still free. There has been an increase in all the states which were reporting dengue except Manipur which showed marginal decrease in 2012; however, in 2011 there were 220 cases in Manipur state. Since 2007, 15 states including north and northeastern states, which were free of dengue, have also reported cases in 2012 (Fig. 6.4b). In 17 states, increase in incidence has been witnessed. Eleven states witnessed cases in thousands and there was many-fold increase as compared to 2007. Overall, there has been about ten times increase in incidence and four times in death due to dengue in 2012 against the year 2007. It signals that DF is going to be a major public health problem in India.

Control of dengue relies basically on vector control through source reduction, which cannot be achieved without the cooperation and involvement of communities. Further, there has been better understanding about the preferred breeding places of *Aedes aegypti* mosquito in peridomestic scraps and household wastes etc. (Arunachalam et al. 2010). Imparting awareness to communities about the breeding places of *Aedes* mosquito, its life cycle, and symptoms of dengue fever, emptying the containers, and seeking health facility as early as possible are still the desired options.

6.3 Way Forward

The analyses of scenario of malaria in the country indicate that the incidence is gradually coming down in hard core areas like northeastern states. Still there are some challenges like estimating actual burden by capturing cases treated by private practitioners (Dhingra et al. 2010; Hay et al. 2010), detection of asymptomatic cases in tribal areas (Ganguly et al. 2013; Karlekar et al. 2012), reaching to inaccessible areas, and acceptance of all interventions by the communities for better compliance (Sharma et al. 2001). The use of polymerase chain reaction (PCR)-based diagnosis in field conditions also looks imminent owing to its efficacy to detect submicroscopic infections (Golassa et al. 2013).

The future scenario of malaria, in spite of various threats like climate change and resistance to antimalarials in *P . falciparum* and *P . vivax* parasites, can be addressed with improved surveillance, identification of foci of persistent transmission, tackling the issue of migration, etc. in addition to ongoing strategies of intervention. The strategic plan of NVBDCP envisages better coverage for more fruitful outcome (http://nvbdcp.gov.in/Round-9/Annexure-2). Of 99 countries having malaria transmission, 9 are in malaria elimination phase, but India has yet to switch over from control to pre-elimination (malaria elimination group). At present also some states like Karnataka, parts of Rajasthan, Maharashtra, Gujarat, Himachal Pradesh, Punjab and Uttarakhand which are having very low API for the last 5 years may be attempted for malaria elimination. There is need of periodic surveys in malaria-free areas at the fringe of endemic areas in Himalayan region to watch the indigenous transmission due to favorable climatic changes.

The future scenario of dengue control in India appears grim due to various confounding factors. As per 12th plan document of planning commission, about 60 % of the growth in urban population is due to natural increase, whereas rural-urban migration has contributed to only 20 %. The number of towns in India increased from 5161 in 2001 to 7935 in 2011. Most of this increase was in the growth of 2532 towns. By 2030, India's largest cities will be bigger than many countries today; therefore, India's urban governance of cities needs an overhaul. Owing to limited water supply, communities have to resort to water storage. Improper disposal of household wastes in peridomestic environment, which has also been found as a preferred breeding ground of *Aedes aegypti* mosquitoes (Arunachalam et al. 2010), warrants careful and advanced planning of water supply, storage, and sanitation facilities. Risk factors for contracting dengue in terms of strata of human density and production of *Aedes aegypti* in urban areas (Padmanabha et al. 2012) also need to be elucidated for targeted intervention. The role of *Aedes albopictus* also needs to be defined in urban or peri-urban environment for devising suitable control strategy.

Since the management through source reduction of breeding grounds of *Aedes* vectors lies mainly at community level, health education should be the mainstay for tackling the problem by improving health-seeking behavior of patients, easily accessible diagnosis and management, and preventive vector control at community level.

References

Arunachalam N, Tana S, Espino F, Kittayapong P, Abeyewickreme W, Wai KT, Tyagi BK, Kroeger A, Sommerfeld J, Petzold M (2010) Eco-bio-social determinants of dengue vector breeding: a multicountry study in urban and periurban Asia. Bull World Health Organ 88:173–184. doi:10.2471/BLT.09.067892

Dhiman RC, Chavan L, Pant M, Pahwa S (2011) National and regional impacts of climate change on Malaria by 2030. Curr Sci 101(3):372–383

Dhingra N, Jha P, Sharma VP, Cohen AA, Jotkar RM, Rodriguez PS et al (2010) Adult and child malaria mortality in India: a nationally representative mortality survey. Lancet 376:1768–74

Ganguly S, Saha P, Guha SK, Biswas A, Das S, Kundu PK, Maji AK (2013) High prevalence of asymptomatic malaria in a tribal population in eastern. India J Clin Microbiol 51(5):1439–44. doi:10.1128/JCM.03437-12. Epub 2013 Feb 20

Golassa L, Enweji N, Erko B, Aseffa A, Swedberg G (2013) Detection of a substantial number of sub-microscopic *Plasmodium falciparum* infections by polymerase chain reaction: a potential threat to malaria control and diagnosis in Ethiopia. Malar J 12(1):352. doi:10.1186/1475-2875-12-352

Hay SI, Gething PW, Snow RW (2010) India's invisible malaria burden. Lancet 376(9754):1716–1717. doi:10.1016/S0140-6736(10)61084-7

Karlekar SR, Deshpande MM, Andrew RJ (2012) Prevalence of asymptomatic *Plasmodium vivax* and *Plasmodium falciparum* infections in tribal population of a village in Gadchiroli District of Maharashtra State, India. Biol Forum Int J 4(11):14:2-142-44

Malaria Elimination Group. http://www.who.int/malaria/areas/elimination/en/

National Vector Borne Disease Control Programme. www.nvbdcp.gov.in

Padmanabha H, Durham D, Correa F, Diuk-Wasser M, Galvani A (2012) The interactive roles of *Aedes aegypti* super-production and human density in dengue transmission. PLoSNegl Trop Dis 6(8):e1799. doi:10.1371/journal.pntd.0001799

Sharma SK, Pradhan P, Padhi DM (2001) Socioeconomic factors associated with malaria in a tribal area of Orissa, India. Indian J Public Health 45:93

Urbanization in India 12th Plan. http://planningcommission.nic.in/hackathon/Urban_Development.pdf

World Health Organization (2003) Climate change and human health – risk and response. Summary. WHO. WHO Publication, pp 1–37. ISBN 9241590815

Chapter 7
Urbanization, Urban Heat Island Effects and Dengue Outbreak in Delhi

Rais Akhtar, Pragya Tewari Gupta, and A.K. Srivastava

Abstract The present paper is the outcome of instigated thought by several studies on urbanization and prevalence of vector-borne disease in tropical areas. It is envisaged and hypothesised to find if any correlation exists between urbanization process, heat island generated due to urbanization (e.g. higher density of road network, buildings and traffic) in the urban areas and the outbreak of vector-borne diseases like malaria and dengue. The paper primarily looked into the temporal data of all metropolitan cities and found that there is an increased incidence of dengue outbreak in all metropolitan cities. There is a lot of variability in the rainfall in Delhi, but all other metropolitan cities have been experiencing an average rainfall over the past two decades. Review of literature led to the construction of hypothesis that there exists a close association between urbanization and the increased incidence of dengue outbreak. Ordinary least square method is used to find if there is any correlation between the urbanization variables and disease outbreak. Regression result shows that there is high possibility between the urbanization and disease outbreak.

Keywords Urbanization • Urban heat island • Vector-borne disease • Rainfall • Temperature

7.1 Introduction

Arthropod disease viruses such as yellow fever, malaria, encephalitides, filariasis and dengue fever are changing the health epidemic scenario mainly in developing countries, especially in the tropical countries. These are the most important arboviral diseases in human both in terms of mortality and morbidity. Dengue fever

R. Akhtar (✉)
International Institute of Health Management and Research (IIHMR), New Delhi, India
e-mail: raisakhtar@gmail.com

P.T. Gupta
National Institute of Urban Affairs, New Delhi, India

A.K. Srivastava
India Meteorological Department, Pune, India

© Springer International Publishing Switzerland 2016
R. Akhtar (ed.), *Climate Change and Human Health Scenario in South and Southeast Asia*, Advances in Asian Human-Environmental Research,
DOI 10.1007/978-3-319-23684-1_7

99

Table 7.1 Dengue fever prevalence in the metropolitan cities in India (no. of patients)

Metropolitan cities	2002	2003	2004	2009	2010	2011	2012
Delhi	45	2882	606	1153	6259	1131	2093
Mumbai	135	328	312	46	32	421	768
Kolkata	0	0	0	254	437	254	3197
Chennai	146	634	571	121	301	1091	941

Source: Indiastat.com

because of its persisting trend in countries can be considered the most common of all arthropod-borne diseases after malaria. Each year an estimated 100 million cases of dengue fever occur worldwide with the greatest risk occurring in Indian subcontinent, Southeast Asia, Southern China, Taiwan, the Pacific Islands, the Caribbean, Africa and Central and Southern America. Though DF is hardly fatal, the disease incapacitates the infected patients in several ways. Over half of WHO member states representing a total of 2000 million are threatened by dengue. It occurs in most of the tropical countries. The dengue infection ranges from asymptomatic through classical dengue to life-threatening dengue hemorrhagic fever with shock (DSS) or without shock (DHF).

7.2   Indian Context

Dengue is a threat in India, especially in the capital city of Delhi, Kolkata and Chennai where the number of reported dengue cases and deaths is higher and trend shows that it may go even higher in the future (Table 7.1). Epidemiologically, dengue fever/dengue hemorrhagic fever is not only changing in disease pattern but has changed its course of manifestation in the form of severe DHF with increasing number of outbreaks. The table also indirectly proves that the disease is more of an urban disease.

Possible cause which has been hypothesised in this paper is the growth of urban structures (buildings, roads and other urban infrastructures) which has been causing rise in temperature by creating urban heat island effect in the main city leading to increased frequency of rain and humidity causing vectors to reproduce at an alarming rate.

7.3   Methodology and Data

The paper tries to assess the series of outbreaks and expected causes of persisting phenomena of disease by finding the climatic variables – average rainfall, average maximum temperature, average annual temperature and some select urbanization-related indicators such as urban density, urban population and vehicular load on roads with the disease data in the metropolitan cities. Procuring the disease data on zone level was difficult, and also no direct data to indicate the urban expansion (until interpretation of satellite images at different points of time) and outbreak of the disease could be utilised. With such limitation, the increased burden of population in the city, urban density and traffic on the roads are considered as the best alternative to analyse urbanization in Delhi.

The paper therefore first looks into the morphology of the urban growth considering urban density and population growth and then further analyses this growth in the context of disease pattern, rainfall and temperature oscillation in Delhi. As already mentioned, the paper hypothesises that the prevalence of dengue is closely linked with the heat entrapped by the urban structure accentuating the temperature and humidity, thus helping *Aedes* mosquitoes to breed faster and in larger number.

Besides plotting the time series data of rainfall, temperature and dengue prevalence from 1996 to 2013 in Delhi, the study also tries to find if there is any changes occurring in some other Indian metropolitan cities. The study tries to statistically estimate the effect of urban growth on disease of dengue by taking the help of ordinary least square method only in the case of Delhi.

7.4 Overview of the Literature: Pragmatic Issues in Urbanization, Urban Heat Island Effect and Health

Literature and studies on climate change and epidemic is fewer but gripping faster to establish climate change and epidemic in tropical countries. Frequent outbreak and consistency in dengue fever attributed to the climate change explored by the researchers at Imperial College London and WHO Kobe Centre in Bangkok in 2008. The report asserts and declares dengue an urban disease (Gubler et al. 2001) because vectors breed mainly in containers such as pots and tyres. There are studies which have reported the climatic change-influencing factors (Napier 2003; Chakravarti and Kumaria 2005; Githeko et al. 2000) such as extreme humidity, extreme temperature, El Niño and human development factors. The current paper envisages that the seasonal or cyclic factors of the disease are combined with the fluctuating humidity and temperature data with the urban density (a proxy variable of urban growth) which creates heat effect in some of the dense area and thus leads to pocketed outbreak of the disease. The "tiger" mosquito *Aedes aegypti* is quoted to be very sturdy and travels up to 500 m. *Aedes aegypti* is extremely common in areas lacking piped water systems and depends greatly on water storage containers to lay eggs.

The mosquito which feeds exclusively on human blood is considered well adapted in the urban surrounding. A study in Delhi metropolitan area says that the disease has changed from being benign to becoming a more frequent and severe form of DHF (Gupta et al. 2006). Delhi has faced seven outbreaks of dengue infection as reported by the study, which may have gone up by two digits considering

2010 and 2013 outbreak years. The study found all four dengue serotypes present in the blood samples, yet in 2003 serotype 3 was more prevalent. A time series study done by Chaosuansreecharoen and Ruangdej in Yala Province, Thailand, found that temperature was positively associated with dengue incidences and relative humidity and monthly rainfall were not positively associated with the dengue incidences in this study area in Thailand (Chaosuansreecharoen and Ruangdej 2014). Another study on the climate change and its impact on vector-borne diseases assessing on continental basis says that the global temperature will rise by 1.0–3.5 °C by 2100 increasing the risk of more outbreak and wider spatial spread of vector-borne diseases by stating that the arthropods can regulate their internal temperature by changing their behaviour but cannot do so physiologically and thus are essentially dependent on climate for their continued existence and reproduction. The authors also propose that the effect of climate change on transmission is more likely to be observed in the extremes of the range of temperature (within the range of 14–18 °C at lower bound and 35–40 °C at the upper bound) (http://www.eliminatedengue.com/faqs/index/type/aedes-aegypti).

In Asia, the temperature has increased by 0.3–0.8 °C across the continent in the past 100 years and is projected to rise by 0.4–4.5 °C by 2070. The authors appear to be fully convinced by citing the incidences in highland Kenya that along with several other factors like seasonal variations, socio-economic status, vector control programmes, drug resistance, climate change and variability are highly likely to influence current vector-borne disease epidemiology. A study in Dhaka found alternate fluctuations in the outbreak of dengue every second year which was observed in the study conducted in Thailand also. The authors give reason where the outbreak was considered urban, and 70–80 % of the population developed herd immunity to the circulating dengue serotypes to thus prevent another outbreak in the following year. The study found some linear relationship with the climatic factors and the outbreak of the dengue fever (Karim et al. 2012). Another study was primary research based on blood samples of 1550 patients in Delhi. In this study, the result showed the large number of serological positive cases coinciding with post-monsoon period of subnormal rainfall and concludes that high rainfall, temperature and relative humidity can be the important climatic factors responsible for an outbreak (Chakravarti and Kumaria 2005).

Sharma and colleagues assert that *Aedes aegypti*, the primary vector of dengue fever, has global distribution, and it is invading areas under urbanization, citing several cities all over the world where different types of vectors have been traced in urban areas (Sharma et al. 2002).

7.5   Discussion

The review of literature and other evidences pinpoints to find the relationship between urbanization and dengue infection prevalence. The first section of the discussion revolves around urbanization, rainfall and temperature variability in Delhi.

This section also briefly looks at the temperature and rainfall distribution in other metropolitan areas in India to find out if the cases of dengue have any correspondence with the flux of rainfall or temperature in the cities.

The second section evolves out of statistical analysis of the data on climatic variables and disease prevalence in the city of Delhi. The variables which have been used as indicators of urbanization are urban density, urban population and vehicular population in Delhi, while for climate variables, average annual rainfall, mean maximum temperature and mean annual temperature are considered to find the climatic influence on the disease. It can be argued that the variables selected for urbanization and climate are insufficient to segregate or decide on the heat island effect and climate influence on the disease prevalence; however, the paper heavily depends on the availability of the secondary data. Several limitations are encountered, e.g. unavailability of zone-wise data of the disease, temperature, and rainfall which could have provided a better relationship between disease, urbanization and climate-related change at micro level. The regression model nevertheless and correlation matrix do good justice to the hypothesis of urbanization, urban heat island effect and dengue prevalence.

7.6 Section I: Urban Expansion of Delhi

Satellite image of Delhi clearly (Fig. 7.1) shows that population expansion in and around the Delhi metropolitan city has increased tremendously. The upper images are a 1974 Landsat MSS (80 m resolution) on the left and a 1999 Landsat Thematic Mapper (28.5 m resolution) on the right. Growth in urban area from 1974 to 1999 is portrayed in the lower maps (areas coloured red). The image clearly shows the growth of the urbanised area has been tremendous in just two decades.

Recent release of the census data provides a clear indication that urban expansion has been tremendous as the density in all districts has reduced except Central and New Delhi (Fig. 7.2). However, a look at the density of Delhi reveals that some of the areas which may have geographically expanded could not possibly reduce the aggregate population density of Delhi which is higher than the 2001 census figures.

7.7 Rainfall and Temperature in Metropolitan Cities

Temperature rise and temperature extremities have been studied at global scale and noted that even rise of temperature as low as 1° may expand the coverage of vector-borne diseases like malaria and dengue in temperate zone as well. There are incidences of increased frequency of flooding of urban areas. With a look at the average annual temperature in all four major metropolitan cities and the bearable limit within one degree of the temperature, then it can be concluded that the temperature

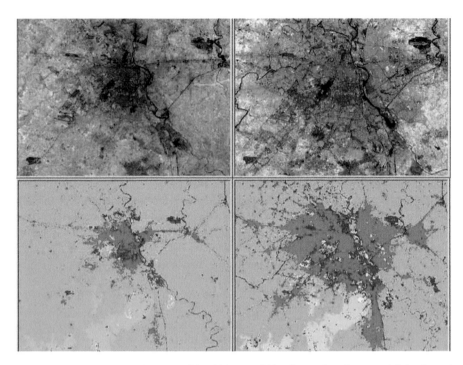

Fig. 7.1 Population expansion in Delhi (1975 and 1999) (Source:http://www.mdafederal.com/
environment-gis/national-security-policy-support/land-use/)

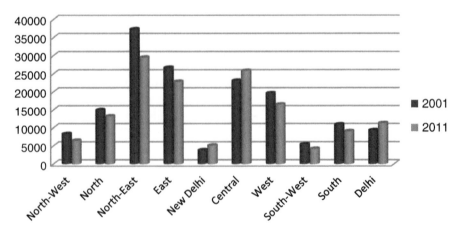

Fig. 7.2 Population density in the districts of Delhi

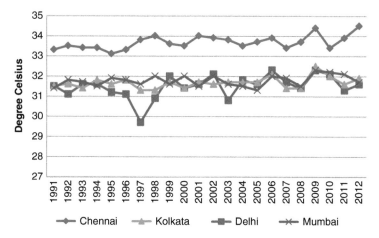

Fig. 7.3 Average annual maximum temperature in the metropolitan cities

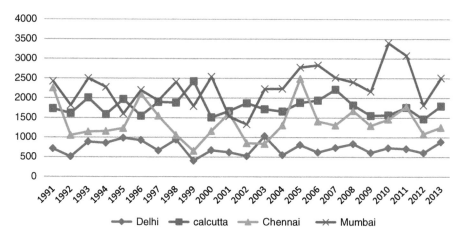

Fig. 7.4 Average annual rainfall in the metropolitan cities (in millimetres)

has been shooting up beyond the recorded average temperature in all cities (Fig. 7.3). Delhi in particular has recorded annual average temperature either below the boundary of 1° (i.e. 1997, 2003) and sometimes above (2002, 2006 and 2009). In recent years, even Kolkata and Mumbai also experienced slightly hot years on an average, whereas Chennai has shown a trend of rising temperature consistently.

The average annual rainfall shows fluctuation especially in Mumbai, Chennai and Kolkata. Delhi however remained within the 500–1000 mm and recorded very regular rainfall for the past two decades (Fig. 7.4). The data on temperature and rainfall is only indicative to measure the climate change or heat island effect; we have to analyse the geographical variability also of rainfall or temperature to conclude the epidemic outcome. Nonetheless, the data still shows the rising trend of fluctuating rainfall and temperature. Further, we are not looking at the monthly variation of rainfall and temperature which is beyond the scope of this paper. This aspect if attempted may reveal some more intriguing aspects of climate change and disease in the urban areas.

7.8   Section II

The cause of heat island effect may be not clearly perceptible but certainly has greatly influenced the span of humidity, longer rain spell thus becoming fertile time for the mosquitoes to breed and increased scope of infecting people also. The studies prove that the longer the duration of humidity and overcast sky leads to increased outbreak of vector-borne diseases especially of different strains. The National Vector-Borne Disease Control Programme and other microbiologists claimed the emergence of strain 2 which is more dangerous than earlier strains. Kolkata has recorded three times more dengue cases than Mumbai and Chennai. In sheer numbers, Delhi is a close second with 2,068 cases of the mosquito-borne diseases in 2012. Kolkata leads among the metros with the highest number of dengue cases (3321). Mumbai and Chennai have recorded 907 and 988 cases of dengue, respectively, in 2012. The number of dengue cases increased in Kolkata which rose from 199 in 2011 to as high as 3,321 in 2012. Mumbai's dengue cases have increased by 283 %, while Chennai has seen over a 100 % rise in dengue cases in 2012. Delhi, which has recorded a 186 % increase in dengue cases in 2012 as compared to 2011, is confronting an outbreak of the most dangerous and virulent form of dengue virus.

Simple time series data however in itself is insufficient to predict the outbreak of dengue, yet the analysis of the dengue prevalence, monthly rainfall and temperature data does indicate that there exists a relationship in the climatic data, temperature and rainfall. Relative humidity is a better measurement to establish the association (if any) between the disease outbreak and the heat island effect. However, data constraints limit the scope of understanding this aspect of hypothesis. Close examination of Figs. 7.5, 7.6 and 7.7 shows a relationship between the disease prevalence temperature and rainfall. Looking at the rainfall and temperature, it is easier to locate more humid year along with all other years; 1996, 2001, 2004, 2006 and 2009 have experienced higher average temperature from June to August, and as observed in Fig. 7.6, which shows an analysis of average monthly rainfall, we find that the rainfall during the month of monsoon has considerably reduced, but its span has increased. Average monthly rainfall in 1993, 1994, 1995, 2003 and 2011 peaked during the monsoon time, but when we see the epidemic outbreak of dengue we find that when the temperature hovers around the average temperature but experience high and longer spell of rainfall disease is more likely to reach a level of outbreak. The paper as discussed does not reveal positive relationship between heat island effect and epidemic but leads to indicate the correlation and explanatory strength of the model.

The paper interprets the relationship existing among all variables by running Pearson's correlation. It can definitely be concluded in two things: firstly, the num-

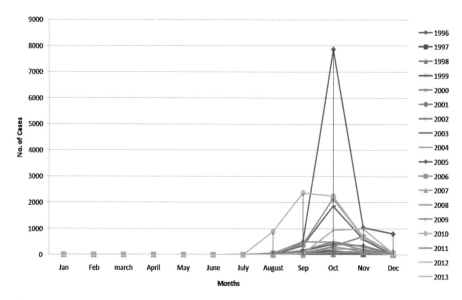

Fig. 7.5 Seasonal variation in dengue prevalence: 1996–2012

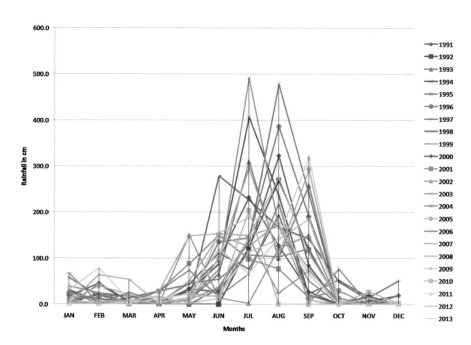

Fig. 7.6 Average monthly rainfall in Delhi {1991–2013}

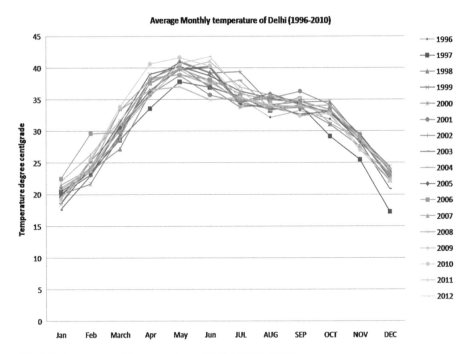

Fig. 7.7 Average monthly temperature of Delhi {1996–2010}

ber of cases reported has a significant positive relationship at 0.05 level with rainfall and, secondly, rainfall has no relationship with the urban density or vehicular population. However, the mean maximum temperature and mean annual temperature show significant positive relationship with the urbanization variables – urban population, urban density or vehicular population (Table 7.2).

Table 7.2 does not determine the strength of the variables. Therefore, the regression was done, and the study is able to explain 57 % of the total model which is reasonably acceptable explanation and shows association between disease prevalence, urbanization and climatic factors. Another important feature is the significant p values where values lower than <.05 are for rainfall and vehicular population. This suggests that the man-made factors such as vehicular density on roads leave higher impact on the disease outbreak and prevalence in the city (Table 7.3).

7.9 Some Points to Ponder on the Rainfall and Temperature Pattern in the Metropolitan Cities

Based on the regression results and the rainfall and temperature pattern in the metropolitan cities including Delhi, it can be easily said that the variability in the minimum temperature has been very low in the past two decades; however, the average

Table 7.2 Correlation matrix of the variables

		Correlations					
	Total cases	Rainfall	Mean max. temp.	Mean annual temp.	Urban population	Urban density	Vehicular
Total cases	1	.496*	0.061	0.052	0.034	0.045	0.164
		0.036	0.809	0.837	0.895	0.86	0.515
Rainfall	.496*	1	−0.287	−0.271	−0.02	0	0.022
	0.036		0.248	0.276	0.939	0.997	0.929
Mean max. temp.	0.061	−0.287	1	.997**	.533*	.533*	0.403
	0.809	0.248		0	0.023	0.023	0.097
Mean annual temp.	0.052	−0.271	.997**	1	.539*	.539*	0.405
	0.837	0.276	0		0.021	0.021	0.095
Urban population	0.034	−0.02	.533*	.539*	1	.997**	.948**
	0.895	0.939	0.023	0.021		0	0
Urban density	0.045	0	.533*	.539*	.997**	1	.935**
	0.86	0.997	0.023	0.021	0		0
Vehicular population	0.164	0.022	0.403	0.405	.948**	.935**	1
	0.515	0.929	0.097	0.095	0	0	

*Correlation is significant at the 0.05 level (2-tailed)
**Correlation is significant at the 0.01 level (2-tailed)

maximum temperature is showing an increasing trend in all metropolitan cities for the same duration (Figs. 7.1 and 7.2). Annual average rainfall in the metropolitan cities is also showing a very high variability which could be due to other global influences, e.g. El Niño impact (Fig. 7.3). Delhi which has expanded remarkably over the last two decades is showing a rising trend of dengue prevalence which is though worrisome yet could be better predicted depending on the urban growth and population density in the particular areas. If rainfall for all other years is not considered except the rainfall in the year of outbreak of dengue, the study noted interesting features. For instance, July has recorded higher rainfall in the outbreak years except year 2010 leading to fertile ground for mosquito breeding in the forthcoming months by increased humidity and temperature (Figs. 7.4 and 7.5). Models have been developed on the basis of entomological, parasitological and also GIS-based meteorological variables to predict the outbreak of vector-borne diseases particularly malaria as both mortality and morbidity have been a cause of concern during malaria outbreak. In one such study based on the meteorological explanation, it is derived that when the mean monthly temperature is between 16.0 and 33.60 and relative humidity does not exceed 61 %, the correlation became stronger to predict the malaria outbreak. With all limitations, the paper quite ably put forth the question of

Table 7.3 Regression and coefficients of the variables

		Model summary		
Model	R	R square	Adjusted R square	Std. error of the estimate
1	.757a	0.574	0.341	2216.317

(a) Predictors: (constant), vehicles, rainfall, mean annual temp., urban density, mean max. temp., urban population

	Coefficients(a)				
	Unstandardised coefficients		Standardised coefficients		
	B	Std. error	Beta	t	Sig.
(Constant)	−37623.393	32423.186		−1.16	0.27
Rainfall	88.469	37.988	0.522	2.329	0.04
Mean max. temp.	7545.605	11782.135	1.755	0.64	0.535
Mean annual temp.	−5261.545	11642.57	−1.247	−0.452	0.66
Urban population	−0.008	0.005	−5.1	−1.483	0.166
Urban density	6.655	6.4	3.143	1.04	0.321
Vehicles	0.003	0.001	1.842	2.413	0.034

a. Dependent variable: total cases

urban planning and disease control especially urban diseases like dengue and malaria. Urban heat island greatly influences the relative humidity and may cause disease prevalence at an alarming rate besides causing other problems like uncomfortable living, boils, depression, etc. The paper also attempts to attract the plight of the poor people especially those who are living at the marginal ends of the city both at economic and geographical location. Advertisements, fuming and DDT spray are becoming ineffective at the time of disease outbreak. What is important is to reduce the cause of heating the core of the cities which is caused by traffic load, lack of green areas and emission of hazardous gases through air conditioning. It is all the more a cause of concern as the dengue epidemic is now becoming more visible in other metropolitan cities too. Another conclusion is that the reporting of the dengue cases is primarily by the hospitalisation cases and there is severe underreporting also by the government departments.

The present paper is a small step to construct the hypothesis of urban influence on dengue which has baffled public health experts and planners to take up preventive measures in the coming years. It is even more important to ascertain the effect of urbanization as the disease has been called an urban epidemic by the WHO and other studies.

Acknowledgement The authors are extremely thankful for helping in procuring the data and their suggestions on the paper to Dr. D. N. Singh (Commissioner, MCD) and Dr. N. K. Yadav (Municipal Health Officer, Municipal Corporation of Delhi).

References

Chaosuansreecharoen P, Ruangdej K (2014) A time series analysis of dengue incidence and weather factors in Yala Province, Thailand. Int J Health Wellness Soc 4. International conference on health, wellness and society, Vancouver, Canada, 14–15 March 2014

Chakravarti A, Kumaria R (2005) Eco-epidemiological analysis of dengue infection during an outbreak of dengue fever, India. Virol J 2:32

Githeko AK, Lindsay SW, Confalonieri UE, Patz JA (2000) Climate change and vector-borne diseases: a regional analysis. Bull World Health Organ 78(9):1136–1147

Gubler DJ, Reiter P, Ebi KL, Yap W, Nasci R, Patz JA (2001) Climate variability and change in the United States: potential impacts on vector- and rodent-borne diseases. Environ Health Perspect 109(suppl 2):223–233

Gupta E, Dar L, Kapoor G, Broor S (2006) The changing epidemiology of Dengue in Delhi, India. Virol J 3:92, http://www.eliminatedengue.com/faqs/index/type/aedes-aegypti

Karim MN, Munshi SU, Anwar N, Alam MS (2012) Climatic factors influencing dengue cases in Dhaka City: a model for dengue prediction. Indian J Med Res 136(1):32–39

Napier M (2003) Application of GIS and modeling of dengue risk areas in the Hawaiian Islands In: Proceedings of the 30th symposium for remote sensing of environment: information for risk management and sustainable development, International symposium on remote sensing of environment, Tucon, AZ, 260–263

Sharma RC, Sharma A, Mathur AC (2002) Forecasting of malaria: a challenge yet to be met. In: Sharma A (ed) Epidemiology, health and population. B.R. Publishing Corporation, Delhi, 209–211

Chapter 8
Urban Industrial Development, Environmental Pollution, and Human Health: A Case Study of East Delhi

R.B. Singh and Aakriti Grover

Abstract The process of urbanization poses threats and opportunities in the urban environment. On the one hand, urbanization leads to development, but on the other as the population increases, the land use/cover changes, pollution increases, and there is alteration of urban heat balance. The combination of heat and pollution acts as a major health threat. This paper, hence, analyzes the historical growth and spread of industries in East Delhi. Further, the impact of growth of industries on human health in East Delhi is examined with the help of primary survey. The research articles on land surface temperature and land use/cover change act as a base to explore the existence of urban heat island in Delhi. It is found that the urban environment of Delhi has undergone rapid shift. This shift is accompanied with conversion of land use/cover and rise in air pollutants. The health hazard in these regions is high and needs urgent attention.

Keywords Land surface temperature • Urban heat island • Industrial growth • Health • Air pollution • Delhi

8.1 Introduction

Industry is key sector in the economic development and prosperity of any nation. Growing industrial sector is crucial for economic development. Ensuring steady industrial growth helps to compliment and sustain continued economic

R.B. Singh (✉)
Department of Geography, Delhi School of Economics, University of Delhi,
Delhi 110007, India
e-mail: rbsgeo@hotmail.com

A. Grover
Department of Geography, Swami Shraddhanand College, University of Delhi,
Delhi 110036, India
e-mail: aakriti.grover@gmail.com

© Springer International Publishing Switzerland 2016   113
R. Akhtar (ed.), *Climate Change and Human Health Scenario in South and Southeast Asia*, Advances in Asian Human-Environmental Research,
DOI 10.1007/978-3-319-23684-1_8

development. Rapid population growth, industrialization, and urbanization accompanied by growing number of vehicles in India are adversely affecting the environment. Though the relationship is complex, population size and growth tend to expand and accelerate these human impacts on the environment (Roy et al. 2011). Associated with this are the land use/cover changes wherein the vegetated land cover and water bodies are converted to concrete areas, thereby changing the heat budget of the city (Singh et al. 2014). The combination of land use/cover change, increased temperature, and varied kinds of air pollutants are sources of microclimatic changes that pose major health problems in large cities of the world.

The impermeable surface, greenhouse gases (GHG), and concretization contribute to increased urban temperature. The phenomenon of creation of island of elevated temperature in comparison to the lower temperature in hinterland is called urban heat island (UHI). Long-term exposure to high temperature and pollutants poses health risk across geographical location. Ebi and McGregor (2008) examined the role of ozone and suspended particulate matter (SPM) in climate change and human health. The ground level ozone is created by reactions occurring between oxides of nitrogen (NOx) with volatile organic compounds (VOCs) in the presence of sunlight and high temperature. The sources of NOx and VOCs are burning of fossil fuels. In urban areas, these are generated in large quantities mainly by vehicles and industries. The increased concentration of ground level ozone and SPM has considerable effect on human health. There is rise in asthma, pneumonia, chronic obstructive pulmonary disease, allergic rhinitis, and other respiratory diseases. Lo and Quattrochi (2003) indicate the interrelationships among land use/cover change, urban heat island, and its implications on health for Atlanta metropolitan area using Landsat data. The land use/cover changes have been analyzed for a period of nearly 25 years from 1973 to 1998. The changes have been calculated for six categories including high-density urban use, low-density urban use, cultivated/exposed land, cropland/grassland, forest, and water. Urban expansion at the cost of forest, rural built up, and cropland is observed. Since land surface temperature (LST) is sensitive to land use changes, the apparent increase in LST is noted in the center of the metropolitan area accompanied with lowering of normalized difference vegetation index (NDVI). Further, the UHI phenomenon and health are correlated with respect to respiratory and cardiovascular diseases that are a result of heat stress and ground level ozone.

Bentham (1992) on the other hand diverts attention to stratospheric ozone depletion and its effects due to increased GHG because of air pollutants released particularly in urban centers. According to him, ozone depletion will increase the temperature and harm human health as the incidence of heat waves will increase. Also, the frequency of climatic disasters will increase leading to loss of lives. The linkage between urban environment and heat waves is well researched. Filleul et al. (2006) established a relationship between temperature, ozone, and mortality in the nine French cities during the 2003 heat wave incidence. The role of UHI on heat waves and human health is presented by Tan et al. (2010). They concluded that heat-related mortality is higher in the inner city as compared to the outskirts, and there is a strong correlation between the thermal conditions of the city and increased mortality events. Parallel results are found by Tomlinson et al. (2011) for Birmingham, UK. Bush et al. (2011) investigated the impacts of climate change on public health in India. They assert that minute increase in GHG, generated by industries and transport, has infused major changes in the climate regime of the country by influencing human health. Heat stress and air pollution are major causes of mortality.

The state of human health globally serves as a key indicator for the conditions of the natural environment and the success of sustainable development (Izhar 2004). While sound development is not possible without a healthy population, lack of development adversely affects the health of many people. Health has direct relation to environmental surroundings (Akhtar 2002). The analysis of WHO has clearly established the interdependence among the factors of health, environment, and development. Polluted air, water, and soil are recognized as major cause for diseases to occur. Harmful effects of air pollution on human health are well recognized and deeply researched. Globally, 1.1 billion people breathe polluted and unhealthy air (UNEP 2002).

It may be recalled that about 80,000 people die every day from diseases related to air pollution exposure across the globe. In addition, air pollution is responsible for 4.6 million lost life years every year (WHO 2005).

Our planet is undergoing a profound demographic shift from rural to urban areas. Nearly 52 % of the world's population is urban. The urban population is projected to reach five billion by 2030. Ninety percent of this growth is expected to take place in developing countries. Cities occupy 2 % of the world's land yet consume 75 % of the world's resources and are responsible for as much as 80 % of global greenhouse gas emissions. Urban air pollution, which has worsened in most large cities in the developing world over the last few decades, imposes a heavy burden on the health of urban populations throughout the developing world. Every year, an estimated 800,000 people die prematurely from lung cancer and cardiovascular and respiratory diseases caused by outdoor air pollution in the world (WHO 2002; CPCB 2008). Other adverse health effects include increased incidence of chronic bronchitis and acute respiratory illness, exacerbation of asthma and coronary disease, and impairment of lung function. Annually, some two million people die from diarrheal diseases. Much of this disease burden is caused by contaminated drinking water and inadequate sanitation (WHO 2002).

The examination of the review suggests that there is urgent need to control population and environmental pollution in the country for better quality of life and health of present and future generation. This is often the case in developing countries, where less attention is paid to environmental protection, environmental standards are often inappropriate or not effectively implemented, and pollution control techniques are not yet fully developed. India is the second most populous and one of the fastest growing countries in the world. All of this development has had a significant impact on the environment, making India an excellent place to investigate the effect of environmental change on human health. Growing populations and high proportions of young people in the Third World are leading to large increases in the labor force. Agriculture cannot absorb them. Industry must provide these expanding societies not only with employment but with products and services for economic growth and development industries cannot be stopped. Also the human health cannot be ignored, and hence it becomes imperative to understand the relationship between industries, environment, and health.

8.2 Rationale for Selecting Delhi as the Study Area

Being the national capital, Delhi is a fast-growing center and hub for in-migration. To suffice the demands and needs of people, both industries and transport are vital. At the cost of development, however, environment shall not be compromised as it has direct as well as indirect ill impacts on health. Delhi is vulnerable to multiple disasters but East Delhi is much sensitive. With the help of climate disaster resilience index tool, Prashar et al. (2012) analyzed the resilience level of the nine districts of Delhi and found that the East Delhi district is least resilient to climatically induced disaster, whereas New Delhi is most resilient. Rahman et al. (2011) explored the quality of urban environment in East Delhi using remote-sensing and GIS techniques. The results reveal environmental deterioration due to unplanned urbanization accompanied with dense haphazard built-up zones and poor physical environment. The shifting of industries to the then periphery of Delhi in East Delhi has manifold impacts on human health. Hence the study aims to understand the industrial development in the study area and develop the linkages between human health and industrial pollution.

8.3 Database and Research Methodology

The analysis is based on primary and secondary sources of data. The secondary data sources include population data from the Census of India, the industrial sector data of Economic Survey of Delhi, and Statistical Abstracts of India. The pollution data was made available from State of Environment Report, CPCB Reports, City Development Plan of Delhi, and Forest Survey of India, and articles on industrial pollution and health were consulted. Apart from these, there are researches related to urban environment and health of Delhi like UHI in Delhi by Pandey et al. (2009), Singh et al. on seasonal variation in UHI, LST of Delhi by Mallick et al. (2008), urbanization and land use/cover change by Rahman et al. (2012), LST and land use/cover change by Singh and Grover (2014), and urban quality of life by Rahman et al. (2011).

Primary survey has been conducted in order to understand the perception of local people on the impact of air pollution on their health. For the purpose, a sample size of 100 persons was taken. The respondents were randomly selected. The questionnaire was divided into three main categories apart from the basic information on the respondent. These sections pertained to status and trend of growth in industries in the study area and changes in environment and health due to increase in pollutants, and the third section focused on the policy control measures and suggestions to be undertaken in order to reduce the ill effects of industrial growth. The questionnaire dealt with questions on changes in air quality, causes of the deteriorating air quality, type of health problems faced by people due to increase in emission, season of highest ill effects on health and solutions suggested to improve the condition of health, and the various initiatives government may incorporate in order to control city pollution. Further, the collected data was coded and tabulated using Microsoft Excel; the expected tables were extracted. Data have been analyzed by using the appropriate pie charts and bar and line graphs, and temporal analysis has been done for the available data sets for different years.

8.4   Study Area

East Delhi is located on the eastern bank of Yamuna River between 28° 34′ 47″ to 28° 40′ 47″ N latitude and 77° 15′ 05″ to 77° 20′ 37″ E longitude (Fig. 8.1), having an area of about 64 km². It is located on fertile-leveled Yamuna Plains that are fast converting into built-up areas. According to the Census of India (2011), the population of East Delhi was 1,707,725. The district has a population density of 26,683 inhabitants per square kilometer. Its population growth over the decade 2001–2011 was 16.68 %. East Delhi has a sex ratio of 883 females for every 1000 males and a literacy rate of 88.75 %.

Delhi has extreme continental climate with annual temperature ranging from 3 °C in winters to 45 °C in June, and average rainfall ranges from 400 to 600 mm. As per the estimates of the Forest Survey of India (2011), the total forest cover and tree cover in Delhi increased to about 176.2 km² (11.88 %) in 2011. Besides, lush green tree cover has grown that cover about 120 km² (8.09 %). The vegetation on the ridge mainly belongs to thorny scrub type, representing semiarid conditions.

Administratively, East Delhi has been divided into three parts – Gandhi Nagar, Preet Vihar, and Vivek Vihar. Gandhi Nagar is a middle-income commercial and residential area characterized by ready-made garments/textile market. The area is one of the most congested colonies with a population of around 3.5 lakh. On the other hand, Preet Vihar and Vivek Vihar are large residential areas in East Delhi. The urban population of East Delhi was over 98 % in 2001 with Vivek Vihar having total urban population of 100 % (Table 8.1).

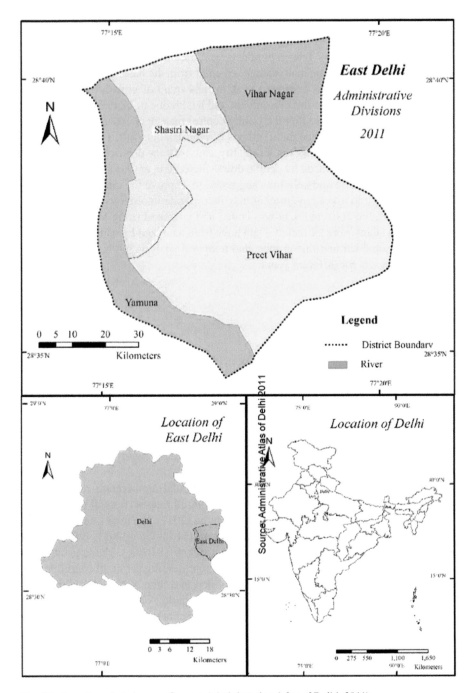

Fig. 8.1 Location of study area (Source: Administrative Atlas of Delhi, 2011)

Table 8.1 Distribution of urban population in East Delhi

Location	Urban population	Percentage
Gandhi Nagar	362,275	98.14
Vivek Vihar	210,055	100
Preet Vihar	873,030	98.69
East Delhi	1,445,360	98.75

Source: Census of Delhi 2001

8.5   Results and Discussion

8.5.1   Industrial Development

As per Economic Survey of Delhi (2001–2002), there were about 129,000 industrial units in Delhi in 1998, against 85,050 units in 1991. An average unit employed 9 workers, while 30 % of the units employed less than 4 workers. The number of industries is continuously increasing from 1981 to 2001. Due to increase in number of industries, a number of employees have also increased except in 1991 when due to court's order, industries were asked to shift to peripheral areas and some most polluted industries were banned. In 2001, again, there was further increase in per unit employee. In 2011, there were 285 food industries, 38 tobacco industries, 682 paper and printing industries, and 293 chemical product industries (Economic Survey of Delhi 2012–2013).

According to the Fifth Economic Census, Delhi was ranked 16th in India (based on the results of 35 states and union territories) with respect to the number of establishments accounting for about 1.80 % of the total establishments in India. The total number of establishments found to be operating during 2005 in the geographical boundaries of the National Capital Territory of Delhi was 757,743.

The density of establishments per square kilometer was highest in Central Delhi (3,223) followed by North East district (1,625) and East Delhi district with 1,492 industries per square kilometer. East Delhi has various kinds of industries like food industry, copper industry, wire industry, electronic industry, printing industry, and paper industry. East Delhi is basically known for medium- and small-scale industries. In East Delhi, there are four industrial areas and three industrial locations. These industrial areas are Shahdara Industrial Area, New Friends Colony Industrial Area Shahdara, Jhilmil Industrial Area, and Patparganj Industrial Area. The industrial locations however are Shahdara, Patparganj, and Jhilmil.

Shahdara is the hub of industries such as paper, printing, wire and cable, food, plastic, chemicals, and metal. There are some famous manufacturing industries; these are DS Automobile Industries, Bakshi Plastic Industries, Ashoka Packaging Industries, Bareja Helmet Industries, Kirdar Chemicals Industries, Shiv Metal Industries, Vikas Dhatu Udyog, and Moon Light Fancy Works. Patparganj Industrial

Area is known as "Functional Industrial Estate." It is basically known for manufacturing industries. There are some examples of manufacturing industries – T. R. Sawhney Motors Private Limited, IG Homoeo Remedies Private Limited (medicine manufacturer), Almonard Private Limited (fans manufacturing and AC repair), industrial chemical, art printing, batteries, and computer hardware. Jhilmil on the other hand is a complex, much systematic, and organized industrial area. This industrial area has been divided in blocks.

The household industry in Delhi consists of agarbatti, aluminum hangers, electrical and electronic gadgets, sewing machines, assembly of hand tools, candles, cane and bamboo products, brushes and brooms, carpentry, cardboard boxes, packaging, dari and carpet weaving, detergent, dairy products, dry cleaning, embroidery, framing of pictures, pencils and pens, hosiery products, leather belts, paper and stationery items, repair of watches, repair of bicycles, stone engravings, sport goods, tailoring, toys, food processing, jewelry, wool knitting, zari, and zardozi.

8.5.2   Urban Environment of Delhi

The studies on urban environment of Delhi reveal that the city has rapidly grown as a megacity. The process of urbanization has inculcated obvious changes in land use/cover of Delhi by replacing the porous land cover with impermeable land use, in order to fulfill the needs of rising population in terms of living space, place of work, and transport network. Associated with the urban development are ill effects reflected in terms of land use/cover changes, pollution, increase in surface temperature, and other environmental changes. Research on these issues has swelled in the recent past due to its inherent impact on human and environmental health.

8.5.2.1   Land Use/Cover Change and LST

Rahman et al. (2012) assessed the land use/cover in North West district of Delhi with the help of *Aster* image. The research concluded that urban expansion and sprawl has altered the land use/cover by reducing forest cover and wasteland by converting them to urban built-up areas and thus is the prime cause for LST rise. The results by Sharma and Joshi (2012) show expansion of built-up areas at the cost of rural hinterland, agricultural zones, and forests. The rapid unplanned growth of the city has pushed the hinterland to peripheral zone and developed dense network of buildings and transport network in the city center. Mallick et al. (2008) have estimated the LST of Delhi using Landsat 7 data. LST is a reflection of physical and environmental processes. The results assert the increasing trend of LST corresponding with concrete land use/cover. The land use/cover change are helpful in understanding LST change and UHI creation.

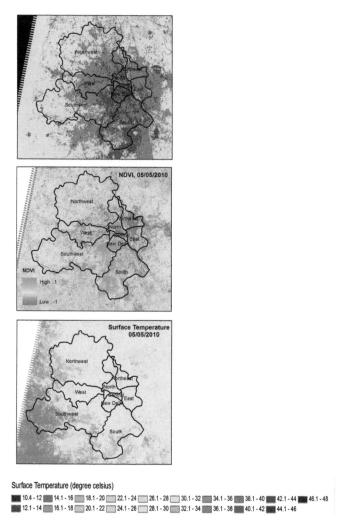

Surface Temperature (degree celsius)

■ 10.4 - 12 ■ 14.1 - 16 □ 18.1 - 20 □ 22.1 - 24 □ 26.1 - 28 □ 30.1 - 32 ■ 34.1 - 36 ■ 38.1 - 40 ■ 42.1 - 44 ■ 46.1 - 48
■ 12.1 - 14 ■ 16.1 - 18 □ 20.1 - 22 □ 24.1 - 26 □ 28.1 - 30 □ 32.1 - 34 ■ 36.1 - 38 ■ 40.1 - 42 ■ 44.1 - 46

Fig. 8.2 Spatial distribution of surface temperature difference (Source: Adopted from Singh and Grover (2014)

8.5.2.2   Changes in LST and Creation of UHI

The recent research by Singh and Grover (2014) correlates land use change and UHI in Delhi. Comparative analysis of a decade of data is done using Landsat 5 image of 2000 and 2010. The surface temperature for the month of May from 2000 to 2010 reflects stark increase all through the city, especially in the western Delhi. The surface temperature difference (Fig. 8.2) is maximum for South West Delhi, which is largely fallow land. Moderate to high change is observed in East Delhi which in the recent past has undergone vast concretization. The study shows that Delhi, unlike many large cities of the world, has UHI creation but of low intensity and extent. Rather, many small UHIs are created in the city. This is due to the fact that concretization in Delhi is much horizontal than vertical and hence the warm air is not trapped due to the design and height of buildings. Added to this is the cooling effect of Yamuna River and large tree cover.

Pandey et al. (2009) presented the study on creation of UHI in Delhi in summer season for day and night using MODIS data. Singh et al. (2014) presented the seasonal variation in UHI creation for Delhi representing summer, winter, autumn, and spring season. This study also unveils similar outcomes. There is a strong correlation between land use/cover type and LST, for instance, vegetation, forest, and water body record low temperature, whereas urban and rural built up have high LST.

8.5.2.3 Air Pollution

The pollution level, especially aerosol, SPM (Pandey et al. 2009), oxides of carbon, nitrogen, and sulfur (Jain and Khare 2008), has increased in Delhi. The main causes of this increase are the high pace of fleet of vehicles added on the roads and the industrial growth. The urban environment owing to combination of factors like land use/cover change, pollution, and LST change has altered to a large extent (Rahman et al. 2011). This is reflected in changes in quality of water, air, noise, living conditions, and overall environment of Delhi (Rahman et al. 2009). The remote-sensing data and GIS techniques are utilized by Rahman et al. (2009) to assess urban environmental issues in Delhi. The air pollution, noise pollution, industrial growth, solid waste generation, and LST have been intelligently correlated.

The Central Pollution Control Board is the primary authority for recording and maintaining data on air quality and pollution level. The industries in 1970–1971 were recorded to be the prime contributor to air pollution, with vehicular pollution and domestic sources having lower share. The following decade experienced increase in vehicular pollution and decrease in industrial pollution. Further, in the 1990s and 2000s, the contribution of vehicles increased constantly, and on the contrary, the industrial polluting units declined (Fig. 8.3). This can be attributed to relocation and closing of many industries as a result of High Court order. By 2000–2001, 72 % pollution was caused by means of transport, 20 % by industries, and only 8 % by domestic sources.

There are multiple forms of pollution and changes in air quality caused by the industries. These include air, water, noise, soil, and solid waste. Most people recognized solid waste as the major component of waste that may be due to its visibility. Nearly 30 % of the respondents pointed out to the toxic gas emissions, and 11 % pointed out the industrial effluents as a form of industrial pollution (Fig. 8.4). These forms of air pollution are toxic gases, bad odor and smell, and suffocation. The assessment of form of degradation of air quality is based on primary survey (Fig. 8.5). It is evident that the emission of toxic gases is much higher in East Delhi than any factor. After toxic gases, suffocation is also responsible for air quality degradation, and other factors are odor and smell. The people of East Delhi believed that effect of toxic gases is much higher in air quality degradation which is almost double in all factors.

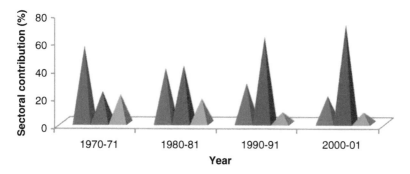

Fig. 8.3 Trend and sources of air pollution, Delhi (Source: CPCB Report 2010)

Fig. 8.4 Constituents of industrial pollution, East Delhi (Source: Primary Survey)

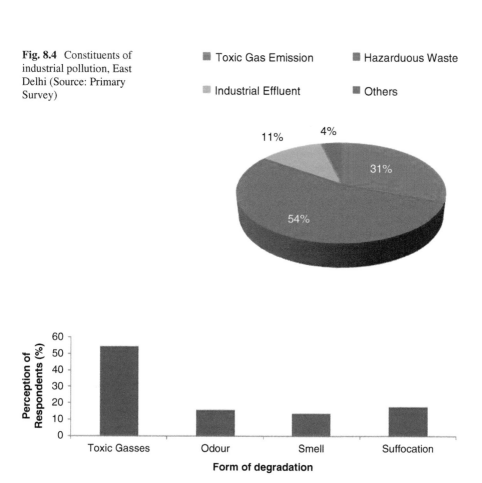

Fig. 8.5 Forms of air quality degradation, East Delhi (Source: Primary Survey)

8.5.3 Air Pollution and Human Health

The gases exhaled from the industries can seriously adversely affect the health of the population and should be given due attention. The gases mentioned below are mainly outdoor air pollutants, but some of them can and do occur indoor depending on the source and the circumstances:

- *Lead:* Prolonged exposure can cause damage to the nervous system, digestive problems, and in some cases cancer. It is especially hazardous to small children.
- *Radon:* A radioactive gas that can accumulate inside the house, it originates from the rocks and soil under the house, and its level is dominated by the outdoor air and also to some extent the other gases being emitted indoors. Exposure to this gas increases the risk of lung cancer.
- *Ozone:* Exposure to this gas makes our eyes itch, burn, and water, and it has also been associated with increase in respiratory disorders such as asthma. It lowers our resistance to cold and pneumonia.
- *Carbon monoxide (CO):* CO combines with hemoglobin to lessen the amount of oxygen that enters our blood through our lungs. The binding with other heme-proteins causes changes in the function of the affected organs such as the brain and the cardiovascular system and also the developing fetus. It can impair our concentration, slow our reflexes, and make us confused and sleepy.
- *Sulfur dioxide (SO_2):* SO_2 in the air is caused due to the rise in combustion of fossil fuels. It can oxidize and form sulfuric acid mist. SO_2 in the air leads to diseases of the lung and other lung disorders such as wheezing and shortness of breath. Long-term effects are more difficult to ascertain as SO2 exposure is often combined with that of SPM.
- *SPM (suspended particulate matter):* SPM consists of dust, fumes, mist, and smoke. These particles when breathed in, lodge in our lung tissues and cause lung damage and respiratory problems. The importance of SPM as a major pollutant needs special emphasis as (a) it affects more people globally than any other pollutant on a continuing basis; (b) there is more monitoring data available on this than any other pollutant; and (c) more epidemiological evidence has been collected on the exposure to this than to any other pollutant (Table 8.2).

The ambient air quality as monitored by CPCB during 1999 shows reduction in levels of various air pollutants in ambient air as compared to previous years. The reducing trend was observed with respect to carbon monoxide, nitrogen dioxide, and lead in residential areas. But in comparison to WHO air quality guidelines, the levels of suspended particulate matter are more than the double, and the levels of lead are much more, whereas the levels of sulfur dioxide and nitrogen dioxide are lower.

It is evident that air pollution poses considerable health concerns. There is a large proportion of human mortality taking place due to respiratory diseases (Table 8.3). However, it is not clear that these are due to industrial pollution. Hence, there is a need to develop dose-response function for city conditions. Few studies show that in Delhi air pollution also causes various diseases. Time series analysis of the impact

Table 8.2 Sources and health impacts of common outdoor air pollutants

S. no.	Pollutant	Source	Zone of influence	Health impact
1	Suspended particulate matter	Natural sources, windblown dust, forest fire, volcanic eruption, combustion	Local and surface	Pneumoconiosis, restrictive lung disease, asthma, cancer, etc.
2	RSPM or PM 10	Industries, combustion of fossil fuels, vehicle exhaust	Regional and surface	Respiratory illness like chronic bronchitis and asthma, heart disease, etc.
3	Sulfur dioxide	Power stations, petroleum refineries, industrial boilers	Regional and global	Heart, respiratory problems including pulmonary emphysema, cancer, eye burning, headache, etc.
4	Oxides of nitrogen	Power plants, electric utility boilers, vehicle emission	Regional and global	Lung irritation, viral infection, airway resistance, chest tightness
5	Carbon monoxide	Incomplete combustion of carbon fuels, motor vehicles	Local and surface	Cherry lips, unconsciousness, death by asphyxiation
6	Ozone	Chemical reaction of NO_2 with VOC	Surface	Impaired lung function, chest pain, coughing, eye and nose irritation
7	Lead	Vehicle exhaust, lead smelting, processing plants	Local and surface	Decrease hemoglobin synthesis, anemia, damage to nervous system and renal system, etc.

Source: Adopted from Patankar (2009)

Table 8.3 Percentage of deaths due to respiratory diseases, East Delhi

Year	In percent
1998	11
1999	9
2000	10
2001	4
2002	10
2003	10
2004	5
2005	7
2006	6

Source: CPCB Delhi (2010)

of the particulate air pollution on daily mortality in East Delhi was carried out. Study shows that age group of 15–44 are more affected by air pollution. AIIMS have correlated daily levels of various pollutants with the number of patient visiting the AIIMS. Casualty for aggravation of certain defined cardiorespiratory disorder study proves the point that respiratory symptoms are more frequent among people

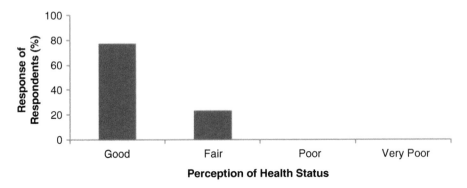

Fig. 8.6 Perception of health status of respondents, East Delhi (Source: Primary Survey)

residing in highly polluted areas. Emergency room visit for asthma, chronic obstructive airway disease, and acute coronary events increased by 21.3 %, 24.9 %, and 24.3 %, respectively, on account of higher than acceptable levels of pollutants.

The research on impact of air pollution from different sources on human health in Delhi is profoundly studied area. Goyal et al. (2006), Ravindra et al. (2006), Jain and Khare (2008), and Nagdave (2004) focused on status and trend of environmental pollution in Delhi. Centre for Science and Environment (CSE) carries out researches on health and environment of Delhi on regular basis. It reveals the outcome of shifting industries to peripheral zones as a wise policy and asserts that fuel quality along with the steep increase in vehicle use has increased air pollution in Delhi. The impact is assessed by correlating respiratory illness and cardiovascular diseases to pollution trends. The impact of air pollution on health in Delhi was carried out by Cropper et al. (1997). They detailed the age, gender, and season-wise incidence of mortality due to air pollution. The trend of pollutants on seasonal basis was also analyzed. The studies reflect that long-term, repeated exposures to air pollution increase the cumulative risk of chronic pulmonary and cardiovascular disease and even death (CPCB 2008).

According to the primary survey, over 70 % people responded that their health is good. None responded on poor or very poor health. Thus, generally, the sample population has fair to good state of physical health (Fig. 8.6).

To understand the correlation between industrial pollution and human health in the study area, people were asked on their perception of causes of their health problems. Many respondents from East Delhi believed that they are much affected by respiratory disease and comparitively less affected by waterborne disease (Fig. 8.7), the reason being that people living there for over 10 years could compare the past and present status of air quality. Many health problems were identified like breathlessness, cough, and cold being common, in general, for all age groups. Some proportion did mention acute respiratory illness like asthma, sinus, and bronchitis to name a few. Over 40 % did respond as having no disease but did mention some minor illness like headache and dizziness that are also well-recognized impacts of emissions.

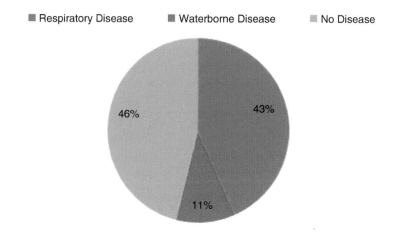

Fig. 8.7 Major health problems due to industrial pollution, East Delhi (Source: Primary Survey)

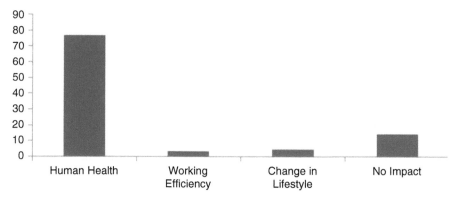

Fig. 8.8 Impacts of industrial pollution on human environment, East Delhi (Source: Primary Survey)

However, most people also recognized that industrial pollution has a major impact on human health and working efficiency. Nearly 80 % acknowledged that their ill health is due to pollution caused by industries in the vicinity (Fig. 8.8). The wind transports the polluted air to nearby residential colonies causing changes in daily lifestyle. Health problems reported most were common cold, cough, headache, and itching in eyes and nose. Across age and gender, these health problems were noted. Some people also responded on asthma, cardiovascular diseases, and respiratory illness. The health of people deteriorated in rainy season. Delhi has developed much space for recreation and parks. Residents, especially the old and women with children, go out to nearby green zones on a regular basis.

This has helped maintain or reduce respiratory problems to a large extent. While moving on two wheelers or three wheelers, people prefer using cloth mask to protect their skin from burning that also saves them from inhaling harmful gases. The respondents recognized that the tree cover in the city has increased over the last few years and hence the air quality is still better. The other impacts of pollution on human health are seen as changes in work efficiency and lifestyle. Only a few percentage of sample population responded on decline in work efficiency or lifestyle changes.

8.6 Conclusion and Suggestions

Rapid, unplanned, and unsustainable patterns of urban development are making developing cities focal points for many emerging environment and health hazards. Unsustainable patterns of transport, industrialization, and urban land use are either the driver or root cause of a number of significant and interrelated environment and health hazards faced by urban dwellers in developing countries. These health and environment linkages cut across a range of policy sectors and thus are often overlooked in policymaking. A growing awareness of industrial pollution and its consequences has led to tighter restrictions on pollution all over the world, with nations recognizing that they have an obligation to protect themselves and their neighbors from pollution.

While industrialization is an essential feature of economic growth in developing countries, like India, industrial practices may also produce adverse environmental health consequences. To provide employment and satisfy the demands of the increasing population, industrial sector is rapidly growing in India. Industries are also the major sources of pollution and hence, the flip side of this development is degrading human health. The considerable magnitude of air and water pollution pulls up the number of people suffering from respiratory and waterborne diseases many a times leading to deaths and serious health hazards.

References

Akhtar R (2002) Urban health in the Third World. APH Publishers, New Delhi
Bentham G (1992) Global climate change and human health. GeoJournal 26:7–12
Bruce N, PerezPadilla R, Albalak R (2002) The health effects of indoor air pollution exposure in developing countries. World Health Organization, Geneva
Bush KF, Luber G, Rani Kotha S, Dhaliwal RS, Kapil V, Mercedes P, Brown DG, Frumkin H, Dhiman RC, Hess J, Wilson WL, Balakrishnan K, Eisenberg J, Kaur T, Rood R, Batterman S, Joseph A, Gronlund CJ, Agrawal A, Hu H (2011) Impacts of climate change on public health in India: future research directions. Environ Health Perspect 119(6):765–770
Central Pollution Control Board (2008) Epidemiological study on effect of air pollution on human health (adults) in Delhi. Ministry of Environment and Forests, Government of India
Central Pollution Control Board (2010) Annual Report. Ministry of Environment & Forests, Government of India
Cropper ML, Simon NB, Alberini A, Sharma PK (1997) The health effects of air pollution in Delhi, India. Policy research working paper 1860. The World Bank, Development Research Group

Department of Planning (2001–2002) Economic Survey of Delhi. Government of Delhi
Department of Planning (2012–2013) Economic Survey of Delhi. Government of Delhi
Directorate of Census Operations (2001) Provisional population totals – NCT of Delhi. Ministry of Home Affairs, Government of India
Directorate of Census Operations (2011) Provisional population totals – NCT of Delhi. Ministry of Home Affairs, Government of India
Ebi KL, McGregor G (2008) Climate change, tropospheric ozone and particulate matter, and health impacts. Environ Health Perspect 116(11):1449–1455
Filleul L, Cassadou S, Médina S, Fabres P, Lefranc A, Eilstein D, Le Tertre A, Pascal L, Chardon B, Blanchard M, Declercq C, Jusot J-F, Prouvost H, Ledrans M (2006) The relation between temperature, ozone, and mortality in nine French cities during the heat wave of 2003. Environ Health Perspect 114(9):1344–1347
Forest Survey of India (2011) State of Forest Report. Ministry of Environment & Forests, Dehradun pp 112–118
Goyal SK, Ghatge SV, Nema P, Tamhane SM (2006) Understanding urban vehicular pollution problem visavis ambient air quality – case study of a megacity (Delhi, India). Environ Monit Assess 119:557–569
Izhar N (2004) Geography and health: a study in medical geography. APH Publishers, New Delhi
Jain S, Khare M (2008) Urban air quality in mega cities: a case study of Delhi city using vulnerability analysis. Environ Monit Assess 136:257–265
Lo CP, Quattrochi DA (2003) Landuse and landcover change, urban heat island phenomenon, and health implications: a remote sensing approach. Photogramm Eng Remote Sens 69(9):1053–1063
Mallick J, Kant Y, Bharath BD (2008) Estimation of land surface temperature over Delhi using Landsat7 ETM+. J Indian Geophys Union 12(3):131–140
Nagdave DA (2004) Environmental pollution and control: a case study of Delhi mega city. Popul Environ 25(5):461–473
Pandey P, Kumar D, Prakash A, Kumar K, Jain VK (2009) A study of the summertime urban heat island over Delhi. Int J Sustain Sci Stud 1(1):27–34
Patankar (2009) Health effects of urban air pollution – a study of Mumbai. Unpublished PhD thesis, Humanities and Social Science Department, Indian Institute of Technology, Bombay
Prashar S, Shaw R, Takeuchi Y (2012) Assessing the resilience of Delhi to climaterelated disasters: a comprehensive approach. Nat Hazards 64:1609–1624
Rahman A, Netzband M, Singh A, Mallick J (2009) An assessment of urban environmental issues using remote sensing and GIS techniques: an integrated approach. A case study: Delhi, India. In: de Sherbiniin A, Rahman A, Barbieri A, Fotso JC, Zhu Y (eds) Urban population: environment dynamics in the developing world – case studies and lessons learned. Committee for International Cooperation in National Research in Demography (CICRED), Paris
Rahman A, Kumar Y, Fazal S, Bhaskaran S (2011) Urbanization and quality of urban environment using remote sensing and GIS techniques in East Delhi, India. J Geogr Inf Syst 3:62–84
Rahman A, Kumar S, Fazal S, Siddiqui MA (2012) Assessment of land use/land cover change in the North West district of Delhi using remote sensing and GIS techniques. J Indian Soc Remote Sens 40(4):689–697
Ravindra K, Wauters E, Tyagi SK, Suman M, Grieken R (2006) Assessment of air quality after the implementation of compressed natural gas (CNG) as a fuel in public transport in Delhi, India. Environ Monit Assess 115:405–417
Roy SS, Singh RB, Kumar M (2011) An analysis of local spatial temperature patterns in the Delhi metropolitan area. Phys Geogr 32(2):114–138
Sharma R, Joshi PK (2012) Monitoring urban landscape dynamics over Delhi (India) using remote sensing (1998–2011) inputs. J Indian Soc Remote Sens 41(3):641–650
Singh RB, Grover A (2014) Spatial correlations of changing land use, surface temperature (UHI) and NDVI in Delhi using Landsat satellite images. In Singh RB (ed) Urban development chal-

lenges, risks and resilience in Asian megacities. Springer, Tokyo, 2015, pp 83–98. ISBN: 978-4-431-55042-6 (Print) 978-4-431-55043-3 (Online)

Singh RB, Grover A, Zhan J (2014) Interseasonal variations of surface temperature in the urbanized environment of Delhi using Landsat thermal data. Energies 7:1811–1828

Tan J, Zheng Y, Tang X, Guo C, Li L, Song G, Zhen X, Yuan D, Kalkstein AJ, Li F, Chen H (2010) The urban heat island and its impact on heat waves and human health in Shanghai. Int J Biometeorol 54:75–84

Tomlinson CJ, Chapman L, Thornes JE, Baker CJ (2011) Including the urban heat island in spatial heat health risk assessment strategies: a case study for Birmingham, UK. Int J Health Geogr. doi:10.1186/1476-072X10-42

United Nations Environment Programme (2002) Integrating environment and development: 1972–2002. United Nations

World Health Organization (2002) The world health report 2002 – reducing risks, promoting healthy life. United Nations, Geneva

World Health Organization (2005) The World Health Report 2005: make every mother and child count. United Nations, Geneva

Chapter 9
Climate Change and Human Health Impact and Adaptation Responses in Nepal

Bimal Raj Regmi, Cassandra Star, Bandana Pradhan, and Anil Pandit

Abstract This chapter aims to understand the human health dimensions of climate change in Nepal. A case study approach was used to describe and quantify the association between climate factors and reported cases of typhoid and other health-related hazards in Nepal along with their impacts and implications. The research findings show that diseases and health-related hazards have increased in the country. There is an association between the incidence of climate-sensitive diseases and changes in temperature and precipitation trends. The study data suggests that climate change is likely to have impacts on the health sector in Nepal; however current adaptation policies, strategies, and response measures in this sector are insufficient to address such impacts. The lack of response measures has resulted in increased risk and vulnerability among the poor and marginalized communities living in both rural and urban areas of Nepal. There is an urgent need to devise policies and strategies to fill the existing information and knowledge gaps and implement an integrated approach for better health planning and research in Nepal, to develop long-term mechanisms of addressing health issues and challenges in urban and rural areas.

Keywords Climate change • Human health • Adaptation • Integrated approach • Health planning

B.R. Regmi (✉) • C. Star
School of Social and Policy Studies, Flinders University, Adelaide, SA, Australia
e-mail: bimalrocks@yahoo.com

B. Pradhan
Institute of Medicine, Tribhuvan University, Kirtipur, Nepal

A. Pandit
Hays Medical Center, 2220 Canterbury Dr, 67601 Hays, KS, USA

© Springer International Publishing Switzerland 2016
R. Akhtar (ed.), *Climate Change and Human Health Scenario in South and Southeast Asia*, Advances in Asian Human-Environmental Research, DOI 10.1007/978-3-319-23684-1_9

131

9.1 Introduction

Nepal is located along the southern slopes of the Himalayas between the Tibetan region of China in the north and the Gangatic plains of India in the south. The country has an area of 147,181 km^2, and it lies between longitudes 80° 4' to 88° 12' east and latitudes 26° 22' to 30° 27' north (Fig. 9.1). The census data shows that Nepal's population is increasing by 1.4 % per annum, the total population reaching 26,620,809 in 2011 (CBS 2011). Agriculture provides a livelihood for three-fourths of the population and accounts for about one-third of the gross domestic product (GDP).

Nepal's geographic and ecological variation makes its vulnerable to climate change. A study by Baidya et al. (2007) shows that the rate of increase in temperature (0.04 °C/year) is higher than the mean global rate. Similarly, Nepal has experienced variability in the rainfall pattern over the last two decades. An analysis of daily precipitation data for a 46-year period from 1961 to 2006 carried out by the Department of Meteorology shows an increasing trend in precipitation extremes such as higher rainfall within a short duration of time (Baidya et al. 2008). Future climate change projection also shows a greater climatic variability and increase in temperature across the country, rising from 0.5 to 2 °C with a multi-model mean of 1.4 °C by 2030 (NCVST 2009).

Weather and climate have a wide range of health impacts and play a role in the ecology of many infectious diseases (Patz et al. 2000). The relationships between health and weather, climate variability, and long-term climate change are complex (McMichael et al. 2006). Changes in climatic elements such as temperature, humidity, precipitation, wind, and pressure have direct and indirect effects on the health of human beings (Regmi et al. 2008).

The fifth assessment report of the Intergovernmental Panel on Climate Change (IPCC) has indicated that climate change impacts on human health will be multifold. A number of adverse public health impacts are expected to worsen in climate-related disasters such as storms, floods, landslides, extreme heat, drought, and wildfires (IPCC 2014). These are highly context specific but range from a worsening of existing chronic illnesses (which could be widespread); to possible toxic exposures via air, water, or food; and deaths, which are expected to be few to moderate in number but could be significantly higher in low-income countries (Keim 2008).

Climate change impacts have been attributed to an increase in the prevalence of both vector-borne and water-borne diseases. A range of vector-borne illnesses has been linked to climate, including malaria, dengue, hantavirus, bluetongue, Ross River virus, and cholera (Patz et al. 2005). The incidence of cholera, for example, has a seasonal variability that may be directly affected by climate change (Koelle and Pascual 2004). Vector-borne illnesses have been projected to increase in geographic reach and severity as temperatures increase (McMichael et al. 2006). According to Lafferty (2009), mosquitoes and other vectors are moving to inhabit areas previously free from such vectors of transmission. Research in the last decade has demonstrated that cholera is also sensitive to climate variability (Rodo et al. 2002; Koelle et al. 2005a; Constantin de Magny et al. 2007; Koelle et al. 2005b).

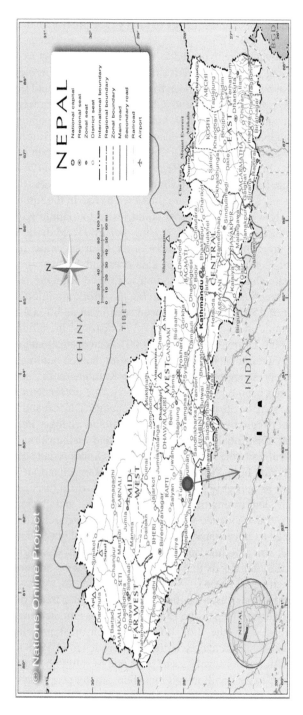

Fig. 9.1 Map of Népal showing the study area

Cholera outbreaks may become more widespread as the climate continues to change due to the projected increase in frequency of heavy precipitation over many areas of the globe (Koelle et al. 2005a).

Extreme climate events are expected to become more frequent as a result of climate change in Nepal (MoE 2010). Historical data show that in Nepal, disasters, famines, and disease outbreaks have been triggered by droughts and floods (NCVST 2009). Due to more extreme temperatures, heat stress is likely to become common, causing heat stroke and other problems for the working-class population in urban areas of Nepal (Pradhan et al. 2013). There is also an increasing trend in other vector-borne and water-borne diseases in both the rural and urban areas of Nepal, often shifting to the higher altitudes (Regmi et al. 2008). Badu (2013) found that with climate change, there are more desirable conditions for water-borne diseases, and diarrheal incidences in particular are predicted to rise in the future.

Malaria is already a problem in most countries in the Hindu Kush-Himalayas (Narain 2008; Bhattacharya et al. 2006; Kristie et al. 2007). Malaria, kala-azar, and arboviral diseases are common particularly in all Tarai and most hill districts of Nepal (Regmi et al. 2008). Outbreaks of kala-azar and Japanese encephalitis are also linked to climate change in Nepal's subtropical and hot regions. There is evidence that malaria vectors such as *Anopheles fluviatilis*, traditionally not found above an elevation of 1500 m, have now been seen in an additional 13 districts of Nepal and are endemic in 65 out of a total of 75 districts, including a few in the mountains (DoH 2008; Dhimal and Bhusal 2009). A cross-sectional entomological survey conducted by Dhimal and Bhusal (2009) after the first outbreak of dengue in Nepal identified the presence of *Aedes aegypti* in Kathmandu in 2009. Previously, *Aedes aegypti* had not been recorded in the country. The presence of this mosquito in these urban areas may be attributed to climate change (Dhimal and Bhusal 2009).

There is a link between disease outbreaks and climate change (Patz et al. 2005). As at the year 2000, the global burden of diarrhea and malnutrition attributable to climate change was already the largest in the world in South Asian countries including Bangladesh, Bhutan, India, Maldives, and Nepal (Ramachandran 2014). A number of recent studies have highlighted the burden of diarrheal diseases in mountain regions. The quantity and timing of runoff from snowmelt and glaciers directly and indirectly influence the incidence and prevalence of water-related diseases (Ebi et al. 2007). The need to store water will increase the risk of water-borne and water-based infections, as will flooding and water contamination. Contagious outbreaks are more pronounced after floods that disrupt sewage systems and contaminate the drinking water (Regmi et al. 2008).

In addition to infectious diseases, there is an increasing trend of injury-based morbidity and mortality in Nepal due to the impacts of climate-induced disasters. Nepal is classified as one of the "hot spots" for geophysical and climatic hazards (NSDRM 2009). By global standards, in the two decades from 1988 to 2007, Nepal was in the 23rd position in terms of loss of lives due to natural hazards (MoHA and DPNet 2009). Due to mass flooding and landslides, many people in both rural and urban areas lost their houses and lands. The data reveal that more than 80 % of property loss was due to climate-related disasters. In 2010, MoHA and DPNet reported that more than 4,000 people had died in the previous 10 years due to climate-induced disasters, causing an economic loss of US\$ 5.34 billion. Besides economic losses, the physical and mental stress on households of losing their family members is high.

Knowledge and information on climate change and its impacts on human health in Nepal are limited. There is also a lack of in-depth studies that investigate the human health implications of climate change on the well-being of people in Nepal. The main objective of the study reported here is to analyze the impacts of climate change on health in selected vulnerable communities. Specifically the study aims to:

- Examine the association between and impact of changing climate factors (temperature, rainfall) and water-borne diseases in urban areas of Nepal
- Assess the impacts and implications of extreme weather events in terms of the health-related hazards in selected rural areas in Nepal

The outcomes of the study are relevant for health-related professionals and policy makers seeking to design both research and development strategies to deal with issues related to human health in Nepal. The empirical evidence provided in this chapter can be used to guide policy makers, researchers, practitioners, and development professionals mostly engaged in implementation in Nepal to develop effective coping and adaptation strategies in the health sector.

9.2 Research Methodology

9.2.1 Research Methods

The research methodology used a case study to document climate change impacts and implications in relation to the outbreak of diseases and health hazards in urban and rural areas of Nepal. A case study approach is useful for providing a cross-sectional and in-depth analysis of a particular problem or context that captures both urban and rural health-related issues (Yin 2003). One limitation of a case study approach is that it can lead to generalization. The authors have endeavored to minimize this limitation by emphasizing the empirical findings.

The first case involved a study of typhoid in urban areas of Nepal, focusing on cases at Patan hospital in Kathmandu. Kathmandu Valley was selected because it is the most populated, fragile, and risk-prone area in terms of climate change impacts and health-related risks and hazards (GoN 2010). Secondary literature, including journal articles, published books and papers (seminar/workshop proceedings), institutional reports, and other publication materials were reviewed to gather data. Primary data and information were collected from consultation with a wide range of stakeholders. Various stakeholders working in climate change were also consulted through a series of meetings and included government agencies. Government officials, including the IPCC national focal point, were consulted to inform them about the study. In addition, a 1-day stakeholder workshop was organized to share the findings of the case study as well as obtain feedback and suggestions from participants.

The second case involved a study of climate change impacts on human health in rural areas. This was carried out involving two Village Development Committees (VDCs) of Pyuthan district in the mid-western region of Nepal (Fig. 9.2). The research sites were selected purposively in order to represent rural locations and investigate climate change impacts at the community level. In-depth interviews and focus group discussions were conducted with selected members of the communities. A total of 128 randomly selected households in the two VDCs were interviewed to map their perceptions of the impacts of climate change on human health. Focus group discussions were held with six community groups and local stakeholders. In addition, semi-structured interviews were undertaken with 28 practitioners and 17 policy makers, including health professionals and other policy makers. The inter-

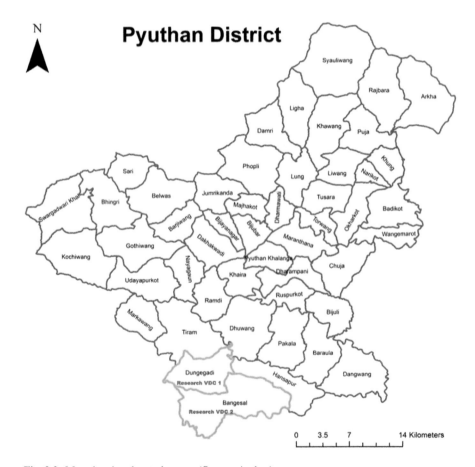

Fig. 9.2 Map showing the study areas (*Source*: Author)

views and discussions with different stakeholders were useful for obtaining insight into the views of policy makers, practitioners (professionals engaged in implementation), and communities on climate change implications in the health sector.

9.2.2   Research Data

Data were collected from a range of sources for the research. Data on typhoid cases were obtained from Patan Hospital, one of three large general hospitals within the Kathmandu metropolitan area. It has 251 beds and provides inpatient and outpatient medical, surgical, pediatric, obstetric, and gynecology services and serves as a primary care facility. Each year Patan Hospital has approximately 250,000 outpatient visits, 30,000 emergency department visits, and 15,000 admissions. Bed occupancy for medical wards typically runs at almost 100 %, and approximately 90 % of the patients are residents of the immediate Kathmandu Valley area. Data on cases of laboratory-confirmed episodes of typhoid fever from 1997 to 2005 were recorded, including both inpatient and outpatient data. The data were retrieved from the microbiology laboratory, where all the blood cultures from inpatients and outpatients are collected and cultured.

Meteorological data (temperature and rainfall) across a 30-year period for Kathmandu Valley and Pyuthan District were obtained from the Department of Meteorology and Hydrology, Nepal Government, and analyzed. The meteorological stations of Khumaltar, Bijuwataar, and Gorahi were selected as sites for analysis of temperature and rainfall trends and variability. In addition, secondary data from the health post of Dhugegadi VDC, the district health office of Pyuthan district, Pyuthan's district administrative office, and Ministry of Home Affairs were collected in order to analyze the health-related implications of climate change at the national and local level.

9.3   Major Findings

9.3.1   Trend of Typhoid Cases in Kathmandu Valley

The aim of this case study was to explore if recent trends in the rising temperature and rainfall variability in the Kathmandu Valley have any association with an increasing number of culture-positive typhoid cases recorded in an urban hospital of the valley. The case study also looked at other external factors that contribute to an increase in climate vulnerability in the health sector in urban areas of the Valley.

Typhoid fever is a systemic bacterial infection caused by *Salmonella typhi*. Typhoid is usually acquired through ingestion of water or food contaminated by the urine or feces of infected carriers, and as such it is a common illness in areas where

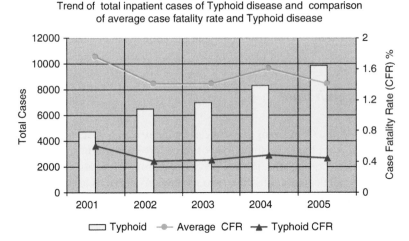

Trend of total inpatient cases of Typhoid disease and comparison
of average case fatality rate and Typhoid disease

Fig. 9.3 Increasing trend of typhoid cases in Kathmandu Valley (*Source*: Regmi et al. 2008)

sanitation is poor and the socioeconomic status of people is low. Worldwide, at least 17 million new cases and up to 600,000 deaths are reported annually. The case-fatality rate of typhoid fever is 10 %, but it can be reduced to 1 % with appropriate antibiotic treatment (WHO 1996). Case fatality rate here refers to the number of patients who die/number of patients who are infected with the disease, expressed in a percentage.

The study found that the number of typhoid cases in urban areas of Nepal has been increasing since 2001 (Fig. 9.3) but that average case fatality rates appear to be generally decreasing across year-to-year fluctuations (DoHS 2005). There are various reasons for the typhoid rising trend in Kathmandu Valley. One of the reasons for the increase can be attributed to rising temperature. The temperature data show that the average temperature in Kathmandu Valley is increasing at a faster rate than other parts of Nepal (0.8 in the period 1968–1998), making it one of the most vulnerable urban areas in Nepal (Baidya et al. 2008; GoN 2010).

Analysis of the data on temperature and typhoid cases for the Kathmandu Valley shows a positive correlation. Figure 9.4 illustrates that the cases of typhoid recorded at Patan Hospital, Kathmandu, tend to rise from the month of May onward, reach a peak in July, and decline in the following months. This pattern coincides with the seasonal rise in temperature and is more closely with the maximum temperature trend. Of particular importance are the months of May and June which fall in the summer season, during which the monsoon rainfall also occurs.

The cases of typhoid can also be linked to the precipitation pattern in the Kathmandu Valley (Fig. 9.5). The rainfall totals rise from mid-May, reach a maximum in July, and decline from the month of September. Coincidently, the cases of typhoid rise during the months of June, July, August, and September. Although more detailed research would be required to establish the true cause and effect

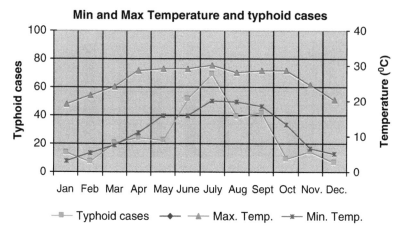

Fig. 9.4 Distribution of temperature and typhoid cases by month (2005), Kathmandu (*Source*: Regmi et al. 2008)

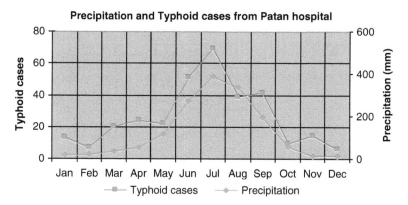

Fig. 9.5 Distribution of rainfall and typhoid cases by month (2005), Kathmandu (*Source*: Regmi et al. 2008)

relationship between temperature and precipitation and typhoid cases, it can be concluded that drinking water sources are more likely to be contaminated with fecal matter during heavy rainfall periods. Overall, there is an increasing trend of typhoid cases from 1997 onward (Fig. 9.4). The typhoid-positive cases were confirmed by blood culture tests that ranged from 6 to 19 % of the total cases cultured (Fig. 9.6).

Apart from temperature and rainfall, other associated factors might have contributed to the increased number of typhoid cases in Kathmandu Valley. One of the important factors identified is poor water quality and sanitation. Table 9.1 indicates that the various sources of water consumed by the people of the valley are not free from fecal contamination, though they are safe within World Health Organization (WHO) guidelines in terms of selected chemical parameters. This has been verified

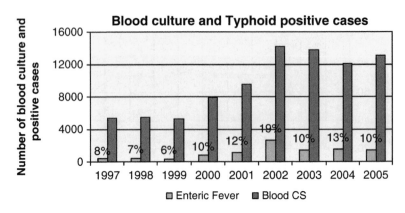

Fig. 9.6 Blood culture and typhoid positive cases (*Source*: Regmi et al. 2008)

Table 9.1 Water quality indicators of different sources in Kathmandu Valley in 2005

Parameter	Water sources				WHO GV
	Pr_Tap	Pu_Tap	Well	S. spout	
pH	6.5–8.2	6.5–7.5	7.5	7.5	6.5–8.5
Iron (mg/l)	ND-0.2	0.2	0.2	0.3	0.3 -3
Chlorine mg/l	ND	ND	ND	ND	0.2
Chloride mg/l	10–30	22–45	26–27	23–45	250
N-NH4 (mg/l)	ND-0.2	0.2	0.2	0.2	0.04-0.4
PO4 – P (mg/l)	0.1	0.1	0.1	0.1	0.4–5.0
Coliform bacteria (source)	+/–	+	+	+	–
Coliform bacteria (consumption point)	+				–
E. coli cfu/100 ml	10–131	3–20	48–200	58	0

Pr_Tap *private tap*, Pr_Tap *public tap*, S. spout *stone spout*, WHO GV *World Health Organization Guideline Value* (*Source*: Regmi et al. 2008)

in the studies undertaken by Pradhan et al. (2005). According to this study, no single drinking water source in the region – including dug wells, shallow wells, deep wells, springs, stone spouts, ponds, river, and pipe water – is 100 % safe from pathogenic bacterial contamination.

According to the government sources, the high occurrence of typhoid cases may also be attributed to the lack of access to clean drinking water sources and the poor sanitation conditions in Kathmandu Valley. Overall, slightly below 30 % of households in Kathmandu, as of 2010, still have no access to water from a pipe source that is considered a "safe" drinking water source. This means they have to depend on other sources, which are relatively less safe because the well and open sources of drinking water are not properly maintained.

Sanitation conditions can be explained in terms of households' access to toilets and sewerage (Pradhan 2004). According to the government sources, more than 10 % of households in Kathmandu still do not have access to toilets. People in the area usually defecate in open fields, river banks, around ponds and lakes, or in the jungle, the waste eventually mixing with water sources such as rivers, ponds, and lakes (Pradhan 2001, 2003). This becomes a serious health risk, particularly during the rainy season.

In addition, the focus group discussion with policy makers revealed that the poor water drainage system in Kathmandu makes the contamination of water more severe. The respondents in the focus group discussions indicated that about 25 % of the houses in Kathmandu often get flooded due to inadequate drainage. Kathmandu's existing drains and sewers are designed as either separate or combined systems. A combined storm water and sewer system, which is 50–70 years old, exists in the core area, which directly discharges into the Bagmati River (Fig. 9.7).

According to the Kathmandu municipality authority, around 17 % of households are being served by the existing system and about 60 % of the depth of the old sewers is still clogged with settled sludge and other debris, lowering the capacity to 40 % (Personal communication with municipal authorities, 2012). The flow from these sewers gets discharged directly into the river during the dry season. This constitutes a serious health hazard, as people are culturally inclined to use river water as holy water which includes use for ritual, bathing, and even drinking.

The results of this case study indicate that with increasing variability in temperature and rainfall and the lack of proper infrastructure for drinking water and drainage systems, health problems in urban areas like Kathmandu will worsen in future.

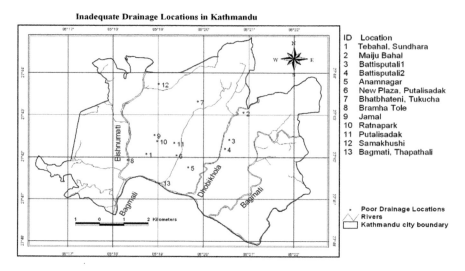

Fig. 9.7 Extent of inadequate drainage in Kathmandu Valley as at 2010 (*Source*: Kathmandu Municipality 2010)

The findings with regard to temperature reported in this paper are consistent with research carried out by Bentham and Langford (1995, 2001), D'Souza et al. (2004), and Kovats et al. (2004) on the impact of increased temperature and precipitation on typhoid prevalence. These studies have demonstrated that temperature and precipitation change have linear relationships with the occurrence of typhoid fever.

The study also looked at the association between rainfall variability and the outbreak of diseases such as typhoid. No other Nepalese studies have investigated the effects of rainfall variability on the occurrence of typhoid fever. This case study found that cases of typhoid fever are positively linked with up to 50 mm of weekly rainfall. When temperature, season, rainfall trend, and the 2002 typhoid outbreak are included in the multivariate statistical analysis, no association is evident. This demonstrates that, in Nepal, temperature has a much stronger relationship with typhoid. But more in-depth study is required to establish the association between rainfall and typhoid cases.

The data reported in this paper do not describe a cause and effect relationship of typhoid with temperature but merely demonstrate a positive relationship between increase in temperature and the number of typhoid cases, as typhoid is usually seen in the summer months of warmer temperatures. During warm temperatures, the growth of typhoid bacilli is favored drastically. It should also be noted that typhoid is not only a water-borne disease but also a food-borne disease. This study thus recommends a more detailed epidemiological study to better investigate the effects of climate factors on typhoid cases in Nepal.

In contrast with the magnitude of likely climate change impacts in urban areas, the municipal health planning response to such change in these areas is almost absent. In addition to increased temperate and greater variability in rainfall, urban populations are facing the problems of poor drainage and sewerage systems, a lack of clean drinking water, and poor health services. The municipalities of Kathmandu, Patan, and Bhaktapur have not yet identified climate change as their priority agenda. Although a few private and nongovernment organizations are working in the climate change sector, their main focus is on pollution control.

The health response at the national level is also lacking. There is still limited knowledge, information, and awareness among health sector communities about the potential impacts of climate change on human health, as climate change is one of several local issues. Climate change seems to be a low priority of the Ministry of Health and Population and a low priority locally. Very few initiatives are carried out at the national level to investigate health impacts and issues resulting from climate change. Although the National Adaptation Programme of Action (NAPA), formulated in 2010, has identified health as a major consideration in climate change responses, government action is almost nonexistent. According to health professionals interviewed, since the NAPA, there has been almost negligible research and development support from the government and other concerned agencies.

According to the majority of the national-level policy makers and practitioners, the health sector is lagging behind in terms of research and development because it is a low priority of government. It was also revealed during the interviews with policy makers that the support from development agencies and NGOs was also not very satisfactory. According to Dhimal (2008), there are a lot of challenges such as limited resources and capacity in conducting research on climate change and health in vulnerable, mountainous countries like Nepal.

9.3.2 Case Study Two: Health Risks and Hazards Due to Variability in Temperature and Rainfall in Rural Areas of Nepal

The second case study examined the incidence of health-related risks and hazards in Bangesaal and Dhungegadi of Pyuthan district, particularly investigating how variability in temperature and rainfall has impacted the health and well-being of households and communities in the rural areas of Nepal.

Analysis of the temperature and rainfall trends in Pyuthan district showed that temperatures are increasing and there is great variation in the rainfall pattern. There was an increase in the mean annual maximum temperature trend from 1975 to 2005 in the study area. The mean annual maximum temperature increased at a rate of 0.055 °C between 1997 and 2005. Analysis of the annual total precipitation between 1975 and 2005 showed a decreasing trend in rainfall by 2.03 mm per year, and there was a decreasing trend in precipitation during the monsoon season in the same time period by 2.0 mm per year (Fig. 9.8).

The impact of increasing temperature and variability in rainfall on households is significant. Climate-induced disasters (floods and landslides) have taken the lives of

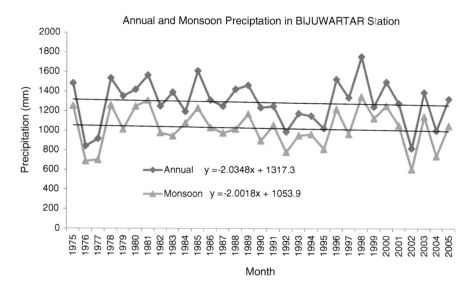

Fig. 9.8 Precipitation trends in Pyuthan district (*Source*: Author)

Table 9.2 Figures from disaster databases on the research and case study sites

	1999–2004 (casualties in numbers)		2005–2010 (casualties in numbers)	
	Death	Injured/lost	Death	Injured/lost
VDCs	15	44	13	143
Dhungegadi Village Development Committee-VDC	1	9	8	15
Bangesaal Village Development Committee-VDC	4	16	3	22

Source: Ministry of Home Affairs, Ministry of Environment report and databases (MoHA 1999–2010; MoHA and DPNet 2009; NHDR 2009; MoE 2010; MoHA and DPNet 2010a, b)

rural people and devastated their livelihoods. The trend of disaster frequency has increased in recent years, with a greater number of casualties and properties lost (MoHA 2009). Taken from databases on casualties and human losses at the study sites, the data in Table 9.2 above show the impact of climate-induced disasters. The death toll in Pyuthan district is comparatively higher. The field data shows that Bangesaal VDC has lost more people through disasters compared with Dhungegadi VDC (Table 9.2). This signifies that the extreme disaster and impact case in the rural area of Nepal is severe. There are some limitations to these data, as many of the losses are not reported and often not documented. The death tolls and number of casualties are more severe at the local level compared to those reported in the national data sources.

A vulnerability assessment carried out in Dhungegadi by the Livelihoods and Forestry Programme of the Pyuthan district showed that more than 200 households (out of 753) were directly impacted by fire, drought, and landslide problems during the period 1999–2009. In the Bangesaal VDC of Pyuthan, in ward number 5, an outbreak of water-borne diseases (cholera and diarrhea) in 2008/2009 took the lives of two villagers. This happened during an extreme flooding season (Rupantaran 2012). A timeline analysis of the disaster trends in Dhungegadi also showed that this is increasing and the impact of disasters has been severe (Table 9.3).

The data on casualties presented in Table 9.3 were not properly reflected in national databases and disaster reports, masking the real urgency needed to address the situation at the local level. In countries like Nepal, central planning depends on relief and rehabilitation efforts that are based on data from the national database of vulnerability. In Nepal, where there are no government systems of systematic data recording at the village level, this might therefore dilute the apparent seriousness of the situation and render it invisible in national scenarios.

The impact of climate-induced disasters on human health is projected to be significant at the national level in Nepal and internationally. Some of the national reports have highlighted examples of major health-related disaster outbreaks in Nepal, leading to injuries, loss of lives, and displacement of people (NCVST 2009). Outbreaks of water-borne diseases are common during the monsoon period. While

Table 9.3 Historical timeline analysis of disaster trends in Dhugegadi (2007–2010)

Year	Major climate disasters	Impact on communities
2010	Drought, landslides, outbreak of diseases, extreme wind	1 person died due to diarrhea – a total of 46 households were affected
		11 livestock were lost due to outbreak of disease
2009	Disease in paddies, drought, landslides, and extreme wind	Agricultural land was lost and production declined – a total of 5 households were impacted
		69 livestock died due to disease outbreak in 20 poor families
		13 houses were damaged by fire outbreak
2008	Drought, fire outbreaks, diseases in livestock, landslides, and hailstones	11 households were directly impacted due to drying up of local water sources
		Agricultural land was damaged
		Fire damaged 6 houses
2007	Drought, fire outbreaks, diseases in livestock	57 ropani of land were left barren due to drying up of local spring
		18 houses were destroyed due to fire outbreak
		150 livestock died due to disease outbreak
		Wheat production declined due to lack of water

Source: Authors' field survey (2012)

there are many factors contributing to such disease outbreaks, increasing temperature and variability in rainfall is one contributor (Regmi et al. 2008).

The findings of this study support that of previous studies. The interviews with communities in the Bangesaal and Dhungegadi VDCs of Pyuthan district revealed that there was a higher degree of uncertainty in terms of projection of the impacts of climate change on human health. More than two-thirds of the respondents (78.6 %) felt that climate change had an impact on human health to some level. A health professional from the Dhungegadi VDC revealed an increasing trend in disease outbreaks and in the number of ill patients, stating that "The children and elderly people visit the health post with illnesses of cholera, diarrhea, coughs and cold during extreme weather events such as heavy rain, chilly days and extremely hot days." He further added that "interestingly in the cold season we are experiencing malaria and diarrhea which is unusual given the history of disease prevalence in our village."

Although the number of malaria cases is said to be low in the far-Western region, the regional and district health data show that the disease is still a major threat to the health of the population in the region. Malaria cases have been slightly increasing in recent years, mostly in the higher altitudes where it had not previously been observed. For example, the Dhungegadi VDC, which is at an elevation of 1000 m, reported that malaria had infected households since 2002. The perception of households also matches the secondary data. The majority of the households interviewed said that the impact of climate change on health is significant. During the focus group discussion, communities revealed that they are now facing issues with mosquitoes that they had not experienced 5 years ago. In the interview with health professionals of Pyuthan district, it was also reported that malaria and Japanese encephalitis are the two most common vector-borne diseases, with an increasing number of cases in the district, both of which are transmitted by the mosquito vector.

While the district and regional health data on malaria and other vector-borne diseases do not show a significant increasing trend, the health assistant of Dhungegadi VDC reported that the number of patients consulting the health post about malaria cases and other diarrheal diseases has significantly increased in recent years. Likewise, the respondent from the district health office stated that, although the hospital records do not indicate a significant increase in the incidence of malaria, such vector-borne diseases may increase their impact through expansion to new regions. This information was validated by some of the interview respondents, who noted that the summer diseases are now prevalent in winter.

The spread of vector-borne disease is also supported by recent national evidence which indicates a migration of the *Anopheles fluviatilis* malaria vector above elevations of 1500 m (Gautam and Dhimal 2009). Malaria is now endemic in 65 districts of Nepal. Incidence of the disease decreased abruptly after a malaria eradication program in 1958. Later, in 1978, the name of the program was changed to the malaria control program. The most common species of malaria parasite occurring in Nepal is *Plasmodium vivax*. Another, more virulent species is *P. falciparum*. The observed increasing trend of this parasite may be due to climate change. Central Nepal has recorded the highest number of malaria-positive cases, whereas mid-Western Nepal had the least number of reported cases in between 1978 and 2009. Central Nepal has more average annual rainfall compared to the mid-western region. Malaria-positive cases are usually at a maximum during wet summers (Regmi et al. 2008; Badu 2013).

In addition to the outbreaks and increasing trends of climate-related diseases, it was found that households and communities are undergoing massive physical and mental stresses due to the impact of climate-related disasters. A variety of health issues induced by climate change were reported by household respondents and health professionals in Pyuthan district. According to the majority of households (92 %), post-disaster recovery is difficult, as residents are mentally shocked due to the loss of loved ones and property. For example, mental illness among two of the parents was recorded after the flooding in Bangesaal VDC in 2009 which killed their family members.

Among other issues, malnutrition is common among children and elderly in the study VDCs. The health professionals reported an increasing trend of cases of malnutrition in children and the elderly in 2012 over previous years (2005–2011), attributed to a lack of proper diet and food. According to the households, the extreme weather events, mostly the variability in rainfall, made cropping more uncertain and risky. The agricultural lands were also impacted due to drought, landslides, and flooding. Communities in Bangesaal VDC reported that more than 30 % of their productive agricultural land is being made unproductive due to extreme climatic events. The anecdotal evidence presented here, however, suggests that the relationship between production, nutrition, and climate change impacts needs to be researched in future.

Low production has implications for the food security situation in the studied VDC. Only 21 % of households in Dhungegadi VDC have food sufficiency, while the remaining 79 % have to struggle. There is not much difference in the food sufficiency status between the two VDCs. However, in the case of Dhungegadi VDC, the data generated from the household survey shows an average of 1 month less food sufficiency (7 months) compared to the VDC profile the government prepared in 2009, which indicates an average food sufficiency of 8 months/year. The government VDC survey data also shows that 28.4 % of the total population has sufficient food for the whole year (DDC-Pyuthan 2008). Likewise, according to the national census data, 15.7 % do not have adequate food to eat (CBS 2011). This implies that climate change will have a significant impact on the health and nutrition of the rural population in coming years.

At the local and district level, in comparison with the severity of the impact, coping and adaptation strategies for health seem to be lacking. Analysis of the local coping strategies in Dhungegadi and Bangesaal VDC shows that at the time of data collection, there was a lack of direct action on dealing with health-related issues. Most of the local practices (more than 80 %) made use of infrastructure, such as constructing drinking water facilities and building check dams (Table 9.4). However, 89.1 % of respondents in both the VDCs perceived that the existing adaptation options – the traditional practices – were ineffective, and they felt unable to address the climate risks and impacts because they lacked information, knowledge, and technology.

The interaction with local health professionals revealed that health workers in the district and local health posts were not aware of climate change. Many of the interviewed respondents said that they had not received any kind of capacity building and resource support from the center to address issues of climate change. As local priorities are immediate and empirical, the lack of support had negative implications for the response measures at the local level. The local and district health

Table 9.4 Local coping and adaptation practices in Dhungegadi and Bangesaal VDCs

Local coping practices	% of respondents ($n = 128$)
Plantation of broom grasses in landslide-prone areas	20
Efficient water management	23
Forest plantation activities	22
Use of local irrigation practices	12
Use of bioengineering practices	17
Others (agriculture and land-use management)	6

Source: Authors' field survey (2012)

infrastructure and human resources also seem to be inadequate. For example, there was only one health worker to provide services to an average of 3000 households residing in a VDC. The district health service was also in a poor condition, with an insufficient number of health workers and doctors to address the service demand.

The NAPA document mentions that because of the poor state of health services in Nepal, public health will indeed be at serious risk from the impacts of climate change (GoN 2010, p.13). The current study also found that access to health services is very limited in the study area. The district report of 2010 indicated that only 18 % of the population can access health post facilities within 30 min distance. The remaining 82 % of the population need to travel more than 1 h to reach the nearest health post. Many VDCs did not even have a health post, so households had to travel hours to access services (DDC-Pyuthan 2010). Improvement in health-related infrastructure and services will be crucial for increasing the resilience of communities to climate change.

9.4 Ways Forward

Climate change impacts on human health are likely to be severe in Nepal. The two case studies discussed in this chapter provide some evidence that temperature increase and rainfall variability can have adverse consequences for public health in both urban and rural areas of Nepal. Extreme changes in temperature and rainfall will very likely lead to a deterioration of the health conditions of households in the future. The findings of the case studies analyzed in this chapter indicate that waterborne diseases like typhoid and other climate-sensitive diseases (e.g., malaria) and natural hazards have increased in Nepal. However, the local and national coping and adaptation strategies in Nepal's health sector are lacking; this sector has not yet responded sufficiently to the challenges posed by climate change to the well-being of communities and households. The lack of timely action to date has made communities' health and well-being more vulnerable to climate-induced risks and disasters.

As there is limited information and knowledge on climate change at the national and local level, projection of the prevalence of climate-induced diseases is difficult and challenging. Climatic factors are one among many factors that contribute to the occurrence of disease and its spread. It was found from the case studies that besides climatic factors, poor hygiene and sanitation, lack of awareness, and poor access to support and health services contribute to an increasing risk and vulnerability of households to diseases and health hazards. The other associated factors that increase vulnerability of households to health hazards are important and need to be controlled for in trying to discern climate change effects. This implies that climate change and health should be dealt with in an integrated manner in order to build effective adaptation responses both at the local and national level.

The existing policy and actions in dealing with climate change issues in Nepal's health sector are inadequate. Even after 4 years of NAPA preparation, there is no project or national initiative focused on addressing public health issues identified in the NAPA. The health sector should have a strategic focus, program, and policy to combat the negative health impacts of climate change. An integrated approach and long-term National Adaptation Plan (NAP) in the health sector must be adopted for health planning and research in different climatic regions in the country.

It is necessary to take both a curative and preventative approach to deal with diseases and health hazards attributed to issues of climate change. More research is needed on epidemiological forecasting and early warning systems for both water- and vector-borne diseases in Nepal. The authors also recommend promoting action in the form of public education programs and other polices for particularly vulnerable groups. In the longer term, new policies may be needed in anticipation of additional cases of disease due to climate change. The NAPA and the Interim Three-Year Plan (2010–2013) have also identified various priorities that include strengthening the health system, awareness and capacity building, and the conduct of research on climate change and health for evidence-based planning. These priorities must be integrated with both short-term and long-term policies and plans in the health sector.

9.5   Conclusion

This study shows that climate-related diseases and other health-related hazards are increasing within the country of Nepal, and there is an association between these increasing trends of water-borne diseases like typhoid due to variability in temperature and rainfall. Water-borne diseases such as typhoid are expected to increase in prevalence and severity with changes in temperature and rainfall. Health-related hazards that lead to loss of life and damage to household infrastructure are placing the health of families at high risk. Consequently, this study has implications for future research work. As the study only found positive associations between climatic factors and the diseases, more detailed epidemiological study is necessary to better investigate the effects of climate factors on other disease in Nepal.

In contrast with the likely future severity of climate change variability and impacts, this study found that coping and adaptation strategies at the local and national level are inadequate. The adaptation policies, strategies, and response measures in the health sector are insufficient to deal with the impacts of climate change. The lack of response measures has resulted in an increased risk and vulnerability of poor and marginalized communities living in both rural and urban areas of Nepal. In addition, there was found to be a lack of systematic information and knowledge on climate change impacts in the health sector. There is also a lack of access to adequate health services and infrastructure to deal with climate change issues at both local and national levels. This has made health professionals and related institutions reluctant to take necessary measures to deal with climate change.

There is an urgent need to devise and implement integrated policies and strategies for better health planning and research in Nepal. These case studies indicate the need for the establishment of a separate research and development department or council for generating information and knowledge on the links between climate change and public health. The government and donor agencies in Nepal also need to invest in capacity building of health professionals and research and academic institutions in light of the changing disease and health risk scenarios associated with climate change. This can be achieved if the Ministry of Health formulates and implements a NAP for the health sector in Nepal.

9.6 Limitations of the Study

This research focused on two case studies conducted in Nepal with reference to a limited time frame of data collection, and hence any projections that can be made regarding future trends are also limited. The data for the case study on typhoid covers only a few years of close study between reported cases in Patan Hospital from 2000–2005. Due to restrictions on time and resources for collecting data, the data series used for the analysis has an end date of 2005. Indications from our analyses of these data that certain disease trends increase with variability in climatic factors does, however, highlight the need to carry out more in-depth studies to clearly establish the relationship between climate change and diseases in Nepal.

Acknowledgments The information for case study one presented in this paper was prepared based on a study carried out by the researchers in 2005–2008 with support from Local Initiatives for Biodiversity, Research, and Development (LI-BIRD) and the International Institute for Environment and Development (IIED). The authors would like to acknowledge the support and contributions of LI-BIRD, IIED, Dr. Sari Kovats, Lalita Thapa, Dr. Anil Pandit, Dr. Pooja Lama, and Emily Collins. We also appreciate the contribution of communities and health practitioners of Dhugegadi and Bangesaal VDC of Pyuthan district for providing valuable information for this chapter.

References

Badu M (2013) Assessing the impact of climate change on human health: status and trends of malaria and diarrhea with respect to temperature and rainfall variability in Nepal. Kathmandu Univ J Sci Eng Technol 9(1):96–105
Baidya SK, Regmi RK, Shrestha ML (2007) Climate profile and observed climate change and climate variability in Nepal. Department of Hydrology and Metrology, Kathmandu
Baidya S, Shrestha M, Sheikh MM (2008) Trends in daily climatic extremes of temperature and precipitation in Nepal. J Hydrol Meteorol 5(1):38–53
Bentham G, Langford IH (1995) Climate change and the incidence of food poisoning in England and Wales. Int J Biometeorol 39(2):81–86
Bentham G, Langford IH (2001) Environmental temperatures and the incidence of food poisoning in England and Wales. Int J Biometeorol 45:22–26
Bhattacharya S, Sharma C, Dhiman RC, Mitra AP (2006) Climate change and malaria in India. Curr Sci 90:369–375
CBS (2011) Preliminary findings of the National Census 2011. Nepal Centre Bureau of Statistics, Kathmandu

Constantin de Magny G, Guegan JF, Petit M, Cazelles B (2007) Regional scale climate-variability synchrony of cholera epidemics in West Africa. BMC Infect Dis 7:20

DDC-Pyuthan (2008) District profile of Pyuthan. Khalanga, DDC Pyuthan: Ministry of Local Development

DDC-Pyuthan (2010) District Profile Khalanga. District Development Committee (DDC), Pyuthan

Department of Health Services-DoHS (2005) Annual report 2004/2005. Department of Health Services, Ministry of Health and Population, Government of Nepal, Kathmandu

Department of Health Services-DoHS (2008) Annual report 2007/2008. Department of Health Services, Ministry of Health and Population, Government of Nepal, Kathmandu, Nepal

Dhimal M (2008) Climate change and health: research challenges in vulnerable mountainous countries like Nepal. Global Forum Health Res 66–69

Dhimal M, Bhusal CL (2009) Impacts of climate change on human health and adaptation strategies for Nepal. J Nepal Health Res Counc 7(15):140–141

D'Souza RM, Becker NG, Hall G, Moodie KBA (2004) Does ambient temperature affect food-borne disease? Epidemiology 15(1):86–92. doi:10.1097/01.doi.0000101021.03453.3e

Ebi KL, Woodruff R, von Hildebrand A, Corvalan C (2007) Climate change-related health impacts in the Hindu Kush–Himalayas. EcoHealth 4(3):264–270

Gautam I, Dhimal MN (2009) First record of Aedes aegypti (L.) vector of dengue virus from Kathmandu, Nepal. J Nat Hist Mus 24(1):156–164

Government of Nepal- GoN (2010) National Adaptation Programme of Action (NAPA). Ministry of Environment, Nepal

IPCC (2014) Climate change 2014: impacts, adaptation, and vulnerability. Working Group II. Contribution to the IPCC fifth assessment report climate: IPCC WGI AR5, Intergovernmental Panel on Climate Change. http://www.ipcc-wg2.gov/AR5. Accessed 15 Aug 2014

Keim ME (2008) Building human resilience: the role of public health preparedness and response as an adaptation to climate change. Am J Prev Med 35(5):508–516. doi:10.1016/j.ampetre.2008.08.022

Koelle K, Pascual M (2004) Disentangling extrinsic from intrinsic factors in disease dynamics: a nonlinear time series approach with an application to cholera. Am Nat 163(6):901–913

Koelle K, Pascual M, Yunus M (2005a) Pathogen adaptation to seasonal forcing and climate change. Proc R Soc Lond B 272(1566):971–977. doi:10.1098/rspb.2004.3043

Koelle K, Rodo X, Pascual M, Yunus M et al (2005b) Refractory periods and climate forcing in cholera dynamics. Nature 436(7051):696–700

Kovats S, Edwards S, Hajat S, Armstrong BG, Ebi KL, Menne B (2004) The effect of temperature on food poisoning: a time series analysis of salmonellosis in ten European countries. Epidemiol Infect 132(3):443–453. doi:http://dx.doi.org/10.1017/S0950268804001992

Kristie LE, Rosalie W, Alexander VH, Carlos C (2007) Climate change-related health impacts in the Hindu Kush–Himalayas. EcoHealth 4(3):264–270. doi:10.1007/s10393-007-0119-z

Lafferty KD (2009) The ecology of climate change and infectious diseases. Ecology 90:888–900. doi:http://dx.doi.org/10.1890/08-0079.1

McMichael AJ, Woodruff RE, Hales S (2006) Climate change and human health: present and future risks. Lancet 367(9513):859–869

Ministry of Home Affairs (MoHA), Nepal Disaster Preparedness Network-Nepal (DPNet) (2009) Nepal disaster report 2008. The hazardscape and vulnerability. Ministry of Home Affairs, Government of Nepal and Disaster, Kathmandu

Ministry of Home Affairs (MoHA) (2009) Annual disaster related losses Nepal (1999–2010). Government of Nepal, Kathmandu

Ministry of Home Affairs (MoHA), Nepal Disaster Preparedness Network- Nepal (DPNet) (2010a) Nepal Disaster Report 2010. The hazardscape and vulnerability. Ministry of Home Affairs, Government of Nepal and Disaster, Kathmandu

Ministry of Home Affairs (MoHA), Nepal Disaster Preparedness Network- Nepal (DPNet) (2010b) Nepal Disaster Report 2009. The hazardscape and vulnerability. Ministry of Home Affairs, Government of Nepal and Disaster, Kathmandu

MOE (2010) Climate change vulnerability mapping for Nepal. Ministry of Environment, Kathmandu

Narain JP (2008) Climate change and its potential impact on vector borne diseases. Health South East Asia. WHO SEARO 14:6

NCVST (2009) Vulnerability through the eyes of vulnerable: climate change induced uncertainties and Nepal's development predicaments. ISET, Kathmandu

Nepal Kathmandu Municipality (2010) Drainage locations in Kathmandu Valley. Kathmandu Municipality, Kathmandu

NHDR (2009) Nepal human development report. State transformation and human development. UNDP, Nepal

NSDRM (2009) National strategy for disaster risk management in Nepal. Ministry of Home Affairs, Kathmandu

Patz J, Engelberg D, Last J (2000) The effects of changing weather on public health. Annu Rev Public Health 21:271–307. doi:10.1146/annurev.pubhealth.21.1.271

Patz JA, Campbell-Lendrum D, Holloway T, Foley JA (2005) Impact of regional climate change on human health. Nature 438(7066):310–317

Pradhan B (2001) Drinking water quality in Kathmandu valley, Nepal. Health Prospect 2(2):10–15

Pradhan B (2003) Sanitation status and soil transmitted Helminthiases in Nepal. Health Prospect 5(1):1–5

Pradhan B (2004) Rural communities' perception on water quality and water borne disease: the case of Bungamati village development committee in Kathmandu Valley, Nepal. J Nepal Health Res Counc 2(1)

Pradhan B, Gruendlinger R, Fuerhapper I et al (2005) Knowledge of water quality and water borne disease in rural Kathmandu valley, Nepal. Aquat Ecosyst Health Manag 8(3):277–284

Pradhan B, Shrestha S, Shrestha R et al (2013) Assessing climate change and heat stress responses in the Tarai region of Nepal. Ind Health 51:101–112

Ramachandran N (2014) Persisting under nutrition in India: causes, consequences and possible solutions. Springer India, New Delhi. doi:10.1007/978-81-322-1832-6_9

Regmi BR, Pandit A, Pradhan B, et al (2008) Climate change and health country report-Nepal, capacity strengthening in the least developed countries (LDC's) for adaptation to climate change (CLACC) Working Paper. Local Initiatives for Biodiversity, Research and Development (LIBIRD), Pokhara, Nepal

Rodo X, Pascual M, Fuchs G et al (2002) ENSO and cholera: a no stationary link related to climate change? Proc Natl Acad Sci 99(20):12901–12906. doi:10.1073/pnas.182203999

Rupantaran N (2012) Climate change vulnerability mapping for Pyuthan District. Rupantaran, Kathmandu

WHO/WMO (1996) Climate change and human health. WHO/WMO/UNEP, Geneva

Yin RK (2003) Case study research: design and methods. Sage Publications, California

Chapter 10
Climate Change and Its Vulnerability for the Elderly in India

S. Siva Raju and Smita Bammidi

Abstract It is amply clear that the two major global drivers of change in the twenty-first century – global warming and population aging – will have points of both convergence and conflict. From the experiences of the developed countries that have experienced them first, we have realized that these phenomena will operate within different frameworks and need to be addressed at various levels, magnitudes, densities, structures, and functions. In the Southeast Asian countries, capital cities like Mumbai have populations above five million and are still growing. Urbanization in Asia is happening at a much higher rate (1.31 % annual average growth rate) than the world average (0.83 % annual average growth rate). Moreover, urbanization and globalization are two processes that catalyzed climate change and brought in further social complications besides the phenomenon of population aging. A large number of elderly now inhabit cities, while their financial and social supports are dwindling. They would soon face serious implications due to the rising rate of urbanization, their growing proximity to the coast that increases their vulnerability to weather/climate risks, and the solutions to which are energy driven (automobiles, elevators, air-conditioning). As climate change poses health, social, and economic risks for the elderly, the discussion on population aging vis-à-vis climate change will focus on ensuring and promoting health and quality of life of the elderly by modifying their built or external environments along with taking up psychosocial interventions. It is thus essential that the governments be open to respond flexibly to these synchronous challenges. In order to gain an understanding about the effects of climate change on the elderly in context of their rising numbers and urbanization, to identify/map the risks or costs involved, to identify possible interventions, to discuss policy support, and to estimate the opportunities and challenges that may arise, efforts for generating knowledge through research in this emerging field are imminent. Following this vein, the current paper contextualizes Mumbai City located in India and its older population, in an attempt to understand the possible impacts of climate change on

S. Siva Raju (✉)
School of Development Studies, TISS, Mumbai, India
e-mail: sivaraju@tiss.edu

S. Bammidi
College of Social Work Nirmala Niketan, Mumbai, India

© Springer International Publishing Switzerland 2016
R. Akhtar (ed.), *Climate Change and Human Health Scenario in South and Southeast Asia*, Advances in Asian Human-Environmental Research,
DOI 10.1007/978-3-319-23684-1_10

them. It also cites few ways of influencing policy making through their inclusion in the discussion and evolving strategies that will come to aid.

Keywords Global climate change • Population aging • Urbanization • Vulnerability • Health impacts • Regional variation • Indian context • Mumbai City • Policy inclusion

10.1 Background

During the twentieth century, the world's population grew almost fourfold, and in the year 2008, it crossed a landmark where more than half of the human population, i.e., about 3.3 billion, became urbanized. It is expected that by 2030, almost five billion people will reside in the urban areas. At the global level, all future population growth is expected to happen in towns and cities, especially in the developing countries. Estimates suggest that about 80 % of the urban dwellers in the world will reside in cities of the developing world, particularly in Asia and sub-Saharan Africa, by 2030. Urban growth now stems more from natural increase (more births than deaths) rather than migration. This is also termed as the "second wave" of urbanization, the first having occurred in Europe and North America in the early eighteenth century. The difference between the two waves is that of its scale. In the first wave, urban population increased from 15 to 423 million between 1750 and 1950. In the second wave, it is expected to have an unprecedented increase from 309 million to 3.9 billion between 1950 and 2030 in the developing world (State of the World Population 2007).

The increase in the world's population coupled with urbanization as mentioned above gave rise to a 12-fold increase in global emissions of carbon dioxide (Haines et al. 2006, p. 2107) primarily due to fossil fuel use and land use change. Similarly, the rise in methane and nitrous oxide within the atmosphere has been the result of increased agricultural production (Intergovernmental Panel on Climate Change 2007, p. 2). The rise in carbon dioxide levels is from approximately 280 parts per million in the 800 years prior to the industrial era to 386 parts per million in 2009 (Commonwealth Scientific and Industrial Research Organization 2011). As a direct consequence of these increases in what are now commonly referred to as "greenhouse gas" emissions, average global air and ocean temperatures have risen. Climate change is defined as a change in the state of the climate that can be identified by changes in the mean and/or the variability of its properties and that persists for an extended period, typically decades or longer (Intergovernmental Panel on Climate Change 2007). Climate change is nothing new. Historically the earth's climate has encompassed variation in temperature and carbon dioxide concentrations; however, there is now significant evidence that recent changes in climatic conditions are occurring much more rapidly. This accelerated warming of global temperatures is also directly attributed to human activity (Intergovernmental Panel on Climate Change 2007; The Global Humanitarian Forum 2009; Commonwealth Scientific and Industrial Research Organization 2011).

10.2   Major Consequences of Global Climate Change

10.2.1   Vulnerability to Climate Change

The Intergovernmental Panel on Climate Change (IPCC) Third Assessment Report defines vulnerability to climate change as "a function of the character, magnitude and rate of climate variation to which a system is exposed, its sensitivity and its adaptive capacity" (Intergovernmental Panel on Climate Change 2001). Ionescu et al. (2006) observe that there are three important caveats in attempting to predict vulnerability that also emphasises the complexity of the exercise:

- The direct effects of climate change will be different across locations.
- There are differences between regions, groups, and sectors in society which determine the relative importance of the direct effects of climate change.
- There are differences in the extent to which regions, groups, and sectors are able to prepare for, respond to, or otherwise address the effect of climate change (Fig. 10.1).

10.2.2   Global Warming and Its Effects

In addressing the impacts of global climate change, the Copenhagen Accord (2009) set the target of limiting global temperature increases to two degrees Celsius relative to pre-industrialized temperatures. However, even a two-degree increase in global

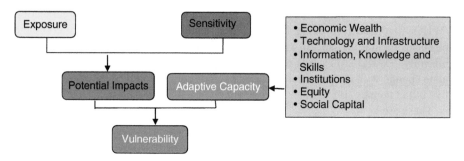

Fig. 10.1 Conceptualizing vulnerability to climate change (*Source*: Intergovernmental Panel on Climate Change, Third Assessment Report (2001))

temperatures will result in significant impacts that are likely to disrupt ecological and social networks alike and such impacts will include:

- Melting of the polar ice sheets resulting in sea-level rise
- Disruption to food supply and water resources
- Damage to physical infrastructure
- Increased public health risks
- Modified global biogeochemical cycles, as well as oceanic and atmospheric circulation patterns

10.2.3 Direct and Indirect Impacts on Human Health

Direct impacts include injuries and death caused by extreme weather events such as flooding, bush fire, and cyclones (Greenough et al. 2001). Haines et al. (2006, pp. 2106) estimate that climate change may have been responsible for around 150,000 deaths (0.3 % of global deaths per year) in 2000. The Global Humanitarian Forum (2009) estimates that approximately 300,000 deaths per year are caused by weather-related disasters and gradual environmental degradation due to climate change and that this figure will rise to 500,000 by 2030. Indirect impacts include changes in incidences of chronic disease and illness, resulting from changes in temperature, food and water supply, pollution levels, as well as the habitat of vectors impacting disease transmission (Haines et al. 2006; Bernard and Ebi 2001). The literature points to the fact that many of these health impacts will particularly affect those with preexisting health conditions or weakened immune and metabolic resistance as a result of age, meaning those who are very young and older will likely be more vulnerable (McMichael et al. 1996, 2006).

10.2.4 Specific Health Impacts

It is important to highlight the definition provided by the World Health Organization (WHO) for "health" as being applied in this context, inferring "a state of complete physical, mental and social well-being and not merely the absence of disease or infirmity" (WHO 2011). It is proposed that the various exposures attributed to climate change with the potential to cause a health impact could be categorized under five broad themes (Bernard and Ebi 2001).

These are:

- Temperature-related morbidity and mortality
- Impacts from extreme weather events
- Impacts on air quality
- Water- and food-borne diseases
- Vector-borne diseases

10.2.5 Psychological and Economic Impacts

Equally, the impact and cost of physiological trauma and stress linked to natural disasters as well as the economic uncertainty induced by climate change is now to be counted (Berry et al. 2008; Morressey and Reser 2007). About 325 million people are currently seriously affected by climate change (projected increase to 660 million people by 2030), and it causes economic losses of US$125 billion per year. The losses include asset values destroyed by weather-related disasters and sea-level rise, lost income due to reduced productivity, and the costs of reduced health or injury (Global Humanitarian Forum 2009, p.19).

10.2.6 Positive Impacts

While most of the literature focuses on the negative consequences, climate change has also brought about positive impacts in some regions and for some industries. For example, agriculture may benefit from changed climatic conditions in terms of temperature, precipitation, plant disease, and pests. This is particularly true for countries further south and north of the equator that are receiving more rainfall and experiencing longer growing seasons and improved productivity (Cairns 2010). In the agricultural sector, changes in climate will also give rise to new technologies and genetically modified plants to improve the resilience (Commonwealth Scientific and Industrial Research Organization 2011).

10.2.7 Geographical Vulnerability

Though the magnitude of these impacts continues to be debated, there is general agreement that low-lying and island nations, as well as those located closer to the equator where temperatures are already hotter, will be most affected by climate change and that the effects will be compounded if the nation's economy and supportive infrastructure is poorly resourced and less able to cope with the fiscal and social consequences. But what also needs to be highlighted is that many of the world's largest cities (and therefore concentrations of economic activity and production) are located in coastal areas, historically linked to seaport development, will therefore also be vulnerable to sea-level rise and storm surge, and represent unprecedented risk to livelihood, property, and urban infrastructure (Organization for Economic Cooperation and Development 2009, p.17). The consequences of these impacts are starting to be felt around the globe.

10.3   Why Are Cities Important in the Climate Change Debate?

Not only is population growth the significant factor in generating primary energy demand, but so too is where and how the world's population will be living (Organization for Economic Cooperation and Development 2008). Recent analysis by the International Energy Agency (IEA) estimated that between 60 and 80 % of the world's energy use currently emanates from cities and urbanized areas (IEA 2010), reflecting the fact that 50 % of the world's population now reside in these environments, coupled with a corresponding concentration of economic activity and production within cities. By 2050 the Organization for Economic Cooperation and Development (OECD) forecasts, based on the current trend of urbanization particularly in the developing economies, that this percentage will increase to 70 %, while in developed economies 86% of the population will be living in cities. In terms of climate change, continued urbanization (again predominately within the developing economies) will increasingly shift the world's reliance from CO_2 neutral energy sources (biomass and waste) to CO_2 intensive energy sources, leading to continued growth in greenhouse emissions from cities (OECD 2009).

10.4   Aging Population and Impact of Climate Change

At the same time that the world is experiencing the impacts of climate change, its population is also aging. The percentage of the elderly within the demographic profile is increasing as overall fertility and mortality rates decline and life expectancy significantly increases (Gavrilov and Heuveline 2003, p. 32). Among the countries currently classified by the United Nations as more developed (with a total population of 1.2 billion in 2005), the overall median age rose from 29.0 in 1950 to 37.3 in 2000 and is forecasted to rise to 45.5 by 2050. The corresponding figures for the world as a whole are 23.9 in 1950, 26.8 in 2000, and 37.8 in 2050. The Oxford Institute of Population Ageing (2016), however, concluded that population aging has slowed considerably in Europe and will have the greatest future impact in Asia. In Japan, one of the world's most quickly aging countries, in 1950 there were 9.3 people under 20 for every person over 65. By 2025 this ratio is forecasted to be 0.59 people under 20 for every person over 65. Because many developing countries are going through faster fertility transitions, they will experience even faster population aging than the currently developed countries in the future without being adequately prepared to address the challenges that may arise. While population aging must be viewed as "a mark of tremendous social achievement and a milestone of human progress" (Zelenev 2008, p. 1), brought about through advances in medicine, improvements in living, work conditions, and lifestyle, it will have considerable social and economic impacts. The potential vulnerability of an aging society to climate change is far greater and the implication this will have for the climate change policy can be juxtaposed for possible actions.

10.4.1   Magnitude of the Shift in Demographic Profile

Globally by 2050, the United Nations predicts that 20 % of the population or one in every five persons will be aged 60 or older, with the percentage of the "oldest of the old" (85 and above) expected to almost double. More staggering is the prediction that the number of persons reaching 100 years of age or older is to increase 14-fold over this same period. There is however significant variation in the pace at which population aging is occurring around the world and in the challenges faced by individual nations as a consequence. As a broad generalization, those nations with developed economies have already started to experience population aging (United Nations 2011). Asia with 1.5 billion people living in cities has the largest urban population in the world. Out of the 75 million born every year the world over, six countries from Asia and Africa – Bangladesh, China, India, Indonesia, Nigeria, and Pakistan – are accounting for almost half of them (United Nations Human Settlements Programme (UN-HABITAT 2006). There are serious implications of such massive populations for these cities, as higher rates of urbanization and their proximity to the coast also enhances their vulnerability to weather/climate risks such as sea-level increases, storm surges, and floods. Asia's densely populated megacities and other low-lying coastal urban areas are described in the Intergovernmental Panel on Climate Change (IPCC) Fourth Assessment Report (AR4) as "key societal hot spots of coastal vulnerability" with millions of people at risk. Mumbai is one of the largest megacities in the world in terms of its population and is currently ranked 4th after Tokyo, Mexico City, and New York (UN-HABITAT 2006). According to the 2011 Census, 121 million persons (9 % of total population) were above 60 years in India and 9.9 million persons in Maharashtra are enumerated to be above 60 years of age (Census of India 2011). Of them, 4.7 million are men and the remaining 5.2 million are women, with nearly half of them residing in urban areas. As the statistics depict, nearly one in ten elderly in the country resides in Maharashtra. Further, in terms of the population composition of the state, over 10 % is comprised of persons aged 60 and above, which is higher than the national average of 8.5 %. Similarly, the proportion of the elderly in the age group 80 and above (oldest-old) in Maharashtra is higher than the national average (Building Knowledge Base on Population Ageing in India Study 2013). Therefore, it is critical for the city to assess the vulnerabilities and devise adaptation and mitigation mechanisms to cope with future climate risks for its elderly.

10.4.2   Diversity in Issues Faced and Propensity for Climate Change Risks

The status of aging in India in coming paragraphs reflects the issues pertaining to socioeconomic and demographic profiles, living arrangements, and services especially of the urban elderly.

10.4.2.1   Poverty

The Ministry of Social Justice and Empowerment, Government of India (1999), in its document on the National Policy for Older Persons, has relied on the figure of 33 % of the general population below poverty line and has concluded that one-third of the population in the 60 plus age group is also below that level. Though this figure may be understated from the older people's point of view, even at this estimate, the number of poor elderly comes to about 23 million. Vulnerable groups like the disabled, fragile elderly and those who work outside the organized sector like landless agricultural workers; small and marginal farmers; artisans in the informal sector; unskilled laborers on daily, casual, or contract basis; migrant laborers, informal self-employed or wage workers in the urban sector; and domestic workers deserve mention here.

10.4.2.2   Health Conditions

It is obvious that people become more and more susceptible to chronic diseases, physical disabilities, and mental incapacities in their old age. Regarding the health problems of the elderly of different socioeconomic statuses, it was found (Siva Raju 2002) that while the older poor largely describe their health problems on the basis of easily identifiable symptoms, like chest pain, shortness of breath, prolonged cough, breathlessness/asthma, eye problems, difficulty in movements, tiredness, and teeth problems, those from upper class, in view of their greater knowledge of illnesses, mentioned blood pressure, heart attacks, and diabetes which are largely diagnosed through clinical examination. In a study by Mutharayappa and Bhat (2008), NFHS-2 data was analyzed to examine the type of lifestyle adopted by the elderly and its effects on their health conditions; it was found that lifestyle adversely affects health and increases the morbidity levels. Lifestyle habits such as alcohol consumption, smoking, and tobacco chewing have adverse effects on one's ability to control diseases.

10.4.2.3   Psychological Health

Psychological changes accompany the passing of years, slowness of thinking, impairment of memory, decrease in enthusiasm, increase in caution in all respects, and alteration of sleep patterns. Social pressure and inadequate resources create many dysfunctional features of old age. Further, it is well known that the incidence of mental illness is much higher among the elderly than among the young. The psychological problems encountered by retired persons are much greater and the impact on the individual is entirely different as compared to those in the unorganized sectors. Attitudes toward old age, degradation of status in the community, problems of isolation, loneliness, and the generation gap are the prominent thrust areas resulting in sociopsychological frustration among the elderly (Mohanty 1989).

10.4.2.4 Changing Values and Attitudes

Due to rapid social change resulting in breakdown of traditional family structures, the elderly may not receive adequate care and attention from their family members. This trend may be fast emerging partly due to the growth of "individualism" in modern industrial life and also due to the materialistic thinking among the younger generation. These changes lead to greater alienation and isolation and the value system, respect, honor, status, and authority, which they used to enjoy in traditional society, have gradually started declining, and in the process the elderly are relegated to an insignificant place in our society (Gupta 2004). Though the younger generation takes care of their elders, in spite of several economic and social problems, it is their living conditions and the quality of care, which differs widely from society to society.

10.4.2.5 Elder Abuse

A review of the few studies that focused on elder abuse indicates that the most likely victim of elder abuse is a female of very advanced age, role-less, functionally impaired, lonely, and living at home with someone primarily their adult child, spouse, or other relatives. Studies in India indicate (Rao 1995; Siva Raju 2002) that more women than men complain of maltreatment in terms of both physical and verbal abuse. The prevalent patterns of elder abuse include mainly psychological abuse in terms of verbal assaults, threats and fear of isolation, physical violence, and financial exploitation. The health profile of the elderly victims indicates that a person suffering from physical or mental impairment and dependent on the caretakers for most of his or her daily needs is likely to be the victim of elder abuse. Though a large section of victims of elder abuse are less educated and have no income of their own, old people with high educational background and sufficient income are also found to be subjected to abuse.

10.4.2.6 Caregiving Needs

For social and familial relations of the elderly, there appears to be a steady change in caregiving from the traditionally secure joint family care of the elderly to extended family care in which care by adult children forms a major part. Scholars cite that if the present trend continues, there will likely be a decrease in elder care by adult children in the future, which will create more demand for old-age homes. India is at a crossroads and has to decide whether to go the family care way or the institutional/community care way. Liebig and Rajan state that for a country like India, the State cannot enter as a major player in elder care in view of the high (prohibitive) cost to the exchequer and the low national priority to elder care. The need to develop models of home or family care may be supplemented by suitably adapting them to a variety of respite services while at the same time suitably adapting them to Indian conditions (Liebig and Rajan 2005).

10.4.2.7 Urban Poor Elderly

Due to industrialization and urbanization and the changing trends in the society, it is the urban elderly who are more likely to face the consequences of this transition as the infrastructure often cannot meet their needs. Lack of suitable housing forces people to live in slums which are characterized by poor physical condition, low income levels, high proportion of rural immigrants, high rates of unemployment and underemployment, and rising personal and social problems such as crime, alcoholism, mental illness, etc., along with total or partial lack of public and community facilities such as drinking water, sanitation, planned streets, drainage systems, and access to affordable healthcare services. With the increasing prevalence of slum dwellers that come to urban areas in search of better opportunities, a significant proportion of them would constitute the elderly.

10.4.2.8 Feminization of Aging

Women are more likely to depend on others, given lower literacy and higher incidence of widowhood among them (Gopal 2006). Hence, the greater vulnerability of women due to their extended life expectancy compared to men and hence higher incidence of widowhood indicates the need to have a special focus on gender-based policy implications and social security needs of women.

This clearly indicates their propensity to be affected due to climate change risks posed. Some important factors among the elderly that decide the risks to them due to climate change are:

- Extent of urbanization and their location in a city with its unique set of features deciding the level of proneness to effects of climate change
- Age, gender composition, and diversity in terms of class, disability, experiences, education levels, current health, culture, religion, attitudes, and information among them
- Infrastructure of cities, their housing environments, facilities, and resource availability
- Policies, acts, financial capacity of governments, and political conviction
- Health and mental health status, functionality, and active aging concepts
- Means of communication, sources of information, and social interaction

The diversity that has emerged in the aging process necessitates research efforts to focus on different aging issues in society that, in turn, is expected to promote development of effective age-related policies and programs.

10.5 Climate Change and Elderly: Contextualizing India and Mumbai City

In the forthcoming paragraphs, the paper provides the profile of the city, specific climate change vulnerabilities, and an overview of the weather/climate risks and identifies the physical, economic, and social vulnerabilities for local communities in Mumbai City. The paper briefly reviews the current status of planning, policy, and implementation efforts to mitigate the identified vulnerabilities in the city and cites some knowledge and research gaps in addressing them. There is scarcity of information on the generic sensitivities of these impacts specifically for older people and hence there is no mode for reaching out to them differently. The concluding paragraph identifies the emerging concerns under climate change and aging that academia and policy makers may require to address seriously.

10.5.1 Profile of the Mumbai Metropolis

Mumbai (formerly known as Bombay) is located on the western seacoast of India on the Arabian Sea at 18053′ N to 19016′ N latitude and 720 E to 72059′ E longitude. The map of Mumbai City, including the location of different administrative wards, is shown in Fig. 10.2. Greater Mumbai Region (referred to as Mumbai in the text) consists of seven islands in the city area and four islands in the suburbs. Mumbai occupies an area of 468 km^2 (sq km) and its width is 17 km east to west and 42 km north to south (MCGM 2007). The entire region encompasses rich natural heritage, such as hills, lakes, coastal water, forests, and mangroves alongside built areas. The coastline of Mumbai has been reclaimed for development purposes; e.g., areas like Cuffe Parade and Mahim Creek were wetlands, later reclaimed for residential and commercial uses. The Municipal Corporation of Greater Mumbai (MCGM) was the first municipal corporation established in India in the year 1882. Since then, the civic body has been responsible for the provision of civic amenities, education, public health, art and culture, and heritage conservation in the city. MCGM holds the distinction of being one of the largest local governments in the Asian continent (MCGM 2007). The Mumbai Metropolitan Region Development Authority (MMRDA), set up in 1975, is responsible for planning and coordination of the development activities of this region. The total area of the MMR, excluding Mumbai City, is 3887 km^2, with a population base of 5.90 million as per 2001 Census (MMRDA 2007). These surrounding areas hold significance for the economy and transportation in Mumbai, as thousands of people travel everyday from these areas into the city for employment.

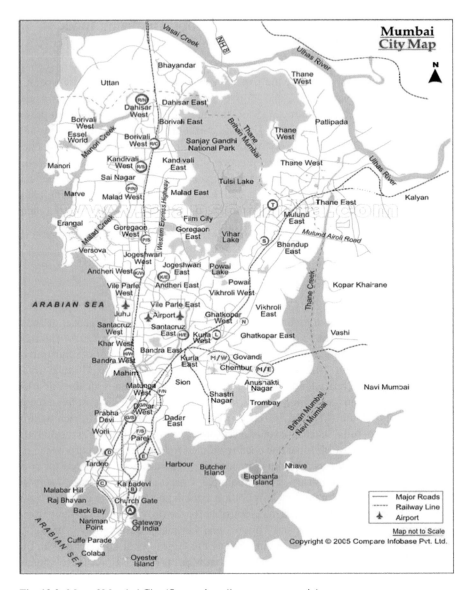

Fig. 10.2 Map of Mumbai City (*Source*: http://www.mcgm.gov.in)

10.5.2 *Physical Vulnerability*

- The city is surrounded on three sides by the sea: Arabian Sea to the west, Harbor Bay in the west, and Thane Creek in the east. The height of the city is just 10–15 m above the sea level. A large part of the city district and suburban district as well as new industrial, commercial, and residential settlements have developed along the reclaimed coastal areas that are low lying and flood prone.

- Being a coastal city, Mumbai is prone to cyclones and gusty winds. There are a number of wards along the coast (Arabian Sea and Thane Creek) that are vulnerable to cyclonic impacts (Govt. of Maharashtra 2007).
- Mumbai also plays host to around 900 industries that are involved in manufacturing, processing, or storage of hazardous goods. The presence of such industries only enhances the vulnerability in case of extreme weather events. The major concentration of such industries is in the Chembur-Trombay belt (Wards M-West and M-East). The area has major chemical complexes, refineries, fertilizer plans, atomic energy establishment, and thermal power plant.
- Mumbai, being on the seacoast, experiences a tropical savanna climate (MCGM 2003; MCGM, MPCB 2005) with a heavy southwest monsoon rainfall of more than 2100 mm a year, and relative humidity is quite high during this season. The flash floods that led to the complete disruption of normal life in Mumbai in July 2005 were the result of an unprecedented rainfall of 944.2 mm on July 26th in the suburban district (Govt. of Maharashtra 2007). Many slum settlements also face the risk of landslides usually occurring during heavy rains with gusty winds.
- Mumbai falls in the seismic zone III that is the Moderate Damage Risk Zone.
- Mumbai, with a large population to cater to, requires basic infrastructure in the form of a large transport network. There has been a massive growth in the number of vehicles (more than 79 %) in recent years between 1991 and 2004. Among the total number of vehicles, the number of two wheelers and three wheelers has gone up by more than 100 % between 1995 and 2004, whereas the number of passenger cars during the same period has gone up by more than 62 % (Govt. of Maharashtra 2005).

10.5.3 Economic and Social Vulnerabilities

Mumbai, being the financial capital of the country with a large industrial and commercial base, attracts a large workforce into the city. The growing population adds to the pressure on basic infrastructure, civic amenities, and housing, which without proper planning of land use and infrastructure development is not sustainable. It also leads to congestion, heavy vehicular traffic, growth in illegal slum dwellings, unhygienic living conditions, and the problem of solid waste disposal. The increasing climate risks lead to health problems, property loss, and disintegration of the social fiber of the city.

10.5.4 Mapping Climate Risks

Revi (2008) has reviewed the climate risks for Indian cities in general in order to highlight the importance of infrastructure investments and urban management and the need to connect these with the official adaptation initiatives. There is a broad consensus among the scientific community on the first-order climate change impacts in India.

- The cities will further face the new hazards in terms of *sea-level rise*. A sea-level rise of 30–80 cm has also been projected over the century along India's coast based on multiple climate change scenarios (Aggarwal and Lal 2001). Such sea-level rises, cyclones, and storm surges could have a devastating impact on a large urban center like Mumbai, which falls into a low elevation coastal zone (LECZ).
- There may be a general *increase in both mean minimum and maximum temperatures by 2–40 °C* depending on the atmospheric GHG concentrations (Sharma et al. 2006). Further, a sea surface temperature rise of 2–40 °C is expected to induce a 10–20 % increase in cyclonic intensity.
- Simultaneously, there might be a decrease in total number of rainy days over much of India, along with an increase in heavy *rainfall days and frequency of such days* in the monsoon season. Mumbai may suffer from acute water shortages due to its heavy dependence on rainfall for water supply in future. Also, droughts might become more common in areas surrounding Mumbai, triggering migrations into the city (Rupa Kumar et al. 2006). A substantial increase in extreme precipitation (e.g., Mumbai floods 2005) is also expected over a large area of the west coast (including Mumbai) and Central India (Rupa Kumar et al. 2006).
- Another important climate risk for Indian cities, in particular Mumbai, is the *onset of waterborne diseases* (diarrhea, cholera, and typhoid) and vector-borne diseases (malaria and dengue). India in future (Bhattacharya et al. 2006) and the city of Mumbai with its large population will be at further risk.
- Vulnerability assessment undertaken by Sherbinin et al. (2007) further suggests that a "bundle" of stresses, such as Mumbai's flat topography, geology, wetlands and flood-prone areas, projected sea-level rise, building conditions including not meeting building codes, squatter settlements, flood-ravaged buildings, poor sanitation and waste treatment, and low incomes, *reducing the ability for disaster preparedness*, will create an enhanced vulnerability for the city.

10.5.5 Adapting to Vulnerability

In December 2005, in the aftermath of the unprecedented Mumbai floods, the Government of India enacted the Disaster Management Act, under which the National Disaster Management Authority and State Disaster Management Authorities have been created. The Act also seeks to constitute the Disaster Response Fund and Disaster Mitigation Fund at national, state, and district levels. In Maharashtra, the state government accordingly has prepared the Greater Mumbai Disaster Management Action Plan (DMAP) in 2007. Under this plan, the risks and vulnerabilities associated with floods, earthquakes, landslides, cyclones, etc., have been identified. The plan further envisages specific relief and mitigation measures for Mumbai:

- Infrastructure improvements: The mitigation strategy seeks to improve the transport, services, and housing infrastructure. These include improvements in road and rail networks, sanitation and sewer disposal system, storm water drainage systems, slum improvements, housing repairs, and retrofitting programs.
- Contingency plan: This strategy includes plans to provide extra transportation if the major transport systems fail, transit camp arrangements, improvements in wireless communication and public information systems, and NGO volunteers' assistance.
- Land use policies and planning: The Draft Regional Plan for Mumbai Metropolitan Region 1996–2011 provides a basic framework for the land use policies for the city. This plan includes strategies like protection of landfill sites, control on land reclamation, shifting of hazardous units from residential areas, and decongestion.

The DMAP looks comprehensive on paper, yet does not provide any specific time frame for achieving the mitigation measures. The experience of the city dwellers in the aftermath of 2005 floods only shows that the city administration and other stakeholders would need more specific strategies and an integrated approach to build resilience of the city to climate risks. The Greater Mumbai Disaster Management Action Plan (DMAP) prepared by the Government of Maharashtra in 2007 has also identified the stakeholders in the government machinery such as the national, state, and district disaster management authorities, local authority (MCGM – specific to Mumbai), MCGM – storm water drainage department, early warning system, public health department, contingency plan, police and fire brigade departments, NGOs and communities, and the specific tasks they need to perform as a part of the mitigation strategies. However, it does not provide any specific time frame for achieving the mitigation measures. No specific attention is given to adaptation strategies that may be more important in the short to medium term, and strategies specific to different age, gender, and other cohorts are not provided, thus overlooking a main way to strategize measures.

10.5.6   Impact for Elderly, Emerging Issues, and Need for Their Inclusion

For each of the climate risks mapped above, while every member of the society has the potential to suffer health and other impacts resulting from climate change, the elderly will be more sensitive and likely to suffer a greater degree of impact than any other age group. While physiological decline and the progression of chronic disease associated with aging are obvious causes of this increased sensitivity, there are other contributing factors that can be bundled under the headings of psychosocial and economic risks during the later stage in their life course. There are benefits in recognizing the high degree of heterogeneity and commonalities that exists within the "senior" cohort. The reality is that there will be a significant variation in the physical, cognitive, social, and financial capacities of elderly and, as such, in the ability of an older person to maintain their independence and well-being. That being said, there is value in considering the range of non-pathological changes and physical as well as some cognitive and behavioral functions as these factors point to the types of sensitivities that older people, as a cohort, will have from the impacts of climate change and also their ability to cope and respond in the face of climate change. There is scarce information about the range of impacts for the elderly in Mumbai, taking into account their diversity.

Some emerging issues to take note are:

- An older person's sensitivity and risk of injury or loss increases in proportion to their level of physical and/or cognitive impairment, level of social isolation, and financial dependency.
- Given preexisting health condition and level of physical and/or cognitive impairment are major determinants of sensitivity to a range of environmental exposures, supporting the elderly to maintain good health and be physically active is a key strategy in building resilience to and reducing vulnerability to climate change.
- The ability for the elderly to obtain assistance and conversely the surveillance can improve their level of preparedness including modifying behavior in response to a threat of an extreme weather event or environmental exposure. As such supporting the elderly to remain socially engaged and active (avoiding becoming socially isolated) can significantly reduce vulnerability to climate change.
- Creating "age-friendly" environments, which promotes positive aging and encourages the elderly to remain socially active, has the potential to deliver the added benefit of also reducing the sensitivity of the elderly to the impacts of climate change and, as such, the overall vulnerability of the community to climate change.
- Adaptation of existing building stock for climate change and supporting aging in place present a more significant issue within the Indian context.
- Older cohorts are more reliant on traditional modes of communication (telephone, newspaper/print media, radio, and television). While this reliance will likely change with the entry of the Baby Boomer cohort into retirement, this needs to be considered when formulating strategies for early warning and information about climatic-induced threats.

Ensuring that strategies are supportive of older people will successfully allow them to mitigate or adapt to climate change. Not only does this make fiscal sense, in helping older people to age in good health and to continue to contribute in a meaningful way to society, but also it has another advantage, i.e., it has the potential to increase the capacity of the community as a whole to cope with the direct and indirect impacts of climate change. In other words, there is a strong cause for considering climate change adaptation in parallel with strategies that take into consideration the needs of the nation's aging population.

10.6 Knowledge and Research Gaps

The state government prepared a Disaster Management Action Plan for the city in the aftermath of the 2005 devastating floods and its shortcomings are cited in the above paragraph. Besides this plan, there are no visible mitigation and adaptation efforts currently underway in Mumbai that would target to reduce vulnerabilities to climate risks. In general, vulnerability reduction and adaptation to the adverse impacts of climate change is an important area for policy formulation at national, regional, and local levels. An interdisciplinary approach is needed to create an information and knowledge base to help identify, develop, and implement effective responses to reduce vulnerability and enhance adaptive capacity. Specific information needs to be built on what the available adaptation options are, under which climatic conditions they will work effectively, anticipated benefits, resource requirements to implement them, requisite institutional structures and processes, and potential spillover effects (Patwardhan et al. 2009). In addition to this, exploratory, descriptive, and experimental studies are required on different adaptation and mitigation options for different cohorts (age, gender, class, location, etc.) as per their vulnerability, requirements for such efforts at all levels, potential performance of these options, and strengthening institutional capabilities to manage adaptations. The research gaps and challenges that need to be addressed for Mumbai in immediate future in order to reduce climate vulnerabilities and build city resilience can broadly be classified into three categories: information, assessment, and knowledge. There is a need to compile *information* regarding different climate-related risks in the city and for different cohorts such as slum dwellers, the elderly, the disabled, young children, women, and unorganized sectors. We further need to *assess* how and where different models and tools can be applied to look at changes in hazards, exposure, and vulnerability. We also need to build on the fundamental *knowledge* about topics where there is inadequate understanding currently, e.g., health impacts of climate change on the elderly, intra-seasonal variability in the monsoon, studies of subsidence and stability of reclaimed lands, etc.

10.7 Future Directions

Future societies will have to adapt and respond to both the "global" issues of climate change and aging, respectively. Yet, little attention has been given to the relationship between the two. One can attempt to explore the potential interactions between these two distinct lines of policy development and find practical solutions to the problems of sustainability in urban living that will harmonize the need to respond to climate change with the needs of a growing older population. It becomes imperative not merely to warn policy makers that the solutions proposed to one set of problems may generate other problems but also, where possible, to create solutions that offer gains across both. Older people, because of a range of physiological, psychological, and socioeconomic dispositions, are more vulnerable to the impacts of climate change and extreme weather events. By understanding the link between climatic exposures on health and other sensitivities of the senior cohort, a range of sustainable interventions can be adopted to reduce vulnerability and therefore increase the resilience of this group. It is hoped that this paper provides a platform to launch further research on this emerging topic in a developing society like India.

References

Aggarwal D, Lal M (2001) Vulnerability of Indian coastline to sea level rise. Centre for Atmospheric Sciences, Indian Institute of Technology, New Delhi

Bernard S, Ebi K (2001) Comments on the process and product of the health impacts assessment component of the national assessment of the potential consequences of climate variability and change for the United States. Environ Health Perspect 109(Supplement 2):177–184

Berry H, Kelly B, Hanigan I, Coates J, McMichael A, Welsh J, Kjellstrom T (2008) Rural mental health impacts of climate change. National Centre for Epidemiology & Population Health, ANU College of Medicine and Health Sciences, Australian National University, Canberra

Bhattacharya S, Sharma C, Dhiman RC, Mitra AP (2006) Climate change and malaria in India. Curr Sci 90(3):369–375

Building Knowledge Base on Population Ageing in India Study (2013) The status of elderly in Maharashtra, 2011. United Nations Population Fund, New Delhi

Cairns E (2010) Climate vulnerability monitor 2010: the state of the climate crisis. Development Assistance Research Associates (DARA), Oxfam

Census of India (2011) Office of the Registrar General of India. Government of India, New Delhi

Copenhagen Accord (2009) http://unfccc.int/meetings/copenhagen_dec_2009/items/5262.php

Commonwealth Scientific and Industrial Research Organisation (2011) Climate change in Australia: technical report. Commonwealth Scientific and Industrial Research Organization, Australia Government, Canberra

Gavrilov L, Heuveline P (2003) Aging of population. Macmillan Reference, New York

Gopal M (2006) Gender, ageing and social security. Economic and Political Weekly:4477–4486

Govt. of Maharashtra (2005) Motor transport statistics. Office of the Transport Commissioner, Mumbai

Govt. of Maharashtra (2007) Greater Mumbai disaster management action plan. Maharashtra Emergency Earthquake Management Programme, Mumbai

Greenough G, McGeehin M, Bernard SM, Trtanj J, Riad J, Engelberg D (2001) The potential impacts of climate variability and change on health impacts of extreme weather events in the United States. Environ Health Perspect 109(2):191–198

Gupta N (2004) Successful ageing and its determinants. Ph.D. thesis, TISS, Mumbai

Haines A, Kovats R, Campbell-Lendrum D, Corvalan C (2006) Climate change and human health: impacts, vulnerability, and mitigation. Lancet 367:2101–2109

International Energy Agency (2010) World energy outlook 2010. International Energy Agency, Paris

Ionescu C, Klein RJT, Hinkel J, Kavi Kumar KS, Klein R (2006) Towards a formal framework of vulnerability to climate change. Environ Model Assess 14(1):1–16

IPCC (2001) In: McCarthy J, Canziani O, Leary N, Dokken D, White K (eds) Climate change 2001: impacts, adaptions and vulnerability. Contribution of Working Group II to the third assessment report of the Intergovernmental Panel on Climate Change (IPCC), Cambridge University Press, Cambridge, U.K./New York

IPCC (2007) Working Group 1: the physical science basis – executive summary. Intergovernmental Panel on Climate Change, Cambridge University Press, Cambridge, U.K./New York

Liebig PS, Irudaya Rajan S (2005) An ageing India: perspectives, prospects and policies. Rawat Publications, New Delhi

MCGM (2003) Environment status report, 2002–03. Municipal Corporation of Greater Mumbai, Mumbai

MCGM (2007) Statistics on Mumbai, Municipal Corporation of Greater Mumbai. Available at http://www.mcgm.gov.in

McMichael AJ, Woodruff RE, Hales S (1996) Climate change and human health. WHO, Geneva

McMichael AJ, Woodruff RE, Hales S (2006) Climate change and human health: present & future risks. Lancet 367(9513):859–869

MMRDA (2007) Basic statistics on Mumbai metropolitan region. Mumbai Metropolitan Region Development Authority. http://www.mmrdamumbai.org/basic_information/html

Mohanty SP (1989) Demographic and socio-cultural aspects of ageing in India: some emerging issues. In: Pati RN, Jena B (eds) Elderly in India: socio-demographic dimension. Ashish, New Delhi, pp 37–45

Morressey S, Reser J (2007) Natural disasters, climate change & mental health considerations for rural Australia. Aust J Rural Health 15(2):120–125

MPCB (2005) Report on environment status of Mumbai region. Maharashtra Pollution Control Board, Government of Maharashtra, Mumbai

Mutharayappa R, Bhat TN (2008) Is lifestyle influencing morbidity among elderly? J Health Manag 10(2):203–217

OECD (2008) Ranking port cities with high exposure and vulnerability to climate extremes: exposure estimates, environment, Working papers no 1. Organization for Economic Cooperation and Development, Paris

OECD (2009) Pensions at a glance: retirement income systems in OECD countries. Organization for Economic Cooperation and Development, Paris

Patwardhan A, Downing T, Leary N, Wilbanks T (2009) Towards an integrated agenda for adaptation research: theory, practice and policy. Curr Opin Environ Sustain 1:219–225

Rao KV (1995) Rural elderly in Andhra Pradesh: a study of their socio-demographic profile. Unpublished doctoral dissertation, Andhra University, Mimeo, Visakhapatnam

Revi A (2008) Climate change risk: an adaptation and mitigation agenda for Indian cities. Environ Urban 20(1):207–229

Rupa Kumar K, Sahai AK, Krishna Kumar K, Patwardhan SK, Mishra PK, Revadekar JY, Kamala K, Pant GB (2006) High-resolution climate change scenarios for India for the 21st century. Curr Sci 90(3):334–345

Sharma S, Bhattacharya S, Garg A (2006) Greenhouse gas emissions from India: a perspective. Curr Sci 90(3):326–333

Sherbinin A, Schiller A, Pulsipher A (2007) The vulnerability of global cities to climate hazards. Environ Urban 19(1):39–64

Siva Raju S (2002) Meeting the needs of the poor and excluded in India, situation and voices. The older poor and excluded in South Africa and India, vol 2. United Nations Population Fund (UNFPA) and The Population and Family Study Center (CBGS), New York, pp 93–110

The Global Humanitarian Forum (2009) Human impact of climate change. Report of the Forum held during 23–24 June 2009 in Geneva, Switzerland

The Oxford Institute of Population Ageing (2016). Accessed from http://www.ageing.ox.ac.uk

UN HABITAT (2006) Urbanization facts and figures, World Urban Forum III, An International United Nations Human Settlement Programme, Event on Urban Sustainability, Canada

UNFPA, State of the World Population (2007) Unleashing the potential for urban growth. United Nations Population Fund, New York

United Nations (2011) United Nations Programme on Ageing. Department of Economic and Social Affairs, Population Division. Retrieved 29 Apr 2011, from http://www.un.org/ageing/popageing.html

WHO (2011) Urban health. Regional Office for Europe, World Health Organization. http://www.euro.who.int/en/what-we-do/health-topics/environment-andhealth/urbanhealth/activities/healthy-cities/who-european-healthy-cities-network/what-is-a-healthy-city/healthycity-checklist. Accessed from 14 June 2011

Zelenev S (2008) The Madrid plan: a comprehensive agenda for an ageing world. Regional dimensions of the ageing situation. UN Department of Economic and Social Affairs, New York

Chapter 11
Heat Effects and Coastal Vulnerability of Population in Thailand

Uma Langkulsen and Desire Tarwireyi Rwodzi

Abstract Excessive heat and flooding are extreme weather events associated with climate change, and they pose serious threats to global health. By conducting a review of literature, we assessed the effect of heat on human health among the Thai population. Our review showed that previous studies focused on mortality, especially among the elderly or persons with preexisting illnesses, and overlooked the idea that physical labor adds to the heat exposure risks via the surplus heat generated due to muscular movements.

We also assessed the demographic profiles and existing public health resources in Bangkok and surrounding provinces as part of the wider project to develop a health and climate change adaptation resilience simulator for coastal cities. This comes following a realization that recurrent flooding makes the Thai capital, Bangkok, and surrounding areas prone to inundation because of the low-lying topography. This has created a need for Thailand to protect its people, natural and man-made resources, and productive capacities in response to the impact of climate change induced floods. Primary and secondary data sources were used to describe demographic characteristics of the population and to assess the adequacy of public health emergency resources. Our study results showed that the study area falls short of the Southeast Asia as well as World Health Organization's set standards for health worker densities and hospital beds.

Keywords Heat exposure • Climate change • Ambient temperature • Resilience • Public health emergency resources

U. Langkulsen, Ph.D. (✉) • D.T. Rwodzi, M.P.H.
School of Global Studies, Thammasat University, Pathumthani, Thailand
e-mail: uma.langkulsen@sgs.tu.ac.th

© Springer International Publishing Switzerland 2016
R. Akhtar (ed.), *Climate Change and Human Health Scenario in South and Southeast Asia*, Advances in Asian Human-Environmental Research, DOI 10.1007/978-3-319-23684-1_11

173

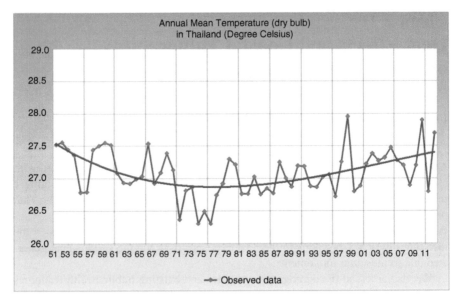

Fig. 11.1 Annual mean temperature in Thailand shows steady rise of temperature since 1975 (Source: Thai Meteorological Department. [Online]. Available: http://www.tmd.go.th/climate/climate.php?FileID=7)

11.1   Effects of Heat on Health among the Thai Population

11.1.1   Introduction

Climate change and rising global temperatures increase concern for the effects of heat on human health and their impact on global health (Tawatsupa et al. 2013). Notwithstanding the growing body of evidence showing that increasing ambient temperatures and extreme weather events are a result of global climate change, researchers and policymakers still pay little attention to the significance of heat effects on human populations worldwide (Ebi and Meehl 2007). In the academic arena, very few studies exist on the health effects of climate change generally, and more specifically, scholarship on developing countries in tropical regions in this study area remains limited. Moreover, most current studies focus on the impact of climate on environmental health, while a few studies approach it from a human health perspective (Langkulsen et al. 2010). Most of them also come from developed countries, and not from developing countries in tropical regions like Thailand.

Developing countries are important, because they may be more sensitive to climate change, and hence, they deserve greater attention now than before (Guo et al. 2012). Thailand saw its temperatures rise significantly over past three decades (see Fig. 11.1). According to a report from the Meteorological Department of Thailand, its monthly mean ambient temperature was above normal from January to November during 2013, especially in the upper region of Thailand. There, the temperature

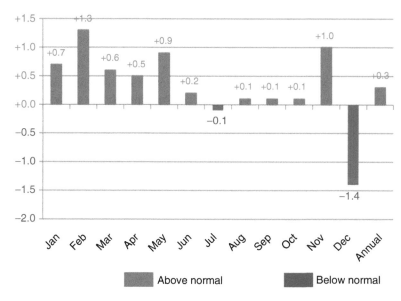

Fig. 11.2 Thailand's monthly mean ambient temperature anomalies (degree Celsius) in 2013 (Source: Annual Weather Summary over Thailand in 2013, Climatological Center, Meteorological Development Bureau, Thai Meteorological Department)

increased by 1 °C during February and November (see Fig. 11.2). Recent temperature forecast models predict temperatures in Thailand will continue to increase by 1.2–1.9 °C during this century (Chinvanno and Southeast Asia START Regional Center 2009; Marks 2011). Elevated environmental temperatures may be a contributing factor to several chronic diseases, such as respiratory and cardiovascular diseases among vulnerable populations, which are also becoming a serious public health issue for Thailand. Both disease categories are among the top five leading causes of death and disabilities in Thailand. Thus, the correlation between temperature and human health is a crucial consideration for improving management of these detrimental health outcomes (Guo et al. 2012).

One important feature of human biology is the maintenance of core body temperature (T_{cb}) near 37 °C, the temperature at which different biochemical and physiological systems function optimally. Basic metabolism, digestion of food, and muscle movement all create "surplus heat," which human body needs to transfer away from it to avoid overheating its T_{cb}, especially in hot environments like the tropics. Adding physical activity to these processes increases the amount of muscle activity that raises internal body heat that the body must remove primarily through the skin. In a hot and humid tropical environment, there is the added risk that the cooling mechanism provided by the physiological process of sweating may be insufficient. Inability to sweat properly may lead T_{cb} increases, because excess body heat cannot be removed effectively. If T_{cb} exceeds 38 °C, then the risk of heat strain increases, producing symptoms of fatigue and lapses in concentration. If T_{cb} goes beyond 39–40 °C, then organ damage may ensue and symptoms may grow worse.

If T_{cb} remains above these temperatures, the brain will be affected resulting in unconsciousness (heat stroke). Death may follow, if the heat excess persists untreated. There are, however, cases of well-acclimatized people who can withstand such high T_{cb} levels for some time. For example, studies in India indicate that workers may continue working efficiently with a T_{cb} up to 39 °C (Kjellstrom et al. 2010). Even so, these are the exception and may not be applicable to other populations, where heat effects may affect human health on a global scale.

11.1.2 Literature Review on Heat Stress Impact in Thailand

To learn the impact of heat effects on human health in developing countries, a review of literature on studies focusing on Thailand was carried out to locate peer-reviewed studies on this topic. A keyword search methodology was utilized incorporating the following keywords (heat stress, heat effects, extreme heat, and hot environments). Searches were performed with the following search engines (Google and Google Scholar). The key search limiter term was Thailand, and based on the search routines conducted from 5 to 10 years. The search found articles meeting the study parameters of heat effects, mortality, and occupation in Thailand. The following articles were identified, reviewed, and reported:

In 2011, Langkulsen cited in *Asian Pacific Newsletter on Occupational Health and Safety* reported that 7 people died from the effects of heat and light, and the primary causes of death included heatstroke, sunstroke, and heat syncope (listing causes of death under the International Classification of Diseases, Tenth Revision (ICD-10) between 2007 and 2009 in Thailand) (Langkulsen 2011).

A cross-sectional, pilot study of multiple worksites ($N=5$) engaged in either agricultural or industrial activities in the Pathumthani and Ayutthaya provinces of Thailand during 2009. The study conducted during the Thai "rainy season" compared different worksites in the industrial and agricultural sectors. Investigators identified 4 of 5 sites with temperatures in the designated heat value danger zone (HI Index >41 °C), and they included industrial activities in the pottery industry, power generation, and knife manufacturing. Of the study sites, only the construction and pottery workers lost productivity due to extreme heat stress and heat-related sicknesses (Langkulsen et al. 2010). The overall loss of productivity related to body heat ranged from 10 to 60 % in these activities. Interestingly, the power plant site had outside temperatures in the HI Index range, but its workers did not experience declines in their productivity. The authors attributed this observation to workers spending more worktime inside air-conditioned structures than outside in the heat. Because the study also took place during the "rainy season" when outside temperatures may be lower than the "dry season," there may be lowering of the impact of temperatures. Thus, heat effects could be more or less depending on the time of season and environmental conditions during the study period.

A second 2010 study from Thailand demonstrated a positive association between heat stress and psychological distress among 24,907 workers surveyed. Of the factors studied, associations between overall health, psychological distress due to reduced work productivity, low income, and disrupted daily social activity, and occupational heat stress were significant and revealed no consistent pattern with age. The adverse effects of heat stress on poor health were worse for females; but males between 15 and 29 years also exhibited a similar association (Tawatsupa et al. 2010).

A more recent study published in 2012 identified an association between occupational heat stress and self-reported doctor diagnosed kidney disease in Thailand. This study found that men exposed to prolonged heat stress increased their odds of developing kidney disease 2.22 times higher than that of men without such exposure. The study also reported that kidney disease in Thailand is a major cause of death among middle-aged adults. Moreover, the number of deaths from renal failure increased from 8895 in 2001 to 11,246 in 2005 to 12,195 in 2007. Whether there is direct correlation between the overall rise in kidney disease deaths and occupationally related heat stress kidney disease remains to be clearly established, but based on the upward trajectory on death rates, the results of this study are crucial to Thailand health sector, and they suggest more studies are needed (Tawatsupa et al. 2012). A peer-reviewed publication in 2013 provided more evidence for a connection between heat stress and occupational injury in tropical Thailand. This study found that one-fifth of workers experienced occupational heat stress, which was strongly and significantly associated with occupational injury (Tawatsupa et al. 2013).

A further two studies published in 2012 provided evidence of an association between temperature and mortality. One population-based study conducted by Guo and others in Chiang Mai province from 1999 to 2008 revealed that exposures to both hot and cold temperatures resulted in a nonlinear relationship between temperatures and increased cause- and age-specific mortality types (Guo et al. 2012). This study also found that hot temperatures have more immediate, acute effects on mortality types than slower cold-related effects. The authors concluded that temperature was an important environmental hazard in Chiang Mai city. A second 2012 study by Tawatsupa and colleagues reviewed temperature and mortality rates for workers in Thailand from 1999 to 2008 (Tawatsupa et al. 2014). Their country-wide study identified a highly significant association between weather variations and mortality in Thai workers, where workers in the central region of Thailand were at the greatest risk for heat-related mortality effects. Based on their findings, the authors cautioned that workers at high risk for heat-related stresses should take safety precautions when working in high heat conditions.

11.1.3 Discussion and Conclusion

Based on the review of studies of heat effects on human health in Thailand conducted from 1999 to 2008, there are health risks that may have local and country-wide impacts, depending on the weather, work, age, and occupation. While the current worldwide scholarship on climate and heat effects remains a work in progress, the studies identified, reviewed, and discussed here show that researchers in Thailand have been at the forefront of such studies. Climate matters, and it is of major concern in tropical, developing countries like Thailand. Past studies published by authors from developed countries on this topic tend to focus on more elderly populations and pay less attention to developing countries. Studies in Thailand suggested that heat effects might be more severe on the health and mortality of young workers, depending on their industrial activity. People carrying out physical labor add to their heat exposure risks via the surplus heat created inside their bodies, especially activities generating muscular movements. Thus, the global impact of heat effects is likely a complex problem that could affect countries and populations differently.

Moreover, and more importantly, heat effects on the health and well-being of populations are a global public health concern, especially in developing countries like Thailand where the general health sector needs to conduct further studies. As noted by Tawatsupa and others, it is very important for policymakers to recognize heat stress-related health outcomes and their effects on economics and production loss. The outcomes of ineffective prevention programs can and will result in ineffectual health services, especially in treating those affected by heat stroke. They may also have a negative impact on social and health-care budgets. Therefore, it is essential to build bridges between general public health services and occupational health and safety services while engaging communities and enterprises in the preventive health activities. Heat effects and climate are important issues in the discussion of the overall advancement in socioeconomic development of countries like Thailand. Tropical countries have to endure excessively long hot seasons, and there is also the link between heat exposures, health protection, work productivity loss, and the community economy. Climate change is a persistent and threatening issue and the public health sector has a major role in analyzing the threats, creating awareness, and finding the best preventive solutions.

11.2 Coastal Cities at Risk: Health and Climate Change Adaptation City Resilience Simulator Development

11.2.1 Introduction and Background

Bangkok is experiencing a need to protect its people, natural and man-made resources, and productive capacities in response to the impact of climate change and increasing numbers of extreme weather patterns and events. In particular, due to the "three waters" of runoff, rain, and sea rise, together with its low-lying topography, much of the capital is prone to inundation. Bangkok and surrounding areas must become resilient to wider factors of risks in order to be prepared for climate change challenges. There is a need to reinvestigate the city as a system, bringing in an urban development approach in order to understand urban risks and promote resilience to climate change challenges. The "Coastal Cities at Risk (CCaR): Building Adaptive Capacity for Managing Climate Change in Coastal Megacities" is a 5-year collaborative research program conducted from 2011 to 2016. Its overall objective is to develop the knowledge base and enhance the capacity of megacities to successfully adapt to and when necessary cope with risks posed by the effects of climate change, including sea-level rise, in the context of urban growth and development. To deal with the shortcomings in existing resilience models and to provide a conceptual basis for establishing baselines for measuring resilience, this project introduces a space-time dynamic resilience measure (ST-DRM). A city resilience system will capture the process of dynamic disaster resilience simulation in both time and space (Angela and Simonovic 2013).

This paper presents the contribution of the School of Global Studies, Thammasat University, to the health component of the City Resilience Simulator Development Project for Bangkok, Thailand. We aimed at describing the demographic characteristics of the population from the study area, describe public sector health care and public health emergency resources, and identify the vulnerable groups in flooding events. As such, our study gives a description of key health system and demographic indicators that have to be considered in planning for the resilience of the city and surrounding areas that are prone to flooding. Selected demographic characteristics of the population from the study area comprising 6 provinces were used to determine the adequacy of public health emergency resources in the event of a flood. Described also is data on vulnerable groups and those who need special care or attention during floods and other climate related hazards in the study area. The data presented in this report is important for determining gaps that should be addressed to increase the resilience of the city to flooding hazards. It also highlights strengths that can be built upon to protect communities threatened by hazards associated with climate change.

11.2.2 Methods

The study area under the health theme comprised six provinces, namely, Bangkok, Nonthaburi, Pathum Thani, Samut Sakhon, Samut Prakan, and Nakhon Pathom. There was severe inundation in these provinces during the 2011 floods in Thailand and this influenced their selection for data collection to boost their resilience. We conducted a descriptive study based on selected secondary health information system data and demographic indicators in relation to public health services, public health emergency system resources, and vulnerable groups in six flood-prone provinces. In-depth interviews were also conducted to gather data from institutions and organizations that were purposively selected. These institutions/organizations include the Local Administration, Ministry of Public Health, Bangkok Metropolitan Administration, and the Department of Disaster Prevention and Mitigation under the Ministry of Interior. Indicators considered for the study were categorized into 4 clusters, namely, demographic data, public health resources, public health

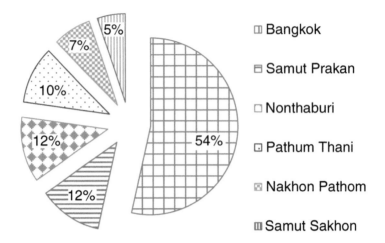

Fig. 11.3 Distribution of households in the study area

emergency resources, and vulnerable groups among the affected population. Quantitative data served a simple descriptive analysis using size, distributions, proportions, and ratios. Processed quantitative data served a gap analysis to identify potential need for adaptation strategies to be included in urban development plans for Bangkok and surrounding areas.

11.2.3 Results

11.2.3.1 Demographic Data

Study Area and Households

Altogether, there are 79 districts and 476 sub-districts within the study area. As of December 2013, a total of 4,848,635 households were reported within the study area. Bangkok had the highest number of households (54 %) within the study area, and Samut Sakhon had the least number of households (5 %), as shown in Fig. 11.3. The median number of members per household across the six provinces was 2.1, with the least (2.0) reported in Nonthaburi province and the highest median of 2.5 reported in Nakhon Pathom.

Population Distribution

As of December 2013, the total population residing within the six provinces under study was 10,538,932 according to data retrieved from the Ministry of Interior. Out of the 6 provinces, Bangkok was the most populous (5,686,252) with its total population being greater than the total of the other 5 provinces combined. The least populated province was Samut Sakhon and its population (519,457) was about a tenth of total population for Bangkok. Overall, there were more females (52 %) than males (48 %) within the study area. The greatest difference in male to female ratio was observed in Nonthaburi province (1: 1.13). The distribution of the population by gender at district and sub-district levels exhibits a similar pattern.

Distribution of Age Groups

Analysis of distribution of population by age groups was restricted to Thai citizens because disaggregated data was not readily available for non-Thai nationals. The population was categorized into 5 distinct age groups relevant to public health and disaster planning. These included children under 5 years, 5–14 years, 15–49 years, 50–64 years, and 65 years and above. Bangkok province had the lowest proportion of people within the 15–49 age group compared to Pathum Thani province which had the highest proportion of 58 % for the same age group. However, Pathum Thani province had the least proportion (7 %) for persons aged 65 and above.

Distribution of Population by Age and Gender

Across the six provinces, proportions for three age categories among males (i.e., under 5, 5–15 years, and 15–49 years) were higher compared to the same categories among females. However, the proportions for females aged 50–64 years and females aged 65 years and above were higher compared to the same age categories among males. Bangkok province had the highest proportion of females aged 65 and above (11 %), while Pathum Thani province had the lowest proportions of people aged 65 and above for both males (6 %) and females (8 %).

11.2.3.2   Public Health Resources

Public Health Resources at Tambon Level (Health Promotion Hospitals)

Table 11.1 shows data on public health resources reported for the year 2012 including the number of Tambon Health Promotion Hospitals (THPHs) and hospital beds. It also summarizes the number of people seen as outpatients, number of visits reported at the outpatients department (OPD), and the average daily outpatients by province. Bangkok province had no health promotion hospitals at Tambon level, and out of the remaining five provinces, no hospital at Tambon level had beds. Nakhon Pathom province had the largest number of THPHs (134), and this is two and half times the number of health centers in Samut Sakhon province (54). Although Samut Sakhon had the lowest number of THPHs, the province had approximately 1.04 health centers per 10,000 population which is a higher ratio

Table 11.1 Tambon Health Promotion Hospitals and out patients department statistics

	Total population (n)	Hospitals (n)	Hospitals per 10,000 population	Beds (n)	Outpatients (n)	OPD visits (times)	Average daily outpatients
Bangkok	5,686,252	–	–	–	–	–	–
Nonthaburi	1,156,271	76	0.657	–	244,398	849,700	2,971
Pathum Thani	1,053,158	78	0.741	–	343,498	1,081,332	3,781
Samut Prakan	1,241,610	68	0.548	–	630,417	1,718,375	6,008
Nakhon Pathom	882,184	134	1.519	–	378,224	1,345,901	4,706
Samut Sakhon	519,457	54	1.040	–	166,897	652,153	2,208
Whole country	63,878,267	9,761	1.528	–	27,514,066	97,839,962	342,287

compared to Nonthaburi, Pathum Thani, and Samut Prakan provinces. Samut Prakan province had the highest number of outpatients, OPD visits, and average daily outpatients reported in 2012.

Public Health Resources at District Level (Community Hospitals)

A hospital is recognized as a community hospital if it is located in the district and had a total number of hospital beds that is less than 200.[1] As shown in Table 11.2, no hospital befits the description of a community in Bangkok and Samut Sakhon provinces. Out of the other four provinces, Samut Prakan had half the number of community hospitals (4) as Nakhon Pathom (8). Nonthaburi and Pathum Thani provinces have 5 and 7 community hospitals, respectively. Even with the least number of hospitals, Samut Prakan had the highest number of inpatient beds (376) at district-level hospitals. The ratios of doctors per bed as well as beds per 10,000 population were all higher for the whole country in comparison with individual provinces. The reported numbers of outpatients, OPD visits, and inpatients are highest in Nakhon Pathom and lowest in Nonthaburi province. Samut Prakan, the province with the highest number of inpatient beds, also had the highest bed occupancy rate of 92 %, followed by Nakhon Pathom, Nonthaburi, and Pathum Thani provinces with bed occupancy rates of 90 %, 83 %, and 63 %, respectively.

---

[1] Bureau of Policy and Strategy, Office of Permanent Secretary, Ministry of Public Health, Thailand.

Table 11.2 Community hospitals, beds, and outpatient and inpatient statistics

	Bangkok	Nonthaburi	Pathum Thani	Samut Prakan	Nakhon Pathom	Samut Sakhon	Whole country
Provincial population	5,686,252	1,156,271	1,053,158	1,241,610	882,184	519,457	63,878,267
Hospitals (n)	–	5	7	4	8	–	759
Hospitals per 10,000 population	–	0.043	0.066	0.032	0.091	–	0.119
Beds (n)	–	180	240	376	360	–	35,763
Beds per 10,000 population	–	1.56	2.28	3.03	4.08	–	5.60
Doctors per bed	–	3	6	6	5	–	9
Outpatients	–	174,424	212,334	210,442	388,498	–	19,178,504
OPD visits (time)	–	568,129	597,805	661,567	1,275,971	–	68,997,747
Inpatients	–	17,732	18,237	30,616	38,519	–	3,219,152
Length of stay (day)	–	54,661	54,752	125,851	118,782	–	10,491,964
Bed occupancy rate (%)	–	83	63	92	90	–	80

Public Health Resources at Provincial Level (General/Referral Hospitals)

Criteria for classifying a hospital as a general hospital was based on the fact that the institution was located in the province/ big district and the number of hospital beds ranged between 200 and 499. The whole country had a total of 68 general/referral hospitals. As shown in Table 11.3, there is no general hospital in Bangkok and Nakhon Pathom provinces. Out of the remaining 4 provinces, only Samut Sakhon province had 2 referral hospitals; the other 3 provinces have 1 referral hospital each. Samut Sakhon, with 2 referral hospitals, had the highest number of inpatient beds (747). All other provinces have doctors per bed ratio lower than that of the whole country, except for Samut Sakhon whose ratio of 8 doctors per bed is equal to that of the whole country. OPD visits and number of inpatients are highest in Samut Sakhon province whereas the reported number of outpatients is highest in Nonthaburi province. Only Samut Sakhon province had a bed occupancy rate (84 %) which is lower than the national average (85 %). Samut Prakan had the highest bed occupancy rate of 107 %, followed by 93 % and 86 % in Pathum Thani and Nonthaburi provinces, respectively.

Public Health Resources at Regional Level (Regional Hospitals Located in the Provinces)

We defined a regional hospital as one that was located in the provinces, had more than 500 inpatient beds, and had medical specialists. With this working definition, only Nakhon Pathom province had a regional hospital out of the six provinces under investigation. Table 11.4 summarizes the statistics for regional hospitals, beds, as well as reported numbers of outpatients and inpatients.

11.2.3.3 Health Personnel

Ratios of health personnel per 10,000 populations as well as the proportions for each group of health personnel by province were calculated. The health worker density, which refers to the number of health workers per 10 000 population, by cadre, is the health workforce indicator that is most commonly reported internationally and represents a critical starting point for understanding the health system resources situation in a country (World Health Organization 2009a). These health personnel included medical doctors, dentists, dental assistants, pharmacists, registered nurses, technical nurses, public health officers, and public health technical officers working at different levels as reported above.

Table 11.3 General/referral hospitals, beds, and outpatient and inpatient statistics

	Bangkok	Nonthaburi	Pathum Thani	Samut Prakan	Nakhon Pathom	Samut Sakhon	Whole country
Total population	5,686,252	1,156,271	1,053,158	1,241,610	882,184	519,457	63,878,267
Hospitals (n)	–	1	1	1	–	2	68
Hospitals per 10,000 population	–	0.009	0.009	0.008	–	0.039	0.011
Beds (n)	–	442	377	385	–	747	23,636
Bed per population	–	2,561	2,712	3,151	–	675	2,179
Beds per 10,000 population	–	3.82	3.58	3.10	–	14.38	3.70
Doctor per bed	–	4	6	5	–	8	8
Outpatients (n)	–	431,053	101,540	137,095	–	246,098	5,307,299
OPD visits	–	674,137	392,608	486,271	–	1,056,331	20,088,277
Inpatients (n)	–	24,687	29,032	31,769	–	63,159	1,650,964
Length of stay (days)	–	138,447	127,948	150,674	–	228,961	7,333,534
Bed occupancy rate	–	86	93	107	–	84	85

Table 11.4 Regional hospitals, beds, and outpatient and inpatient statistics

	Bangkok	Pathum Thani	Samut Prakan	Nakhon Pathom	Samut Sakhon	Whole country
Total population (n)	5,686,252	1,053,158	1,241,610	882,184	519,457	63,878,267
Hospitals (n)	–	–	–	1	–	28
Bed (n)	–	–	–	670	–	19,660
Hospitals per 10,000 population	–	–	–	0	–	0
Beds (n)	–	–	–	670	–	19,660
Bed per population	–	–	–	1,299	–	3,269
Beds per 10,000 population	–	–	–	7.59	–	3.08
Doctor per bed	–	–	–	6	–	5
Outpatients (person)	–	–	–	152,637	–	3,571,742
OPD visits (times)	–	–	–	768,315	–	16,632,996
Inpatients (person)	–	–	–	49,858	–	1,466,463
Length of stay (days)	–	–	–	240,670	–	7,145,179
Bed occupancy rate (%)	–	–	–	98	–	100

Health Personnel at Tambon Health Promotion Hospitals

Across the whole country, there are no medical technologists and medical scientists, neither are there medical technician assistants at Tampon Health Promotion Hospitals. Table 11.5 shows the numbers and ratios per 10,000 population for the available health personnel at THPH level by province and for the whole country. The ratio per 10,000 population for public health technical officers and public health officers is highest in Nakhon Pathom (2.54/10,000 population and 1.72/10,000 population, respectively) and is even higher than the ratios observed for the whole country.

Medical Personnel at Community Level Hospitals

Again at community-level hospitals (located in the district, number of beds < 200), Bangkok province had none of the above-mentioned health professionals. As shown in Table 11.6, the ratios of health personnel per 10,000 population were highest for registered nurses, followed by medical doctors across the provinces under investigation. For registered nurses, only Nakhon Pathom province had a ratio greater than the average for the whole country (5.86 versus 5.21). For Samut Sakhon province, data is only available for medical technologists and medical technician assistants whose ratios are 0.04 per 10,000 population. There are no medical scientists in Bangkok, Samut Prakan, Nakhon Pathom, and Samut Sakhon.

Table 11.5 Number and ratios per 10,000 population for health personnel at Tambon level

Province	Public health Technical officers		Dental assistants		Public health officers		Registered nurses		Technical nurses	
	Number	Ratio	Number	Ratio	Number	Ratio	Number	Ratio	Number	Ratio
Bangkok										
Nonthaburi	105	0.91	21	0.18	64	0.55	71	0.61	–	–
Pathum Thani	134	1.27	25	0.24	89	0.85	65	0.62	–	–
Samut Prakan	194	1.56	37	0.30	77	0.62	77	0.62	1	0.01
Nakhon Pathom	224	2.54	23	0.26	152	1.72	72	0.82	–	–
Samut Sakhon	78	1.50	11	0.21	85	1.64	29	0.56	–	–
Whole country	14,365	2.05	1,478	0.21	9,341	1.33	7,097	1.01	8	0.00

Table 11.6 Ratios of health personnel per 10,000 at community hospital level

Province	Medical technologists	Medical scientists	Medical technician assistants	Medical doctors	Dentists	Pharmacists	Registered nurses	Technical nurses
Bangkok	–	–	–	–	–	–	–	–
Nonthaburi	0.05	0.01	0.14	0.54	0.42	0.32	2.78	0.05
Pathum Thani	0.02	0.01	0.13	0.40	0.35	0.29	3.53	0.09
Samut Prakan	0.02	–	0.06	0.52	0.27	0.32	2.27	0.07
Nakhon Pathom	0.02	–	0.18	0.86	0.48	0.48	5.86	0.09
Samut Sakhon	0.04	–	0.04	–	–	–	–	–
Whole country	0.09	0.01	0.16	0.60	0.35	0.45	5.21	0.11

Medical Personnel at General/Referral Hospitals

At general hospitals (located in the province or big district, number of beds 200–499), we did not find data for two of the provinces under investigation, that is, Bangkok and Nakhon Pathom. However, for the provinces where data was available, registered nurses had the highest ratios per 10,000 population followed by medical doctors as shown in Table 11.7. Samut Sakhon province had the highest ratio of registered nurses and medical doctors per 10,000 population, 9.49 and 1.89, respectively. Noteworthy is the fact that for all the provinces where data was available, the ratios per 10,000 population for registered nurses, medical doctors, pharmacists, and dentists were higher at provincial level compared to ratios for the whole country.

Medical Personnel at Regional Level Hospitals

A regional hospital, by working definition, referred to one that was located in the province, had more than 500 beds, and had medical specialists. Out of the six provinces, only Nakhon Pathom province had a regional hospital, and the numbers and proportions of different health personnel are shown in Table 11.8.

At the regional hospital in Nakhon Pathom, 70 % of health personnel were registered nurses and that the same average was observed for regional hospitals in the whole country. However, the ratio per 10,000 population for registered nurses is higher in Nakhon Pathom (6.4) compared to the national average (2.45). For all health personnel at this level, the density (ratio per 10,000 population) is higher in Nakhon Pathom compared to the whole country.

11.2.3.4   Other Public Health Resources Available at Hospitals

From Tambon to provincial levels, all hospitals reported that professional tasks were being completed primarily through electronic systems, and a variety of electronic communication modalities were being used in exchanging information with other health service providers. All hospitals in the study area reported that they maintained registers for chronic conditions. Not only did they conduct outreach activities to deliver primary health-care services, but they also provided specialized programs (JHCIS) for vulnerable populations. With the exception of Sam Khok, a district-level hospital, all other hospitals at different levels were providing after-hour services to practice populations. At all levels, hospitals in the study area had collaborative care arrangements with other health-care organizations as well as other providers beyond the health sector such as housing, justice, police, and education.

Table 11.7 Ratio of health personnel per 10,000 population at provincial level (general hospital)

Province	Medical technologists	Medical scientists	Medical technician assistants	Medical doctors	Dentists	Pharmacists	Registered nurses	Technical nurses
Bangkok	–	–	–	–	–	–	–	–
Nonthaburi	0.08	0.05	0.10	0.88	0.20	0.34	4.00	0.11
Pathum Thani	0.06	0.02	0.12	0.61	0.14	0.22	3.32	0.06
Samut Prakan	0.03	0.02	0.17	0.57	0.13	0.28	2.59	0.29
Nakhon Pathom	–	–	–	–	–	–	–	–
Samut Sakhon	0.12	0.08	0.27	1.89	0.71	0.73	9.49	0.60
Whole country	0.07	0.02	0.11	0.40	0.09	0.17	2.83	0.18

Table 11.8 Ratio of health personnel per 10,000 population at provincial level (general hospital)

Health personnel	Nakhon Pathom			Whole country		
	Number	% (out of health workers at province)	Ratio per 10,000 population	Number	% (out of health workers at province)	Ratio per 10,000 population
Medical technologists	13	2 %	0.15	374	2 %	0.05
Medical scientists	4	0 %	0.05	88	0 %	0.01
Medical technician assistants	21	3 %	0.24	550	2 %	0.08
Medical doctor	106	13 %	1.20	3734	15 %	0.53
Dentist	16	2 %	0.18	416	2 %	0.06
Pharmacist	29	4 %	0.33	1013	4 %	0.14
Registered nurse	565	70 %	6.40	17,224	70 %	2.45
Technical Nurse	48	6 %	0.54	1176	5 %	0.17

11.2.3.5   Available Public Health Emergency Resources at Hospitals

Primary data regarding the status of public health emergency resources was gathered through in-depth interviews that were conducted with a number of agencies or organizations that were purposively sampled. These included 2 hospitals at Tambon level, 2 at district level, and 2 general/referral hospitals. In addition we purposively selected Nonthaburi and Pathum Thani Provincial Health Offices, Disaster Prevention and Mitigation Provincial Offices (Nonthaburi and Pathum Thani Provincial Offices), as well as the Bangkok Fire and Rescue Department.

Hospitals at all levels confirmed that they had in place either policies or plans that included emergency preparedness and response. One district-level hospital (Sam Khok) indicated that they did not have formal emergency preparedness and response coordination mechanisms in place. However, at all levels, hospitals had roles and responsibilities and lines of authority that were clearly laid down in the policy frameworks. None of the 6 hospitals (from Tambon to provincial level) had an emergency preparedness and response plan that included standard operational procedures (SOPs). Hospitals at all levels indicated that their policies provided budget allocations for emergency preparedness and response plans. Only Sam Khok, a hospital at district level, indicated that they will have the budget allocated when flood or disaster strikes. At all levels, hospitals confirmed that delegation of authority for the EPR focal point was supported by administrative procedures. With the exception of one hospital at district level (Sam Khok hospital), the other five hospitals at different levels confirmed that they conducted risk and vulnerability assessments that reflect specific public health concerns. It was also clear from the interviews that at most hospitals, not all staff were trained on emergency preparedness and response, and there were differences on the type of staff to be trained, even at hospitals at the same level. Only Sam Khok (a hospital at district level) did not have the essential supplies for health response prepositioned in strategic locations.

In addition, inventory systems for essential supplies and equipment for health were operational at all other hospitals except Sam Khok.

All the other 5 hospitals (except for Sam Khok) indicated that health systems-related EPR information was in place and was made available across and within various levels/sectors. Only two hospitals (Pranangklao and Sam Khok hospitals) indicated that resources meant for vulnerability assessment and risk mapping were neither developed nor utilized at all levels. In completing tasks, hospitals at all levels primarily use electronic systems. Risks from existing hazards were assessed at all key health facilities at 5 hospitals except Sam Khok. However, hospitals at all levels had emergency plans in place outlining emergency management, mass casualty management, as well as evacuation procedures. It was only at Sam Khok hospital that knowledge of public health threats in emergencies was not integrated in the existing disease surveillance system. In addition, it was only at Sam Khok hospital where emergency surveillance and response needs were not assessed, and no measures were taken.

11.2.3.6 Available Public Health Emergency Resources at Non-hospital Institutions

Qualitative data was gathered from a total of 14 non-hospital institutions within the study area. These comprised 4 institutions under Local Administration, 5 institutions under Ministry of Public Health, 4 institutions under the Ministry of Interior's Department of Disaster Prevention and Mitigation, and Bangkok Metropolitan Administration. Table 11.9 summarizes the availability of public health emergency resources at surveyed non-hospital institutions. Policy frameworks were available across all the 14 surveyed non-health institutions under the four different types of administration. Less than half (43 %) of the 14 surveyed non-hospital institutions reported that they had a developed plan for emergency preparedness and response which included standard operational procedures. Under the Ministry of Public Health, out of the 4 institutions surveyed, only Pathum Thani Provincial health office (20 %) had the plans in place. All 14 non-health institutions reported that emergency preparedness and response focal points, units, or persons responsible were available at their institutions. However, 14 % of the non-health institutions stated that their policies did not provide budget allocations for emergency preparedness and response plans.

All surveyed institutions reported that risk and vulnerability assessments conducted by their organizations reflected specific public health concerns. In addition, any assessed risks to public health were reflected in institutional preparedness and response plans. Training for response and preparedness was reported to be in place for only 50 % of the surveyed institutions. However, all surveyed institutions reported that training was in place and focused on first aid in emergencies. All surveyed institutions under the Ministry of Public Health, Ministry of Interior, and BMA indicated that financial resources were allocated for essential response sup-

Table 11.9 Availability of public health emergency resources at non-hospital institutions

	Local Administration (%)	Ministry of Public Health (%)	Ministry of Interior (%)	Bangkok Metropolitan Administration (%)	Out of 14 surveyed institutions (%)
Policy framework					
Policy/plans include emergency preparedness and response (EPR)	100	100	100	100	100
Formal EPR coordination mechanisms	100	100	100	100	100
Coordination mechanism is developed and institutionalized	100	100	100	100	100
Roles, responsibilities, and lines of authority are defined	100	100	100	100	100
Disaster preparedness plans and standard operational procedures					
A plan for emergency preparedness and response is developed including standard operational procedures (SOPs)	50	20	50	100	43
Drills and simulation exercises are conducted at various levels of the health system	100	80	100	0	86
Public health emergency resources					
Policies provide budget allocations for emergency preparedness and response plans	100	60	100	100	86
EPR focal points/units/ persons responsible are identified	100	100	100	100	100
Delegation of authority for EPR focal points is supported by administrative procedures	75	100	100	100	93
Plan for mitigation, preparedness, and response					
Risk and vulnerability assessments reflect specific public health concerns	100	100	100	100	100

(continued)

Table 11.9 (continued)

	Local Administration (%)	Ministry of Public Health (%)	Ministry of Interior (%)	Bangkok Metropolitan Administration (%)	Out of 14 surveyed institutions (%)
Preparedness and response plans reflect assessed risks to public health	100	100	100	100	100
Response and preparedness capacity					
Training is in place of staff	50	20	100	0	50
Training is in place focusing on first aid and their role in public health interventions in emergencies	100	100	100	100	100
Basic supplies for public health and mass casualty care are in place for trained staff	75	60	75	0	64
Simulation exercises and drills based on preparedness plans are conducted	50	60	75	0	57
Capacity for emergency provision of essential services and supplies					
Plans list essential services, supplies, and logistics requirements for emergency response to health needs	50	60	100	0	64
Financial resources are allocated for essential response supplies and equipment for health needs	50	100	100	100	86
Essential supplies for health response are prepositioned in strategic locations	75	80	100	0	79
Inventory system for essential supplies and equipment for health is operational	50	60	100	0	64
Transport and distribution arrangements are identified and plans are in place	75	80	100	100	86

(continued)

Table 11.9 (continued)

	Local Administration (%)	Ministry of Public Health (%)	Ministry of Interior (%)	Bangkok Metropolitan Administration (%)	Out of 14 surveyed institutions (%)
Suppliers and transportation means are identified and plans are in place for distribution of emergency supplies	75	80	100	100	86
Advocacy and awareness					
Awareness materials on EPR and public health issues are developed and disseminated widely to populations at risk	75	80	100	100	86
Health systems-related EPR information is in place and made available across and within various levels/sectors	75	100	100	100	93
Identify risks and assess vulnerability					
Resources for vulnerability assessment and risk mapping are developed and utilized at all levels	75	40	75	100	64
Repository of information from vulnerability assessments and risk mapping includes information specific to the health sector	33	100	100	100	78
Incident command system					
Primarily use electronic systems to complete their tasks	75	100	100	100	93
Use a variety of electronic communication modalities in the exchange of information with others	75	100	100	100	93

(continued)

Table 11.9 (continued)

	Local Administration (%)	Ministry of Public Health (%)	Ministry of Interior (%)	Bangkok Metropolitan Administration (%)	Out of 14 surveyed institutions (%)
Health facilities resilience					
Risks from existing hazards are assessed in all key health facilities	75	60	100	100	79
Assessed risks in health facilities are prioritized and essential problems are mitigated and reduced	100	100	100	100	100
Health facility maintenance staff is trained in mitigating the nonstructural risks	25	80	100	100	71
Emergency plan is in place which outlines emergency management, mass casualty management, and evacuation procedures	50	80	100	100	79
Early warning and surveillance systems for public health					
Disease surveillance system is in place with regular reporting	50	100	0	0	50
Functional response mechanism integrated in disease surveillance system at all levels	25	100	0	0	43
Knowledge of public health threats in emergencies is integrated in the existing disease surveillance system	25	60	0	0	29
Emergency surveillance and response needs are assessed and measures taken	25	100	0	0	43
SRRT are available	25	100	0	0	43
Staff is trained in risk communication	25	60	50	100	50

plies and equipment for health needs. However, under BMA, neither were essential supplies for health response prepositioned in strategic locations, nor were inventory systems for essential supplies operational. Almost all institutions (93 %) reported that information technologies were being used to manage public health information, with the exception of only Nonthaburi Provincial Administration Organization under the Local Administration.

Almost all (93 %) surveyed non-hospital institutions reported that they primarily used electronic systems to complete their tasks. In addition, 93 % of the institutions reportedly used a variety of electronic communication modalities in the exchange of information with others. Under the Local Administration, only 25 % of the institutions reported that health facility maintenance staff was trained in mitigating non-structural risks. Fifty per cent (50 %) of the surveyed institutions under Local Administration and 80 % of institutions under Ministry of Public Health indicated that emergency plans were in place and outlined emergency management, mass casualty management, and evacuation procedures. None of the institutions under Ministry of Interior and under BMA had early warning and surveillance systems for public health in place except for trained staff in risk communication, which was reported in 50 and 100 % of surveyed institutions under Ministry of Interior and BMA, respectively. Disease surveillance systems with regular reporting were in place at 50 % of surveyed institutions under Local Administration, 100 % on institutions under Ministry of Public Health, and none under Ministry of Interior and BMA. Less than half (43 %) of all surveyed institutions reported that they had functional response mechanisms integrated in disease surveillance systems at all levels. Only 29 % of the surveyed institutions indicated that knowledge of public health threats in emergencies was integrated in existing disease surveillance systems.

11.2.3.7   Vulnerable Groups Among the Affected Population

Disabled Persons

Four out of the six provinces had centers/homes for disabled persons. As of 2014, data was not made available on disabled persons in Samut Sakhon and Nakhon Pathom provinces. Centers/homes/foundations for people living with disabilities within the four provinces provided care and support for persons of different age groups and various disabilities. These included mentally disabled children and babies, autistic persons, as well as disabled men and women in general, as shown in Table 11.10.

Population Under 5 Years

Children under the age of 5 depend on adults for their daily needs and safety, and they are more vulnerable to vaccine preventable diseases, injury, and death during floods. This group includes breastfeeding children who require special feeding arrangements during and after disaster to prevent malnutrition and diarrheal

Table 11.10 Number and proportion of disabled persons by home, center, or foundation

Province	Home/center/foundation	Age group (years)	Male (n)	Male (%)	Female (n)	Female (%)	Total
Bangkok	The Autistic Thai Foundation (AU-Thai Foundation)[a]	12+	47	90	5	10	52
Nonthaburi	Pak Kret Home for Mentally Disabled Babies (Fueang Fa Home)	0–7	241	58	171	42	412
	Pak Kret Home for Mentally Disabled Children (Girls) (Rachawadee Home)	7–18	0	0	529	100	529
	Pak Kret Home for Mentally Disabled Children (Boys) (Rachawadee Home)	7–18	603	100	0	0	603
	Pak Kret Home for Children with Disabilities (Nonthaburi Home)	7–18	252	60	168	40	420
	Nonthaburi Service Center for Autistic Persons	13–25	100	96	4	4	104
	Vocational Development Center for Disabled Persons [a]	18+	6	23	20	77	26
Pathum Thani	Half-Way Home for Men	18+	484	100	0	0	484
	Half-Way Home for Women	18+	0	0	577	100	577
Samut Prakan	Phra Pradaeng Home for the Disabled People	18+	238	50	242	50	480
	Phra Pradaeng Vocational Rehabilitation Center[a]	15–40	75	70	32	30	107

[a]Includes disabled persons who stayed overnight

Table 11.11 Distribution of population under 5 years by gender

Province	Male	Male (%)	Female	Female (%)	Total	Total under 5 as % of provincial population
Bangkok	135,343	51	127,802	49	263,145	5
Nonthaburi	30,859	52	28,984	48	59,843	5
Pathum Thani	32,647	52	30,572	48	63,219	6
Nakhon Pathom	25,677	52	23,982	48	49,659	6
Samut Prakan	37,145	52	34,624	48	71,769	6
Samut Sakhon	16,077	51	15,164	49	31,241	6

Table 11.12 Distribution of population above 60 years by gender

Province	Male	Male (%)	Female	Female (%)	Total	Total 60+ as % of provincial population	Ratio: male to female
Bangkok	342,548	42	475,305	58	817,853	14 %	01:01.4
Nonthaburi	72,451	44	93,033	56	165,484	14 %	01:01.3
Pathum Thani	50,115	43	65,763	57	115,878	11 %	01:01.3
Nakhon Pathom	50,727	42	69,364	58	120,091	14 %	01:01.4
Samut Prakan	63,211	43	84,674	57	147,885	12 %	01:01.3
Samut Sakhon	27,472	43	36,278	57	63,750	12 %	01:01.3

diseases. The proportions of persons under the age of 5 in relation to the whole provincial population were within the range of 5–6 % across the 6 provinces. Table 11.11 gives a summary of the gender proportions for persons under 5 years within each province.

Population Above 60 Years

The UN agreed cutoff is 60+ years to refer to the older population.[2] A common feature among the six provinces is that females constituted greater proportions among persons aged 60+ as shown by the gender proportions as well as the male to female ratios in the Table 11.12.

[2] World Health Organization. Definition of an older or elderly person. http://www.who.int/health-info/survey/ageingdefnolder/en/. Accessed on 05/10/2013

Table 11.13 Elderly persons in homes/foundations by province and by gender

Province	Homes/centers/ foundations (n)	Male (n)	Male (%)	Female (n)	Female (%)	Total
Bangkok	4	84	21	310	79	394
Pathum Thani	1	36	36	65	64	101
Samut Prakan	1	36	100	0	0	36
Nakhon Pathom	2	27	25	80	75	107

Table 11.14 Number of prison and correctional institutions by type

	Facilities for male only	Facilities for females only	Facilities for both males and females	Total number of prison facilities
Bangkok	4	2	2	8
Nonthaburi	2	0	0	2
Pathum Thani	2	1	3	6
Samut Prakan	0	0	2	2
Nakhon Pathom	0	0	2	2
Samut Sakhon	0	0	1	1

Elderly Persons in Homes/Foundations by Province and by Gender

As of October 2014, homes or foundations for elderly persons aged 60 and above were available in 4 out of the 6 provinces, i.e., Bangkok, Pathum Thani, Samut Prakan, and Nakhon Pathom. The highest number of homes (4) was in Bangkok province, followed by Nakhon Pathom province with 2 homes, and Pathum Thani and Samut Prakan provinces had a center each. All homes/foundations for the elderly in the study area housed persons aged 60 years and above. As shown in Table 11.13, Bangkok province had a total of 394 elderly persons housed in homes/foundations, and 79 % of these were females. Pathum Thani on the other had 101 elderly persons in homes or foundations (36 % males and 64 % females). The single center for the elderly in Samut Prakan province catered only for males.

Prisoners

There were 21 prisons and correctional facilities/institutions within the study area. Table 11.14 shows the number and types of prison and correctional institutions within the six provinces under investigation. As of February 2015, Bangkok Metropolitan Province had 8 prisons located in three districts. Out of the 8 facilities in Bangkok, 2 facilities accommodated both male and female while 4 and 2 facilities housed male and female prisoners separately. In Nonthaburi, the only two prison facilities available accommodated only males. In Samut Prakan, Nakhon Pathom, and Samut Sakhon provinces, there were no separate facilities/institutions for male and female prisoners.

Table 11.15 Number and proportion of prisoners by gender

	Male (n)	Male (%)	Female (n)	Female (%)	Total (n)
Bangkok	29,338	81	6879	19	36,217
Nonthaburi	6665	100	0	0	6665
Pathum Thani	15,721	84	2968	16	18,689
Samut Prakan	5985	85	1030	15	7015
Nakhon Pathom	3837	86	643	14	4480
Samut Sakhon	2810	84	542	16	3352
Total	64,356	84	12,062	16	76,418

Table 11.16 Orphans in residential care/orphanages

Province	Number of orphanages	Range of ages	Boys	Proportion of boys (%)	Girls	Proportion of girls (%)	Total	Ratio of boys to girls
Bangkok	2	5–18 years	125	26	360	74	485	1 : 2.9
Nonthaburi	4	0–18 years	578	68	277	32	855	1 : 0.5
Pathum Thani	3	0–18 years	357	61	226	39	583	1 : 0.6

Table 11.15 shows the number and proportions of prisoners by gender. The data includes numbers for convicted prisoners, prisoners during appeal, inquiry and investigation, as well as young prisoners in youth detention centers. Only one prison, Nonthaburi provincial prison, accommodated prisoners aged less than 15. Across all the provinces, female prisoners constituted less than 20 % of the total prisoners. Out of a total of 76,418 prisoners, Bangkok had the highest number (36,217), followed by Pathum Thani (18,689), and Samut Sakhon had the least (3352).

Orphans in Residential Care/Orphanages

As of July 2013, there were 2 homes for orphaned children in Bangkok with a total of 125 boys and 360 girls. Nonthaburi province had 4 homes with a total of 855 children, whereas Pathum Thani province had 583 orphaned children from 3 homes. Table 11.16 gives a summary of the ratios and gender proportions of orphaned children housed at orphanages within the 3 provinces.

Schools and School Students

In total, there were 461 schools in the study area comprising 165 government boarding schools and 296 government non-boarding schools. At government boarding schools, the ratios of boys to girls were 1:1.16 for Bangkok, 1:1.02 for Pathum Thani, and 1:0.93 for Nonthaburi province. A total of 894 private schools were also identified in the study area. Out of the 531 students at private boarding schools in Nonthaburi province, 85.5 % were girls, giving a boy to girl ratio of 1:5.9.

Government Schools

In total, there were 461 schools in the study area comprising 165 government boarding schools and 296 government non-boarding schools. Table 11.17 summarizes the number and gender proportions of students at government boarding schools.

Private Schools

A total of 894 private schools were identified in the study area. This included seven schools that were unreachable and one school that was closed at the time of data collection. The majority (858) of the private schools were non-boarding. Out of the 531 students at private boarding schools in Nonthaburi province, 85.5 % were girls, giving a boy to girl ratio of 1: 5.9. Similarly, there were more girls enrolled at private boarding schools in Bangkok province (63 % for girls versus 37 for boys) as shown in Table 11.18. However, in Pathum Thani 64.7 % of the private boarding students were boys.

Table 11.17 Students at government boarding schools

Province	Number of boarding school	Number of boys	Proportion of boys (%)	Number of girls	Proportion of girls (%)	Total number of students	Ratio of boys to girls
Bangkok	75	5181	46.37	5993	53.63	11,174	1:1.16
Nonthaburi	26	687	51.81	639	48.19	1326	1:0.93
Pathum Thani	64	3378	49.56	3438	50.44	6816	1:1.02

Table 11.18 Students at private boarding schools

Province	Number of boarding schools	Number of boys	Proportion of boys	Number of girls	Proportion of girls	Total number of students	Ratio of boys to girls
Bangkok	23	1517	37	2611	63	4128	1:1.7
Nonthaburi	3	77	14.5	454	85.5	531	1:5.90
Pathum Thani	2	123	64.7	67	35.3	190	1:0.54

11.2.3.8   Vector-Borne Diseases

The annual incidence for both dengue hemorrhagic fever (DHF) and malaria was calculated per 10,000 population for three provinces – Bangkok, Nonthaburi, and Pathum Thani for the period from 2008 to 2012. Annual incidence for the period under review ranged from a low of 3.90 cases/10,000 population for Pathum Thani in 2009 to a high of 23.55/10,000 for Nonthaburi in 2008. Annual incidence for malaria is significantly lower than DHF in all the three provinces under study. All 3 provinces recorded less than 1 case per 10,000 population. The highest (0.33/10,000) for the 5 years was recorded for Pathum Thani in 2008, while the lowest (0.04/10,000) was recorded in Nonthaburi in 2011.

11.2.4   Discussion and Conclusion

Bangkok province had a population greater than that of the other five provinces combined. Overall, there were more females than males in each of the provinces. While there are no universal standards for assessing the sufficiency of the health workforce, it had been estimated that countries with less than 23 health-care professionals (counting only physicians, nurses, and midwives) per 10,000 population will unlikely be able to achieve adequate coverage rates for the key primary health-care interventions prioritized by the Millennium Development Goals (World Health Organization 2009b). All six provinces fall way too short of this WHO standard, and the health workforce densities are even lower than the Southeast Asia regional estimates of 5 physicians per 10,000 and 12 nurses per 10,000 (World Health Organization 2009b). Hospital beds on the other hand are used to indicate the availability of inpatient services, but again, there is no global norm for the density of hospital beds in relation to total population (World Health Organization 2009b). The numbers of available hospital beds too are below the Southeast Asia regional estimates.

   We observed that none of the six hospitals (from Tambon to Provincial level) had an emergency preparedness and response plan that included standard operational procedures (SOPs). In addition, training of staff on emergency preparedness and response targeted different cadres at different hospitals. It is true that discrepancies on the types of public health emergency resources exist both between hospitals at the same level and hospitals at different levels. In addition, a considerable number of non-hospital institutions in the study area were not fully equipped with the necessary public health emergency resources to enable them respond effectively to disaster situations. However, hospitals at all levels had emergency plans in place outlining emergency management, mass casualty management, as well as evacuation procedures. Conclusively, it can be inferred from this preceding discussion that the six provinces under study may not be capacitated well enough to effectively and efficiently meet the public health needs of their target population in the event of an unprecedented catastrophe.

Acknowledgments This work was supported by the National Research University Project of Thailand Office of Higher Education Commission, the International Development Research Centre (IDRC), the Canadian Institutes of Health Research (CIHR), the Natural Sciences and Engineering Research Council of Canada (NSERC), and the Social Sciences and Humanities Research Council of Canada (SSHRC), Ottawa, Canada. The research team on behalf of Thammasat University's School of Global Studies would also like to express their sincere gratitude to the following organizations/institutions for their unwavering support and participation:

Pranangklao Hospital
Pathum Thani Hospital
Bang Kruai Hospital
Sam Khok Hospital
Plai Bang Tambon Health Promotion Hospital
Bann Gew Tambon Health Promotion Hospital
Local Administration
Ministry of Public Health
Ministry of Interior
Bangkok Metropolitan Administration

Conflicts of Interest We declare that we have no conflicts of interest.

References

Angela PA, Simonovic AP (2013) Excerpt: "Coastal Cities at Risk (CCaR): generic system dynamics simulation models for use with city resilience simulator". The University of Western Ontario, Dep. of Civil and Environmental Engineering, Ontario

Chinvanno S, Southeast Asia START Regional Center (2009) Future climate projection for Thailand and surrounding countries: climate change scenario of 21st century. Regional Assessments and Profiles of Climate Change Impacts and Adaptation in PRC, Thailand and Viet Nam: Biodiversity, Food Security, Water Resources and Rural Livelihoods in the GMS:4

Ebi KL, Meehl G, Bachelet, D, Twilley R, Boesch DF (2007) Regional impacts of climate change: four case studies in the United States, Pew Center on Global Climate Change Arlington, VA

Guo Y, Punnasiri K, Tong S, Aydin D, Feychting M (2012) Effects of temperature on mortality in Chiang Mai city, Thailand: a time series study. Environ Health 11(36):10.1186

Kjellstrom T, N LB, Nuntavarn V-V, Sasitorn T, Langkulsen U, Balakrishnan K, Ayyappan R, Rajkumar P, Mohan Raj S, S G (2010) Climate change, increased human heat exposure, and impacts on occupational health in South-East Asia -- a preliminary analysis based on data from India and Thailand. WHO/SEARO

Langkulsen U (2011) Climate change and occupational health in Thailand, Asian-Pacific Newsletter on Occupational Health and Safety. Retrieved from http://www.ttl.fi/Asian-PacificNewsletter

Langkulsen U, Nuntavarn V-V, Taptagaporn S (2010) Health impact of climate change on occupational health and productivity in Thailand. Global Health Action 3

Marks D (2011) Climate change and Thailand: impact and response. Contemp Southeast Asia 33(2):229–258

Tawatsupa B, Dear K, Kjellstrom T, Sleigh A (2014) The association between temperature and mortality in tropical middle income Thailand from 1999 to 2008. International journal of biometeorology 58(2): 203–215

Tawatsupa B, Lim LL, Kjellstrom T, Seubsman S-A, Sleigh A (2012) Association between occupational heat stress and kidney disease among 37,816 workers in the Thai Cohort Study (TCS). J Epidemiol 22(3):251–260

Tawatsupa B, Lim LL, Kjellstrom T, Seubsman S-A, Sleigh A, Thai Cohort Study Team (2010) The association between overall health, psychological distress, and occupational heat stress among a large national cohort of 40,913 Thai workers. Global Health Action 3

Tawatsupa B, Yiengprugsawan V, Kjellstrom T, Berecki-Gisolf J, Seubsman S-A, Sleigh A (2013) Association between heat stress and occupational injury among Thai workers: findings of the Thai Cohort Study. Ind Health 51(1):34–46

World Health Organization (2009a) Toolkit on monitoring health systems strengthening. Human Resources for Health, Geneva

World Health Organization (2009b) World Health Statistics 2009: health workforce, infrastructure. Essential Medicines, Geneva

Chapter 12
Climate Change and Health Impacts in Bangladesh

K. Maudood Elahi

Abstract There is increasing evidence that global climate change will have adverse effects on human health, mainly among the poorest population of the developing countries. Bangladesh may experience some of the more severe impacts because of its characteristic climatic and geographical conditions, coupled with high population density and poor health infrastructure. Many of the climatic events would make the climate change-induced health impacts worse as a result of newer environmental threats, such as changes in microclimate, erratic climatic behavior, and salinity intrusion in soil and water. This paper identifies some of the possible direct and indirect impacts of climate change on health condition of Bangladesh. In identifying such impacts, secondary sources of information have been widely reviewed. It has been seen that even though climate change-induced health impacts have been gaining importance in Bangladesh, there is still a lack of research and capacity in this field and its ever-increasing level of vulnerability of the people. Linkage between climate change and the increased incidences of disease, rate of mortality, and availability of safe water has not yet received the proper focus it requires.

Keywords Climate change • Contamination • Drought • Food security • Heat stress • Salinity intrusion

12.1 Introduction

Today, the world is already experiencing various impacts of climate change. These are reflected through melting glaciers on high hills and subarctic areas to perch lands, floods, and cyclone disasters. The changing climates are changing all forms of lives in many countries – and perhaps forever. Therefore, climate change

K.M. Elahi (✉)
Department of Environmental Science, Stamford University Bangladesh,
Dhaka 1209, Bangladesh

Geography and Environment, Jahangirnagar University, Dhaka 1342, Bangladesh
e-mail: iamelahi.km@gmail.com

© Springer International Publishing Switzerland 2016   207
R. Akhtar (ed.), *Climate Change and Human Health Scenario in South and Southeast Asia*, Advances in Asian Human-Environmental Research,
DOI 10.1007/978-3-319-23684-1_12

is considered as the single most important challenge of this century. This is more so for countries like Bangladesh, Sri Lanka, India, Maldives, Solomon Island, Vanuatu, and other small low-lying island states and coastal communities around the world. Bangladesh will be severely affected by inundation of land and resultant displacement of population.

The scenario for Bangladesh is very scary to say the least. It is the most densely populated deltaic country in the world with low-lying coastal zone, which has been repeatedly devastated by tropical cyclone. For Bangladeshis – who are habitually geared for tragedy – the experiences of 2007 floods, followed by devastating cyclone, *Sidr*, are still very vivid in their memories. Historically, Bangladesh used to endure a big flood every 10 years or so. Now, such devastating floods and cyclones are likely to be regular event every year due to global warming and changes in the climatic pattern. With these, a range of health-related problems is bound to follow as a set pattern for this country (Zaman et al. n.d.).

About 23 % of the country's area is critically vulnerable due to sea-level rise and will be permanently flooded, according to many experts, by 2050 displacing an estimated 25 million people. At the same time, the northwestern region will be subject to scarcity of water leading to drought and changing in hydrological regime creating untold misery, including affecting food security and agricultural production. Bangladeshi scientists estimate that approximately 40 % of crop yield will be reduced by 2050 due to climate change or variability. Both fishery and forestry will likewise be immensely affected. Climate change is also likely to have wide-ranging adverse impacts on human health and well-being. Decreased availability of potable water due to raising salinity will be responsible for increased illness and death, while many infectious diseases, including malaria and dengue, could rise due to climate change. In sum, these cumulative factors, associated with weak infrastructure, poor governance, and lack of resources, will lead to greater risks in the future, to say the least. A recent World Bank study (cited by Islam 2008) revealed that about 4 % of the GDP is being eroded by environmental degradation in Bangladesh annually.

12.2   Climate Change, Sea-Level Rise, and Bangladesh

One of the reasons for defining disasters in Bangladesh in terms of health issues is that Bangladesh has emerged as one of the world's predicted hot spots for potential tropical diseases due to climate change. Bangladesh is particularly vulnerable to sea-level rise because of the compounding effects of global warming and land subsidence due to removal of groundwater and the weight of accumulation of huge quantities of river sediments in the Ganges-Brahmaputra-Meghna (GBM) river system, estimated at between 0.5 and 2 billion metric tonnes per year (Ahmed 2006). The most recent projections by the Intergovernmental Panel on Climate Change (IPCC) estimate global sea-level rises within the range of 0.18 to 0.59 m (IPCC 2007). These estimates are considerably lower than that of the first IPCC assessment in 1990 (70 cm by 2100) but are more consistent with recent findings by Church and White (2006) and Church et al. (2004, 2006a, b). Much more cata-strophic projections (exceeding a rise of 1 m by the end of 2100) by a group of non-IPCC scientists are based on computer simulations of accelerations of glacier melting (Munro 2006; Hansen 2007). Even with a conservative estimate of 70 cm of sea-level rise (the maximum projection of the first IPCC assessment), Bangladesh is likely to experience a sea-level rise of up to 1.9 m by 2100, largely because of the remaining contribution by land subsidence (Houghton 2004).

Since the 2007 IPCC assessment projects about 10 % lower sea levels than in its first assessment, thus prorated, Bangladesh is still expected to experience a sea-level rise of about 1.7 m by the end of this century. This amounts to a rate of 1.7 cm or 0.66 in. of rise per year.

Expressed in this manner, this may not sound alarming, but in a low-lying coastal environment with extremely gentle slopes, about one-fifth of the land area of Bangladesh is at risk of inundation (Houghton 2004). Estimates of the number of people that would be displaced by inundation of one-fifth of Bangladesh range from 15 million to as high as 25–35 million (Nicholls and Mimura 1998; Rabbani et al. 2010). Notwithstanding such divergence in estimates, large-scale population dis-placement and associated socioeconomic and health problems due to sea-level rise is a disaster-in-waiting for Bangladesh as it is the most anticipated impact of climate change in this low-lying country. However, the health issues as an inevitable conse-quence of the climate change and sea-level rise has not so far been addressed to in-depth in the existing literature. The National Adaptation Program of Action (NAPA) of the Government of Bangladesh does not shed much light on this particular prob-lem either (MoEF 2010).

12.3   Climate Change, Sea-Level Rise, and Health Issues

There is increasing evidence that global climate change will have adverse effects on human health, mainly among the poorest population of the developing countries. Bangladesh may experience some of the more severe impacts because of its charac-teristic climatic and geographical conditions, coupled with high population density and poor health infrastructure. Many of the climatic events would make the climate change-induced health impacts worse. For example, in 2004, about 50 % of the total land mass of the country was inundated for nearly 2 months, while in 1998, flood affected approximately 30 million people in 52 out of 64 districts of the country. At the same time, a large number of people are annually affected by coastal cyclones leading them to lead economic deprivation affecting their nutritional level and gen-eral health. Thus, such high degrees of "vulnerability" to the impacts of climate change make the population particularly susceptible to adverse health impacts and threaten development achievements (Khan et al. 2011a, b).

Such vulnerability conditions are directly linked with some human health prob-lems as a result of newer environmental threats, such as changes in microclimate, erratic climatic behavior, and salinity intrusion in soil and water. This paper identifies some of the possible direct and indirect impacts of climate change on health condition of Bangladesh. In identifying such impacts, secondary sources of information have largely been reviewed.

12.3.1   Climate Change and Health: Some Generalizations

As indicated above, climate change affects human health both directly and indirectly as people get exposed to some of the key weather elements (temperature, precipitation, air mass, etc.) directly and indirectly through changes in the quality of water, soil, given ecosystems, agriculture, industry, human settlements, and nature of health infrastructure. Such exposures in varying degrees can cause disability and sufferings – even death. Further, health problems can increase vulnerability and reduce the capacity of individuals and/or groups to adapt to climate change. At present the impacts of climate change are rather subtle to observe, but the IPCC has projected a progression in health hazards for all regions of the world. The IPCC (2007 quoted by Rahman 2008) has observed some emerging evidence of climate change affecting human health as:

- Altered the distribution of some infectious disease vectors
- Altered seasonal distribution of some allergic pollen species
- Increased heat wave-related deaths

Systematic reviews of empirical studies provide the best evidence for the relationship between health and weather or climate factors, but such formal reviews are rare. The evidence published so far indicates that:

- Climate change is affecting the seasonality of some allergic species as well as the seasonal activity and distribution of some disease vectors.
- Climate plays an important role in the seasonal pattern or temporal distribution of malaria, dengue, tick-borne diseases, cholera, and other diarrhea diseases.
- Heat waves and flooding can have severe and long-lasting effects.

Most importantly, waterborne diseases will pose a major public health concern in the country with changes in climate factors in terms of yearly maxima and minima as observed by a study conducted by Bangladesh Center for Advanced Studies (BCAS) and National Institute of Preventive and Social Medicine (NIPSOM) with support from Climate Change Cell of the Environment Directorate of the Government of Bangladesh. The study was conducted in three climatic zones, i.e., drought prone (Rajshahi), flood prone (Manikganj), and coastal zone (Satkhira) (Climate Change Cell 2009). According to the study, climatic factors are associated with malnutrition problems. Incidence of diarrhea was found to have positively correlated with annual rainfall in Rajshahi and Satkhira districts. Total monsoon rain was also found to have positively correlated with incidence of rainfall. In contrast, dry season rainfall was found to have positive correlation with Manikganj area. Skin diseases and malnutrition were also found to be positively correlated with temperature differentials in both Rajshahi and Satkhira and negatively correlated in Manikganj area. The study also shows that the climatic factors of Satkhira are sensitive to diarrhea, skin diseases, and malnutrition as they are positively correlated with at least one of the climatic variables. Rainfall variations came next as the main cause of such diseases and are followed by natural hazards and other disorders (UNB 2009). However, these findings should be viewed with caution as simple statistical correlation does not imply any causal relationship between variables under consideration as there are other elements that could have interventions, including access to health infrastructure, with human health and diseases in a given area.

12.4   Climate Change and Diseases Scenario

In the early 1990s, there was little awareness of health risks posed by global climate change. Broadly a change in health conditions can have three kinds of health impacts:

- Impacts on human health directly by weather extreme (death, injury by flood, storm)
- Indirect impacts through changes in the range of disease vector-borne and water-borne pathogens
- The health consequences of various processes of environmental and ecological disruption that occur in response to climate change

Over the years it has been suggested that changes in the incidences of diseases and other health problems related to climate and environmental changes have occurred. Vector-borne and waterborne diseases have been on the rise along with the increased shortage of freshwater supply and sanitation. Even though it has been recognized that climate change affects the health sector as much as the other sectors, climate change-related health problems have been virtually ignored. Research in this area has been very limited, and there has been no study to find out the extent of impact climate change has had on health-related issues (IIED and BCAS 2008).

According to the World Bank (2000 quoted by Rahman 2008), Bangladesh is vulnerable to outbreaks of infectious, waterborne, and other types of diseases. Available information shows that the incidence of malaria has increased from 1,556 cases in 1991 to 15,375 in 1981 and from 30,282 cases in 1991 to 42,012 in 2004 (Elahi and Sultana 2010). Other diseases, such as dengue, diarrhea, dysentery, and other gastroenteric diseases, are also on the rise especially during summer months when humidity is high and monsoon rains break in (MoEF 2005) (Table 12.1). Other high-temperature and rainfall-related diseases on the increase are dehydration, malnutrition, and heat-related morbidity mainly among children and aged population groups. These diseases are also linked to some environmental conditions, like water supply, sanitation, and seasonal food habits.

Table 12.1 Incidence of some climate-induced diseases in Bangladesh

Diseases	Period	Total cases	Average annual cases
Diarrhea	1988–2005	48,302,636	2,842,273
Skin diseases	1988–1996	23,697,833	2,623,092
Malaria	1974–2004	1,018,671	33,956
Mental disorder	1988–1996	201,881	22,431
Dengue	1999–2005	19,830	3,305

Source: WHO (2006), Director-General, Health, GoB (1996, 1997), MoEF (2005) (After Rahman(2008))

Streatfield (2012) followed the framework for Climate Change-Health Research of the WHO projects' impacts of climate change that indicated increase of global temperature (relative to preindustrial) ranging from 0 to 5 °C and identified five major areas, i.e., food, water, ecosystems, extreme weather events, and risk of abrupt and major irreversible changes. Consequently, with above 3 °C changes in temperature, each of the above areas will lead to falling crop yield particularly in developing regions and also falling yields in many developed regions, and many major cities will be threatened by sea-level rise, extinction of increasing number of species, rising intensity of climatic events, and increasing and abrupt risks of large-scale shifts in the climate system, respectively (Streatfield 2012). Most of these will directly impact on human health conditions in Bangladesh. Such a situation is now already emerging in Bangladesh in terms of increased incidence of infectious diseases which may be termed as the *direct impacts* of climate change. These are briefly discussed below:

Malaria About 15 million people are now at risk of malaria attack, and according to the Ministry of Health and Family Welfare (GoB 2011), 37 deaths were reported, but the "actual cases may be 5,000 to 10,000" (Streatfield 2012). The most affected areas are in the northeastern and southeastern parts of Bangladesh (Elahi and Sultana 2010). Many parasites prefer 24–26 ° C and mosquitoes (*Plasmodium falciparum*) like similar range (above 20 °C). A study was conducted in Rangamati, Sylhet, and Faridpur districts over the period 1972–2002 to observe the impacts of climate change on public health especially on malaria. The climate-related variables included were temperature, rainfall, and relative humidity. It was observed that with the rise of yearly average maximum temperature, yearly total rainfall, and yearly average humidity, malaria prevalence in Rangamati has increased, and with the rise of yearly average maximum and minimum temperature in Sylhet and Faridpur, malaria prevalence was also increased. Yearly average minimum temperature in Rangamati, yearly total rainfall in Sylhet and Faridpur, and yearly average humidity in Faridpur were found to be negatively correlated with malaria prevalence (Amin et al. 2011).

Dengue The rise in average temperature and increased instances of waterlogging due to floods has increased the population of mosquitoes in some tropical countries. The Bangladesh National Adaptation Program of Action (NAPA) noted the increasing trend and variation of dengue outbreaks are consistent with the corresponding trend and variation of temperature, which indicates that the anticipated future warming in Bangladesh might enhance the dengue occurrence (MoEF 2005). Dengue is a viral febrile illness that is also transmitted by mosquito vectors (*Aedes aegypti*). Outbreaks of dengue have become frequent in recent years for a number of reasons. Almost all ages and both sexes are susceptible to dengue. Dengue transmission occurs during the rainy season of the year. Since July 2000 there had been four outbreaks of dengue in the Dhaka City, and cases had also been reported from other big cities of different parts of the country (Dhaka, Chittagong, Khulna, Rajshahi) (IIED and BCAS 2008). Dengue fever is showing a fluctuating trend – 6,000 cases with 58 deaths in 2002 and 500 cases with 0 deaths in 2007. But there was a trend of dengue outbreaks in every 2 years. An increase of 3° to 4 ° C of temperature may double the reproduction of *Aedes aegypti* mosquito – the carrier of dengue – and there is no vaccine yet to treat this disease.

Kala-azar In Bangladesh kala-azar had almost disappeared in the late 1970. But it reappeared sporadically and an increase of kala-azar cases reported during 1982–1988. During the last few years, cases have been increasing but the reasons for this are not clear. The disease pattern is extremely localized with most cases reported from rural areas, a familial and contiguous household clustering pattern among the lower socioeconomic groups (IIED and BCAS 2008). There was an incidence of kala-azar in only 8 upazilas (out of 509) in 1981–1985, it increased to 105 in 2011, and a total of nearly 9,000 cases were detected causing 15–30 deaths per year. As the surveillance is weak, the estimated cases will be about 45,000 (Streatfield 2012). Female sand flies (*Phlebotomus argentipes*) carry the parasitic protozoa (*Leishmania donovani*) whose favorite habitats are river embankments and crevices in mud houses and in weeds/vegetation around houses. The prevalence of kala-azar is expected to rise with temperature as well as erratic rainfall due to global warming affecting this part of the world.

Diarrhea and Cholera Diarrhea, the most common disease in Bangladesh, accounts for about 16 % of total morbidity. The transmission of disease is associated with poor hygiene and inadequate access to water and sanitation and a warmer and humid environment. In the summer months, the incidences increase. Diarrhea and cholera epidemics are often reported in Bangladesh, and especially the children are affected by excessive heat, floods, waterlogging, lack of safe drinking water, etc. The proportional morbidity rate due to diarrhea disease has decreased in Bangladesh ranges from 17.5 % (1992) to 14.2 % (1996). The most vulnerable groups to diarrhea which ranged are children under five. Of them, infants had relatively high proportional morbidity due to diarrhea (IIED and BCAS 2008). A warmer water body is favorable to the growth of cholera pathogen in blue-green algae and copepods which explains the endemic nature of the disease in the Ganges floodplain in Bangladesh. A Bangladesh study found that abundance of *V. cholera 01* increases with copepods which feed on phytoplankton in coastal waters (Colwell 1996;

Pascual et al. and Rodo et al. *Science*, 2000 as quoted by Streatfield 2012). No wonder cholera is quite endemic along a wide belt of coastal Bangladesh.

Drought-Related Ailments There has been an increasing trend of erratic climatic behavior in Bangladesh reflected through shifts in seasonal pattern in terms of change in rainfall regime and localized drought condition. The effects of drought on health include malnutrition, various infectious diseases, and respiratory diseases (Menne and Bertillini 2000). Because of the loss of productivity of land and loss of crops due to drought, it will reduce dietary diversity and negatively affect overall food intake and may therefore lead to malnutrition and/or nutrition-related diseases. A study by Aziz et al. (1990) in Bangladesh has found that drought and lack of food were related with an increased risk of mortality from diarrhea illness.

Heat Stress Globally cities are experiencing "urban heat island effect" with higher temperature than the surrounding areas. Most major cities of Bangladesh are also experiencing the same effect. Again, some parts of northwestern (i.e., Rajshahi, Naogaon) and southwestern (Kushtia, Satkhira) Bangladesh experience extremely high summer temperature – often reaching a maximum of around 38° to 42 °C. The heat index analysis from 1961 to 2010 portrays significant amount of change in both temperature and relative humidity in the past 20 years. Table 12.2 summarizes observed change in temperature over Bangladesh. Both the temperature and humidity are showing increasing trends in almost all parts of Bangladesh except that the SWR is showing decrease of humidity. It is decreasing up to 0.16 %. In the NWR, the increase of maximum temperature in recent years is quite considerable. On the other hand, comparatively small increase of temperature is observed in the very region in case of mean temperature. The changes of mean maximum temperature in SWR (+0.650 °C), mean temperature in ER (+0.550 °C), and humidity in NWR and CR (+4.3 % and +4.2 %, respectively) between the period range of 1961–1990 and 1991–2010 are very significant (Rajib et al. 2011).

It is believed that the global warming is responsible for the remarkable increase of maximum and mean temperature during summer in Bangladesh (and the adjoining areas in West Bengal in India). The mean heat index value ranges from 42° to 59 °C in different parts of the country. Besides adverse climatic condition, environmental pollution and the erratic pattern of rainfall result in noticeably rising trend of humidity in northwest, central, and eastern parts of the country. Consequently, both heat and humidity levels play significant role to physical discomfort and diseases for many people (Rajib et al. 2011). But most observable health ailments are reflected through dehydration, heat stroke/exhaustion, and aggravation of cardiovascular diseases in elderly people and reduced work capacity and productivity.

Water and Food Contamination With the change in the pattern of temperature and precipitation with climate change, the waterborne and food-borne diseases will increase. In general, increased temperature results in higher pathogen replication, persistence, survival, and transmission for bacterial pathogens and has mixed effects on viral pathogens but often reduces the overall transmission rate (MoEF n.d.).

Table 12.2 Observed change in temperature and humidity by region in Bangladesh

Years	NWR	SWR	CR	ER
Mean maximum temperature				
1961–1990	32.970C	32.80C	31.760C	31.330C
1991–2010	33.470C	33.450C	32.280C	31.730C
Change	**+0.50 °C**	**+0.650 °C**	**+0.520 °C**	**+0.40 °C**
Mean temperature				
1961–1990	28.370C	28.80C	27.90C	27.320C
1991–2010	28.520C	29.210C	28.240C	27.870C
Change	**+0.150 °C**	**+0.410 °C**	**+0.340 °C**	**+0.550 °C**
Humidity (%)				
1961–1990	81.50	85.78	80.40	84.97
1991–2010	85.00	85.64	83.80	85.10
Change	**+4.3 %**	**−0.16 %**	**+4.2 %**	**+0.15 %**

Source: Rajib et al. (2011)
NWR northwestern region, *SWR* southwestern region, *CR* central region, *ER* eastern region

Higher temperature produces a greater number of waterborne and food-borne parasitic infections, and overall increased rainfall is associated with increased burdens of disease for bacteria, viruses, and parasites though the causes of these increases differ by pathogen and ecological setting. On the other hand, changes in rainfall patterns are likely to compromise the supply of safe water, thus increasing the risk of waterborne diseases. They are also associated with floods and waterlogging that increase the incidence of diarrhea, cholera, skin, and eye diseases (Rahman 2008).

Apart from the above diseases, other conditions having bearing on human health are related to displacement of population due to various climatic disasters associated with global warming and sea-level change triggering off, what may be called, "climate refugees." Such a situation would cause social disruption and conflicts, unemployment, and economic uncertainty leading to anxiety syndrome and mental instability (Khan 2010). Thus, there will be a number of *indirect impacts* or slow onset of a number of health conditions due to climate change. Some of these are:

Physical Injury and Chronic Stress Extreme climate events and inevitable impacts of climate change, in the form of flood, torrential rain, cyclone, and tropical storms – often with salinity intrusion into coastal and its inland fringe areas – are likely to cause physical injury to loss of lives. Very recently, the super cyclone – *Sidr* – killed thousands of people and made millions homeless rendering them to experience physical stress and mental agony. Acute and chronic disorders, like post-traumatic stress disorders (PTSD), may occur in which a person experiences emotional numbness or withdrawal syndromes, flash back of traumatic events, extreme form of anxiety, chronic depression, etc. Such conditions have been observed in the cyclone-affected areas and in areas prone to riverbank erosion. A report by the Climate Change Cell of Department of Environment, GoB, indicates that the annual rate of mental disorder was 22,431 in Bangladesh (quoted by Khan 2010).

Domestic Violence Any environmental disaster is likely to cause frustration and anger due to loss of economic and social status. The climate change-induced disaster, like flood, drought, and cyclone, causes food scarcity, hunger, and malnutrition and leads to PTSD mostly among household heads, women, and female-headed members due to mental agony and depression as a result of financial hardship. These are manifested through domestic violence, family breakups, and often suicide.

Health Hazards due to Groundwater Depletion The risks of human health in countries of climate change exposure are multifarious. Importantly, drainage congestion and standing water will increase the likelihood of potential outbreaks of cholera and other waterborne and diarrhea diseases in the floodplain areas of Bangladesh. The pressure on the availability and access to safe water, in particular during the dry season or in draught condition, and the increasing dependence on groundwater are an additional threat.

The Global Agriculture Information Network claims that increasing HYV rice (Boro) is causing Bangladesh's water table to drop by 4–5 ft annually. Climate change is likely to bring more extreme events, possibly including a failure of the monsoon in South Asia, while IFPRI simulates an extended drought beginning in 2030 and continuing through 2035. Such a situation will hamper food production in the face of increasing population, demand for more food, and accelerating food price (Rahman 2012). The inevitable result of this situation will be a widespread malnutrition among the middle- and low-income groups of population in the country resulting increased morbidity situation. Further, the pressure irrigation for agricultural production will lead to water crises particularly in the drought-prone areas in northwestern Bangladesh. Such a situation may very well negatively impact upon food security.

Impact on Air Quality Climate change can contribute to some air quality problems. Respiratory disorders may be exacerbated by frequency of smog (ground-level ozone) events and particulate air pollution particularly in dry seasons (IPCC 2007). This can damage lung tissue and is especially harmful for those with asthma and other chronic lung diseases (MoEF n.d.)

Health Hazards Due to Salinity Intrusion Sources of safe water supply in coastal Bangladesh have been contaminated by varying degrees of salinity intrusion from rising sea levels, cyclone, storm surges, and transgression of tidal water through river inlets and upstream withdrawal of freshwater. On the other hand, in southern and coastal Bangladesh, a widespread intrusion of salinity into water and soil has already affected crop production and cropping practices to a great extent. The situation has further been aggravated with the increasing withdrawal of groundwater further inland. It is thought that this salinity intrusion problem will aggravate health condition related to hypertension and pre-eclampsia in pregnancy (Streatfield 2012).

Water salinity data for Dacope, a rural area in coastal Bangladesh, were collected by the Environment and Geographic Information System (EGIS) during 1998–2000, and information on drinking water sources from 343 pregnant women through 24-h surveillance of urine samples, blood pressure monitoring during dry season (2009–2010), and hospital-based prevalence of hypertension in pregnancy for 969 women (2008–2010). The average estimated sodium intakes from drinking water ranged from 5 to 16 g/day in dry season compared with 0.6 to 1.2 g/day in rainy season. Average daily sodium excretion in urine was 3.4 g/day (range, 0.4 to 7.7 g/day) (Khan et al. 2011a, b). It may be mentioned that as per WHO and FAO, the current dietary intake of sodium is 2 g/day (<85 mmol/day) (Nishida et al. 2004), but in coastal Bangladesh, the mean sodium intake in pregnant women is well above this level.

The annual hospital prevalence of hypertension in pregnancy was higher in the dry season (odd ratio, 12.2 %) that in the rainy season (odd ratio, 5.1 %) with high confidence level. Evidently these health-related conditions are likely to be exacerbated by climate change-induced sea-level rise. Hypertension in pregnancy is associated with increased rates of adverse maternal and fetal outcomes, both acute and long term, including impaired liver function, low platelet count, intrauterine growth retardation, preterm birth, and maternal and perinatal deaths (Sibai 2002), The adverse outcomes are substantially increased in women who develop pre-eclampsia (see above). At the same time, with the increased density and distribution of salinity, cholera germs are getting favorable habitat and spreading in the coastal area (MoEF n.d.).

Food Security and Health In Bangladesh, food security is increasingly being adversely affected by extreme climatic events. Such a situation develops with the fall of income and rise in expenditure, loss of assets and often land through repeated disasters (mainly flood, coastal inundation, and bank erosion), and burden of loan repayments at household level. As such most low-income households fall into a process of pauperization as they reduce regular consumption of food, and the eroded livelihoods expose people to increased health risks. While the impoverishment is exposing the poor more to the adverse impact of climate change, this also bars people from prioritizing their health needs (Alamgir et al. 2011). On the other hand, some of the major riverbeds experience excessive sedimentation. This further aggravates the impact of sea-level rise upstream and increases the risk of local flooding and increased salinity moving further inland upstream resulting in restricted drinking water supply and crop failure in many areas. Such conditions lower the nutritional status as well as various health problems among the affected population.

12.5 Concluding Remarks

Bangladesh is already vulnerable to vector-borne and waterborne diseases, and over the years, their prevalence has shown an upward trend. Thus, diseases like malaria, diarrhea, and cholera are also on the increase especially during the summer and monsoon/rainy months. Climate change is also bringing about additional stresses like dehydration, malnutrition, and heat-related morbidity especially among children and elderly populations. These problems are thought to be closely interlinked with water supply and sanitation issues as well. Climate change has already been linked to salinity intrusion from coastal areas to inland and subsequently leading to land degradation, groundwater contamination, and biodiversity loss and ecosystem damage. Changes in the above factors have direct impact on human health as well. Development of a nation is highly dependent on the health of the people. Bangladesh already carries the burden of high population and is characterized by natural disasters, diminishing and limiting natural resources, and the further burden of increased health problems will push back its development achievements.

Even though climate change-induced health impacts have been gaining importance in Bangladesh, there is still a lack of research and capacity in this field and its ever-increasing level of vulnerability of the people. Linkage between climate change and the increased incidences of disease, rate of mortality, and availability of safe water has not yet received the proper focus it requires. On the whole climate change is expected to present increased risks to human health in Bangladesh, especially in the light of the country's overall socioeconomic infrastructure and development (World Bank 2012). With the increased health risks, the expenditure on public health infrastructure will have to be enhanced by the government from now on.

In view of the above, the impacts of climate change on health would most likely depend upon the success to adapt to various aspects of climate change. And in particular, they would depend upon improved health infrastructure including health care and delivery systems, supply of safe drinking water, and improved sanitation that has to be included within the national climate change adaptation policy in a substantial way right now.

References

Ahmed AU (2006) Bangladesh climate change impacts and vulnerability: a synthesis. Government of Bangladesh, Department of Environment, Climate Change Cell, Dhaka

Alamgir F, Nahar A, Collins AE, Ray-Bennett NS, Bhuiya A (2011) Climate change and food security: health risks and vulnerabilities of the poor in Bangladesh. Int J Clim Change 1(4):37–54

Amin MR, Tareq SM, Rahman SH (2011) Impacts of climate change on public health: Bangladesh perspective. Global J Environ Res 5(3):97–105

Aziz KMA, Hoque BA, Huttly S, Minnatullah KM, Hasan Z, Patwary MK, Rahman MM, Cairncross S (1990) Water supply, sanitation and hygiene. Sanitation report – 1. World Bank, Washington, DC

Church JA, White NJ (2006) A 20th century acceleration in global sea-level rise. Geophys Res Lett 33, L01602. doi:10.1029/2005GL024826

Church JA, White NJ, Coleman R, Lambeck K, Mitrovica JX (2004) Estimates of the regional distribution of sea-level rise over the 1950 to 2000 period. J Clim 17:2609–2625

Church JA, White NJ, Hunter JR (2006a) Sea level rise at tropical Pacific and Indian Ocean islands. Glob Planet Chang 53:155–168

Church JA, Hunter JR, McInnes KL, White NJ (2006b) Sea level rise around the Australian coastline and the changing frequency of extreme sea levels. Aust Meteorol Mag 55:253–260

Climate Change Cell (2009) Climate change and health impacts in Bangladesh: BCAS and NIPSOM

IPCC (Intergovernmental Panel on Climate Change) (2007) Climate change 2007: synthesis report. In: Core Writing Team, Pachauri RK, Reisinger A (eds) Contributions of working group I, II and III to the fourth assessment report of the Intergovernmental Panel on Climate Change. IPCC, Geneva

Elahi KM, Sultana S (2010) Resurgence of malaria in Bangladesh. In: Akhtar R, Dutt AK, Wadhwa V (eds) Malaria in South Asia. Springer, New York, pp 107–122

GoB (2011) Bangladesh health statistics. Ministry of Health and Family Welfare, Dhaka

Hansen J (2007) Scientific reticence and sea level rise. Environ Res Lett 2:024002

Houghton J (2004) Global warming: the complete briefing, 3rd edn. Cambridge University Press, Cambridge, UK

IIED, BCAS (2008) Climate change and health in Bangladesh. CLACC, Dhaka

Islam MA (2008) Climate change and development risk: local perspective. The Daily Star (Dhaka), 17 March 2008

Khan ZR (2010) Climate change on mental health. Ecademics 286:325. Accessed 28 Dec 2012

Khan A, Xun W, Ahsan H, Vineis P (2011a) Climate change, seal level rise & health impacts in Bangladesh. Environment: Science and Policy for Sustainable Development, Sept.–Oct

Khan AE, Ireson A, Kovats S, Mojumder SK, Khusru A, Rahman A, Vineis P (2011b) Drinking water salinity and maternal health in coastal Bangladesh: implication of climate change. Environ Health Persp 119(9):1328–1332. (online. Doi: 10.1289/ehp.1002804)

Menne B, and Bertillini R (2000) The health impacts of desertification and drought. Down Earth 14:4–6

MoEF (2005) National adaptation programs of action (NAPA). GoB, Dhaka

MoEF (2010) National adaptation programs of action (NAPA). GoB, Dhaka

MoEF (n.d.) Climate change and health in Bangladesh. DFID/DANIDA, Dhaka

Munro M (2006) Climate change to redraw map. Winnipeg Free Press, March 24, 2006

Nicholls RJ, Mimura N (1998) Regional issues raised by sea-level rise and their policy implications. Clim Res 11:5–18

Nishida C, Uauy R, Kumanyika S, Shetty P (2004) The joint WHO/FAO expert consultation on diet, nutrition and the prevention of chronic diseases: process, product and policy implications. Public Health Nutr 7:245–250

Rabbani G, Rahman AA, Islam N (2010) Climate change and sea level rise: issues and challenges for coastal communities in the Indian Ocean region. In: Michel D, Pandya A (eds) Coastal zones and climate change. The Henry L. Stimson Center, Stimson Pragmatic Steps for Global Security, Washington, DC, pp 17–29

Rahman A (2008) Climate change and its impact on health in Bangladesh. Reg Health Forum 12(1):16–26

Rahman MM (2012) Food security in the realm of climate change. The Daily Star, 17 November

Rajib MA, Mortuza MR, Selmi S, Ankur AK, Rahman MM (2011) Increase of heat index over Bangladesh: impact of climate change. World Acad Sci Eng Technol 58:402–405

Sibai BM (2002) Chronic hypertension in pregnancy. Obstet Gynecol 100:369–377

Streatfield PK (2012) Climate change and health in Bangladesh. In: National conference on community based adaptation to climate change. ICDDRB, Dhaka

UNB (2009) Climate change to cause more health hazards in Bangladesh. The Daily Star, 27 November

World Bank (2012) Bangladesh: towards accelerated, inclusive and sustainable growth – opportunities and challenges. World Bank, Dhaka

Zaman M, Rasid H, Elahi KM, Khatun H (n.d.) Climate change, disasters and development in Bangladesh: introduction and overview. (Draft)

Chapter 13
Health Adaptation Scenario and Dengue Fever Vulnerability Assessment in Indonesia

Budi Haryanto

Abstract Climate change is a serious challenge we are currently facing, as the impacts have already been occurring. Indonesia's geographic and geological characteristics are also easily affected by climate change, natural disasters (earthquakes and tsunamis), and extreme weather (long drought and floods). Its urban areas also have high pollution levels. Much evidence can be seen, ranging from an increase in global temperature, variable season changes, extremely long droughts, high incidences of forest fires, and crop failures. This paper describes some evidences of climate change in Indonesia to weather disasters such as floods, landslides, and drought, burden of vector-borne diseases, air pollution from transportation and forest fires, and reemerging and newly emerging diseases. An evidence of health related to climate is based on a vulnerability assessment for dengue fever. To respond to its negative impacts to human, the health adaptation strategy and efforts undertaken in Indonesia nowadays include the following: to increase awareness of health consequences of climate change, to strengthen the capacity of health systems to provide protection from climate-related risks and substantially reduce health system's greenhouse gas (GHG) emissions, and to ensure that health concerns are addressed in decisions to reduce risks from climate change in other key sectors.

Keywords Climate change • Health impact • Vulnerability assessment • Adaptation strategy

13.1 Introduction

Located between 06° 08' North–11° 15' South and 94° 45'–14° 05' East, Indonesia is known as the biggest island country in the world which consists of five main islands and 30 smaller island groups including 17, 500 islands. The total area of Indonesia comprises 3.1 million km² of ocean (62 % of the total area) and

B. Haryanto, MSPH, M.Sc., Ph.D. (✉)
Research Center for Climate Change, University of Indonesia, Depok 16424, Indonesia
e-mail: bharyant@cbn.net.id

© Springer International Publishing Switzerland 2016   221
R. Akhtar (ed.), *Climate Change and Human Health Scenario in South and Southeast Asia*, Advances in Asian Human-Environmental Research,
DOI 10.1007/978-3-319-23684-1_13

approximately 2 million km^2 of land (38 % of the total area), with a coastline that extends to 81,000 km. If the economic exclusive zones of 2.7 million km^2 are included, the total jurisdiction area of Indonesia is no less than 7.8 million km^2.

Indonesia's geographic and geological characteristics are also easily affected by climate change, natural disasters (earthquakes and tsunamis), and extreme weather (long drought and floods). Its urban areas also have high pollution levels.

Climate change is a serious challenge we are currently facing, as the impacts are already happening. Much evidence can be seen, ranging from an increase in global temperature, variable season changes, extremely long droughts, high incidence of forest fires, and crop failures. In Indonesia, climate change can be detected from an increase in temperature at a rate of 0,03 °C and an increase in rainfall of 2–3 % per year (Hulme and Sheard 1999), as well as a change in *ENSO* (El Niño Southern Oscillation) to 2–5 years from what was normally 3–7 years (Ratag 2001). This affects not only the environment but also public health in a global scale, both directly and indirectly. A drop in rainfall due to climate variability or seasonal change combined with a rise in temperature has significant effects on water supply (El Niño effect). A rise in seawater temperature has contributed to the spread of diseases such as malaria, dengue fever, diarrhea, cholera, and other vector-related diseases (La Niña years). A change in temperature, humidity, and wind speed is also contributing to the increase in vector population, increasing their life-span and also widening their spread. This in turn may intensify the occurrence of vector-related communicable diseases such as leptospirosis, malaria, dengue fever, yellow fever, schistosomiasis, filariasis, and plague.

Extreme occurrences influenced by climate change in Indonesia, such as floods, hurricanes, tidal waves, landslides, droughts, and forest fires, are happening more often than before. There are 300 events of extreme occurrences from January to August 2008, resulting in 263 deaths, 1,927 critically injured, 66,988 with mild injuries, 7 missing, and 92,210 refugees. Those refugees are susceptible to easily spreading communicable diseases, even worsened by unpredictable climate.

The health risks to the community due to climate change include the spread of diseases such as malaria, dengue fever, diarrhea, cholera, and other vector-related diseases. There is an increase of risks for certain populations who are more susceptible to damages caused by climate change, such as the poor and the very poor (45.2 % from 105.3 million), coastal community (64 % of people in Java Island live in coastal areas), children and the elderly, farmers, traditional community, and people living in small islands (17,500 islands). Climate change is definitely causing negative impacts on the community, with the worse to expect (Ministry of Health, Republic of Indonesia 2011).

The availability of relevant hydrometeorological, socioeconomic, and health data is limited, and available data are often inconsistent and seldom shared in an open and transparent manner. Furthermore, there is insufficient capacity for assessment, research, and communication on climate-sensitive health risks in Indonesia, as well as insufficient capacity to design and implement mitigation and adaptation programs. However, in Indonesia, the urgent need to incorporate health concerns into the decisions and actions of other sectors has been recognized since the last couple years, in which a scenario to mitigate and adapt to climate change, to ensure that these decisions and actions also enhance health, had been developed.

13.2 Climate Health Impacts in Indonesia

The current and emerging climate change-related health risks in Asia and the Pacific include heat stress and water- and food-borne diseases (e.g., cholera and other diarrhea diseases) associated with extreme weather events (e.g., heat waves, storms, floods and flash floods, and droughts); vector-borne diseases (e.g., dengue and malaria); respiratory diseases due to air pollution; aeroallergens, food, and water security issues; malnutrition; and psychosocial concerns from displacement (WHO-SEARO 2007). Change in climate temperatures and rainfall in Indonesia is shown in Figs. 13.1 and 13.2:

The temperature in Indonesia tends to increase since 1950 to 2000 and is predicted to increase up to the year 2100 (Fig. 13.1).

The precipitation trend in Indonesia during 1950 to 2000 is generally increased as shown in Fig. 13.2.

The occurrences related to climate change and its health impacts in Indonesia are the following:

13.2.1 Weather Disasters: Floods, Landslides, and Droughts

Climate change is likely to have major effects on human health via changes in the magnitude and frequency of extreme events: floods, landslides, windstorms, and droughts. The effects of drought are primarily associated with food security and

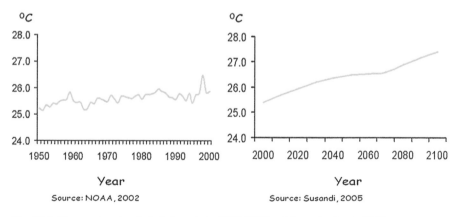

Fig. 13.1 Temperature trends in Indonesia in 1950–2000 and its projection to 2100

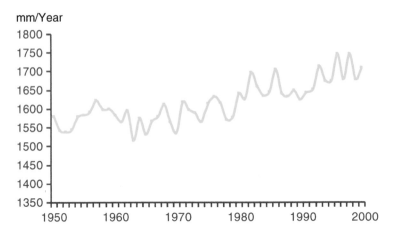

Fig. 13.2 Precipitation trend in Indonesia in 1950–2000 (Source: NOAA-CIRES (2005))

increasing waterborne disease. Weather disasters affect human health by causing considerable loss of life. Extreme weather events cause death and injury directly. Following disasters, deaths and injuries can occur as residents return to clean up damage and debris. The nonfatal effects of natural disasters include physical injury; reduced nutritional status, especially among children; increases in respiratory and diarrhea diseases because of crowding of survivors, often with limited shelter and access to potable water; effects on mental health that may be long lasting in some cases; increased risk of water-related diseases from disruption of water supply or sewerage systems; and exposure to dangerous chemicals or pathogens released from storage sites and waste disposal sites into floodwaters.

Floods occurred almost in all regions and cities in Indonesia, contributing so many health effects and economic lost. Landslides had affected thousands of population life in West Sumatra, South Sumatra, Central Java, Sulawesi, and Halmahera. Droughts had made severe the lives of people living in Kalimantan, Sulawesi, East Nusa Tenggara, Papua, and Maluku.

There were more than 1,400 disasters in Indonesia in the period of 2003–2005 which were about 53 % related to hydrometeorology (34 % was flooding and 16 % landslide). During El Niño years (1994, 1997, 2002, 2003, 2004, and 2006), eight reservoirs in Java had produced electricity below normal capacities. Specifically, El Niño 1997 had caused serious problems to coral reef ecosystems where 90–95 % of coral reefs at the depth of 25 m have experienced coral bleaching (Bucheim 1998).

The availability of water is very dependent on the climate, due to the limited supply of water (only covers about 37 % of urban population and 8 % of rural population) causing people and industries to use deep groundwater resources, in which then causes land subsidence that creates areas vulnerable to flood and saltwater intrusion. Water scarcity is an additional issue as a result of global and regional climate change in which between 2010 and 2015 the country is predicted to experience a major clean water shortage, and this is expected mainly in urban areas (Boer et al. 2007).

13.2.2 Burden of Vector-Borne Diseases

Changes in climate that can affect the potential transmission of vector-borne infectious diseases include temperature, humidity, altered rainfall, soil moisture, and rising sea level. The factors responsible for determining the incidence and geographical distribution of vector-borne diseases are complex and involve many demographic and societal as well as climatic factors. Transmission requires that the reservoir host, a competent vector, and the pathogen be present in an area at the same time and in adequate numbers to maintain transmission.

The important vector-borne diseases related to climate change in Indonesia are dengue hemorrhagic fever (DHF), malaria, and leptospirosis. South Sumatra, Java, Bali, Central Kalimantan, and North Sulawesi are endemic areas for DHF. Meanwhile, Papua, Nusa Tenggara Timur, Nusa Tenggara Barat, and Maluku are endemic areas for malaria. Flood areas in almost all of Indonesian regions are at high risk for leptospirosis.

Dengue hemorrhagic fever has spread to all cities throughout Indonesia since 1968. In 1968 IR of DHF reported cases was 0,05/100.000 population with a case fatality rate (CFR) of 41,3 %; thereafter, outbreaks frequently occur in several areas. In 1998 an outbreak with 72.133 cases, and a mortality rate of 2 %, was the most severe outbreak that ever happened since the first DHF case reported in Indonesia. In 2004 a national outbreak occurred, in 40 districts and cities in 12 provinces with a number of 28.077 cases, with mortality of 381 cases (CFR 1.36 %). DHF increases continuously for the period of 1999–2007 reaching an IR of 71,78 per 100.000 population, even if there had been a case of decrease in 1998. Throughout 2007, 11 provinces suffered from DHF outbreaks, namely, West Java, South Sumatra, Lampung, DKI Jakarta, Central Java, East Kalimantan, Central Sulawesi, East Java, Banten, and DI Yogyakarta. Its number of cases was 156.767, with an IR of 71,18/100.000 population and fatality of 1,570 cases (CFR 1,00 %). The total number of dengue hemorrhagic fever (DHF) cases reported in 2008 is 136,333 cases with 1,170 deaths (CFR = 0.86 % and incidence rate = 60.06 per 100,000 population). The highest incidence rate is in DKI Jakarta, which is 317.09 per 100,000 population, and the highest CFR recorded is in Jambi (CFR = 3.67 %). In 2009, 154,855 cases of DHF were reported with 1,384 deaths (CFR = 0.89 %). In 2010, 159,322 cases of DHF were also reported with 1,317 deaths (CFR = 0.87 %) (Ministry of Health, the Republic of Indonesia 2008a). The increasing number of cases over the years and its spread in Indonesia are shown in Fig. 13.3.

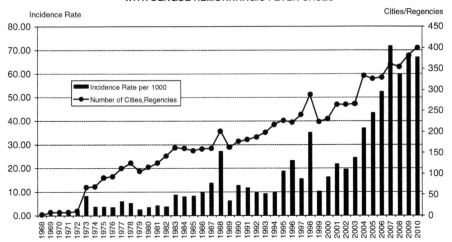

Fig. 13.3 Incidence of DHF per 1,000 population by number of cities/regencies infected

In Indonesia, cases of malaria, dengue, diarrhea, and cholera are predicted to increase as temperatures rise and water becomes contaminated, affecting scores of poor populations that do not have the resource to cope.

In 2010, a study reported by ICCSR (Indonesian Climate Change Sectoral Roadmap) generated maps of DHF's risk caused by climate change across the country (Fig. 13.4) (ICCSR 2010).

It is estimated that 45 % of Indonesian people are living in the high-risk malaria endemic areas. About 1.14 million of clinical malaria cases were reported in 2009, and outbreaks of malaria were reported to occur in 20 villages spreading in 11 provinces. The total number of malaria cases during the outbreaks was 869 cases with 11 deaths (case fatality rate, 1.23 %). It is estimated that more than 150 million people are living in malaria endemic areas, compromising the productivity of the productive age group and the health status of pregnant women and children under five years old (Ministry of Health, the Republic of Indonesia 2008b).

In 2007 the number of positive malaria was 311.789 cases. Furthermore, malaria in Indonesia reemerges and is being influenced by malaria termination program intensity and several environmental factors. Cases of malaria in Java and Bali, expressed as annual parasite incidence (API) for the period of 1995–2000, increase rapidly from 0.07 % in 1995 to 0.81 % in the year 2000 (DG-CDCEH 2010).

The ICCSR (Indonesian Climate Change Sectoral Roadmap) maps on malaria's vulnerability and malaria's risk caused by climate change across the country in 2010 are shown in Fig. 13.5 (ICCSR 2010).

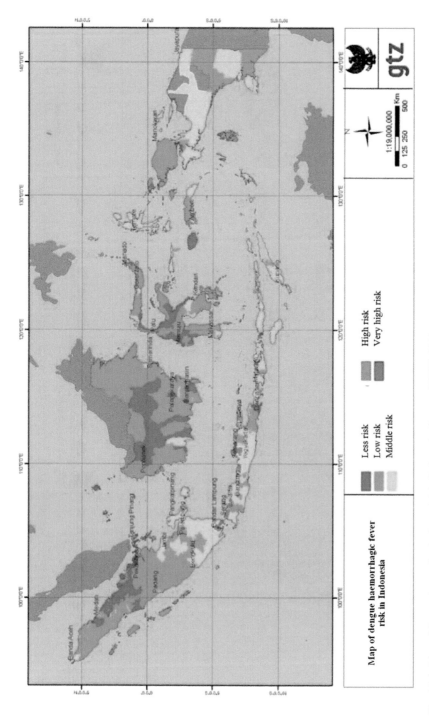

Fig. 13.4 Map of dengue hemorrhagic fever risk in Indonesia (Source: ICCSR 2010)

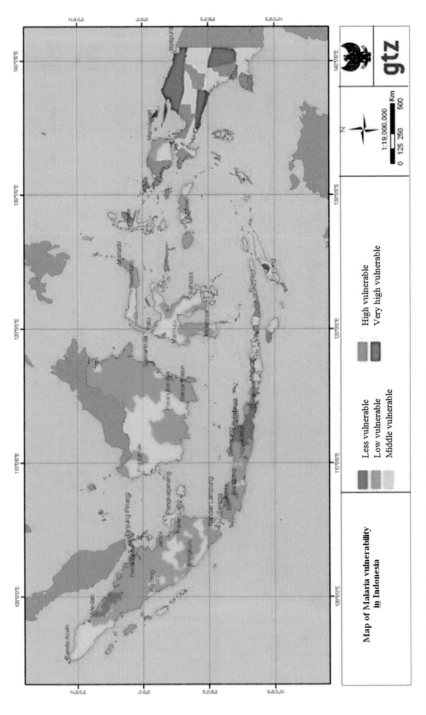

Fig. 13.5 Map of malaria vulnerability in Indonesia (Source: ICCSR 2010)

13.2.3   Air Pollution Sources: Transportation and Forest Fires

Weather conditions influence air quality via the transport and/or formation of pollutants (or pollutant precursors). Weather conditions can also influence air pollutant emissions, both biogenic emissions (such as pollen production) and anthropogenic emissions (such as those caused by increased energy demand).

Exposure to air pollutants can have many serious health effects, especially following severe pollution episodes. Long-term exposure to elevated levels of air pollution may have greater health effects than acute exposure. Current air pollution problems are greatest in Indonesia as it caused 50 % of morbidity across the country (Haryanto and Franklin 2011).

Transportation source in big cities contributes 60–80 % of air pollution (Ministry of Environment 2005). Kerosene is the cooking fuel used by 45 % of the households sampled (National Bureau for Statistics 2005). Fuelwood is used by 42 % of the households sampled (in 12 provinces >50 % households used fuelwood for cooking).

Air pollution from forest fire source in Sumatra and Kalimantan has had affected millions of human health in Sumatra, Kalimantan, and neighborhood countries, Singapore and Malaysia.

The index of air pollution in Jakarta and Surabaya in 2007 (using air quality monitoring system) shows the status of good day only in 72 days and 62 days, respectively. Air quality monitoring using non-AQMS in 30 cities shows high concentration for NO_2 (0–30 ppm) and SO_2 (0–50 ppm). The number of vehicles used on the road increases annually with an average of 12 % (motorcycle 30 %) which is in line with the increase of fuel consumption. Emission test in Jakarta in 2005 found that 57 % of vehicles did not pass the test. Meanwhile the traffic jams among cities continue and worsen (Ministry of Environment 2009).

The prevalence of acute respiratory infection exceeds the national prevalence (25.5 %) in 16 provinces. The prevalence of cough in 2007 is 45 % and flu 44 % without any significant difference between urban and rural. The prevalence of pneumonia exceeds the national prevalence (2.18 %) in 14 provinces. The prevalence of tuberculosis (TB) exceeds the national prevalence (0.99 %) in 17 provinces. In 2007, a number of 232,358 cases found out of 268,042 TB cases (86.7 %). The prevalence of asthma exceeds the national prevalence (4 %) in nine provinces (Ministry of Health 2008c). Pneumonia is overall the number one killer disease for infants (22.3 %) and children under 5 years of age (23.6 %) and is among the top ten diseases that result in deaths among the adult population. The WHO in 2002 estimates acute lower respiratory infection (ALRI) deaths attributable to solid fuel use (for children under 5 years) in Indonesia at 3,130 population, while chronic obstructive pulmonary disease (COPD) deaths attributable to solid fuel use (for 30 years old and more) was estimated at 12,160 population.

13.2.4   Reemerging and Newly Emerging Diseases in Indonesia

The World Health Organization's 2007 report stated that climate change would bring severe risks to developing countries such as Indonesia and has negative implications for achieving the health-related Millennium Development Goals (MDGs) and for health equity. Infections caused by pathogens transmitted by insect vectors, in Indonesia, are strongly affected by climatic conditions such as temperature, rainfall, and humidity.

The reemerging and newly emerging diseases in Indonesia include malaria with more than 100 million people, out of a population of about 230 million, at risk and 700 deaths in 2007, dengue fever or dengue hemorrhagic fever with 1,570 deaths in 2007 (but many more unrecorded), tuberculosis, polio, HIV/AIDS, severe acute respiratory syndrome (SARS), and avian influenza. More than two-thirds of the 576 districts in Indonesia have been classified as malaria endemic areas (MOH 2010).

13.3   Dengue Fever Vulnerability Assessment

Conducted by the RCCC – UI (Research Center for Climate Change – University of Indonesia) in collaboration with the Directorate of Environmental Health of the Ministry of Health and ICCTF (Indonesia Climate Change Trust Fund), a study on dengue fever vulnerability had been carried out with the aims to reveal the relationship of dengue hemorrhagic fever (DHF) to climate change or variability and to develop models and vulnerability map for the distribution of communities and regions prone to climate-induced DHF in 2038, in 20 districts/cities in five provinces: West Sumatra, Jakarta, East Java, Bali, and Central Kalimantan.

The 20 cities/districts were selected based on the availability of monitoring station of the Indonesian Agency for Meteorology, Climatology, and Geophysics (BMKG). Research designs used in this study are the following: (1) ecological time-series study to prove the statistical relationship between climatic factors with the incidence of DHF and malaria; (2) vulnerability analysis of the IPCC 2001 used to develop a model of community vulnerability in accordance to the character of DHF and malaria diseases based on variables in the component such as exposure, sensitivity, and adaptive capacity; and (3) geographic information system (GIS) with its following extensions used to map the distribution of DHF by year, patient's address, flight range of *Aedes aegypti* mosquitoes from its original breeding place, land use, hydrology, contours, roads, and socioeconomic data.

The data revealed that a number of DHF patients per 100,000 population (incidence rate, IR) in almost all study areas show a similar pattern since 2001, which was likely to increase significantly with the peak in year 2010. Then IR decreased sharply in 2011, but was followed by a tendency to increase in subsequent years. Correlation and regression analysis showed that higher rainfall was associated with

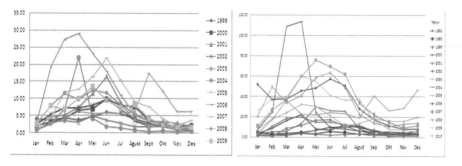

Fig. 13.6 Incidence rate per 100, 000 population of DHF in Surabaya in 1999–2009 and Central Jakarta in 1992–2010

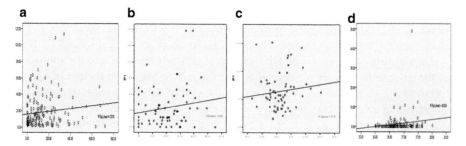

Fig. 13.7 Association between IR of DHF and precipitation in Central Jakarta in 1992–2010 (**a**) and in Bukittinggi in 2006–2011 (**b**). Association between IR of DHF and temperature in Padang in 2006–2010 (**c**) and in Palangkaraya in 1999–2009 (**d**)

an increased incidence of DHF ($p < 0.05$), with a weak level of correlation ($r < 0.25$) which was found in 11 districts/municipalities in 4 provinces, Central Jakarta, North Jakarta, Tangerang City, Tangerang Regency, Denpasar, Jembrana, Badung, Banyuwangi, Kotawaringin Barat, Kotawaringin Timur, and Palangkaraya. This research also found that in Central Jakarta, Denpasar, and Surabaya, the increasing temperature was associated with an increased incidence of DHF ($p < 0.05$) with a weak level of correlation ($r < 0.25$). Other districts/municipalities did not show any correlation between the increased rainfall and air temperature with the incidence of DHF (Figs. 13.6 and 13.7).

The vulnerability component of exposure includes land use (settlement, offices, business, schools, etc.) and population density. The component of sensitivity includes breeding places and resting areas of *Aedes* mosquitoes; pupa and adult density; incidence of DHF by sex, age, occupation, education, and address; and population mobility. The component of adaptive capacity includes availability of health services, such as hospitals, clinics, and public health centers, treatment management and skilled providers, implementation of DHF intervention program, community participation and involvement on DHF prevention program, and personal protection behavior.

The DHF vulnerability in almost all districts/municipalities (17 regencies/cities) indicated a very serious condition, with very high coping range index (CRI) (level 5) in the last few years since 2005. Very high CRI level was shown in the city of Padang in 2005, 2007, 2008, 2009, and 2012; in Padang Pariaman Regency in 2008, 2011, and 2012; in the city of Padang Panjang in 2007 and 2008; in the city of Denpasar in 2006, 2009, and 2010; in Jembrana Regency in 2007; in the city of Badung in 2007, 2009, and 2010; in the city of Surabaya in 2007, 2008, 2009, 2011, and 2012; in Malang Regency in 2007, 2008, 2009, 2011, and 2012; in the city of Pasuruan in 2007, 2008, 2009, 2010, and 2011; in Sumenep Regency in 2007, 2008, 2009, 2011, and 2012; in the city of Central Jakarta in 2005, 2006, 2007, 2008, 2009, and 2012; in the city of North Jakarta in the year 2006–2012; in the city of Tangerang in 2007–2012; in the city of Palangkaraya in 2006, 2008, and 2012; in Kotawaringin Barat Regency in 2005–2008 and in 2012; in Kotawaringin Timur Regency in 2008, 2010, and 2011; and in Barito Utara Regency in 2008. High CRI (level 4) also dominated the community vulnerability to dengue incidence in the study area during those years of very high CRI levels (Figs. 13.8, 13.9 and 13.10).

Distribution of patients with DHF in almost all study sites showed the spreading of the disease to a larger residential area from year to year. The pattern of spread followed the main road network. Mapping the risk of dengue disease transmission by providing an imaginary boundary with a radius of 200 m from the DHF patients each year showed the widespread areas of risk and that there was almost no safe area for people to get infected with dengue risk in each study area (Fig. 13.11).

Some important findings of this assessment are the following: (1) Increased incidence of DHF was associated with the increase in the amount of rainfall and with the increase of temperature in several study areas and in particular years. (2) Community vulnerability to dengue was dominated by very high coping range index (CRI) and high vulnerable conditions each year with the high level of exposure and sensitivity but low level of adaptation capacity. (3) The maps of patient distribution and the risk of dengue disease transmission showed that there is almost

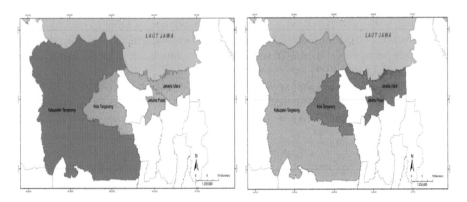

Fig. 13.8 Map of coping range index of population vulnerability to DHF in Jakarta and Banten in 2011 and 2012

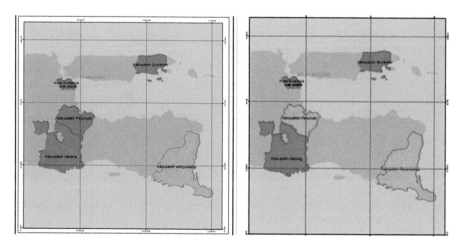

Fig. 13.9 Map of coping range index of population vulnerability to DHF in East Java in 2011 and 2012

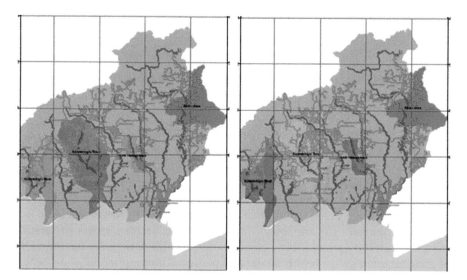

Fig. 13.10 Map of coping range index of population vulnerability to DHF in Central Kalimantan in 2011 and 2012

no safe place without the risk of being transmitted. The findings suggest that health vulnerability assessments and mapping of the potential human health impacts of climate variability and change are needed to inform the development of adaptation strategies, policies, and measures to lessen projected adverse impacts.

Revitalization of dengue disease eradication program as a whole at every stage with closed- monitoring implementation is urgently needed. Technical guidance

234   B. Haryanto

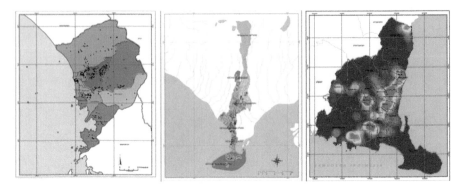

Fig. 13.11 Distribution of DHF cases and its buffering risk area (100 m radius) in Padang, Badung, and Banyuwangi in 2006–2010

and increased skills of health officers are indispensable. Socialization of program activities and increased capacity and participation of community on management measures could be a joint action in preventing the increase in dengue disease associated with climate change.

13.4 Indonesian Strategy on Health Adaptation from Climate Change

The spread of every single disease-related climate change is unique with the differences of community vulnerability and its location. Its characteristics and complexity pose as baseline information to be explored to support the accurate policy strategy development. The national assessments of the potential effects of climate change on human health are needed to better understand current vulnerability and to evaluate the country's capacity to adapt to climate change (Kovats et al. 2003).

The Government of Indonesia's response to the health impacts of climate change is currently prioritized to protect human health. It includes development of national strategy on health adaptation plan and training modules on health adaptation developed by the Ministry of Health in 2009 involving other sectors, such as the Ministry of Environment, Ministry of Public Works, Ministry of Agriculture, Ministry of Forestry, academicians, and NGOs with support funding from WHO-Indonesia office. The key contents of health adaptation strategy are based on the recommendation of WHO Adaptation Focus 2007, such as health security, strengthening health systems, health development, evidence and information, delivery, and partnership. Indonesia's health adaptation strategy is set to focus on infectious disease surveillance, health action in emergencies, safe drinking water, integrated vector management, environmental health capacity building, and healthy public policy (healthy housing, school, forest, industry, city). However, the implementation of the products has not yet been done due to the limited research evidences and institutionalization.

The Indonesian academicians and public health professionals at a parallel meeting of COP 13 2007 in Bali, Indonesia, developed a strategy for public health adaptation in Indonesia, which was consisted with the following steps:

1. Empowering ecological disease surveillance system and developing public health early warning system
2. Developing response to disaster effects of climate change
3. Enhancing capacity building for the government, private sector, and civil society on managing prevention and controlling climate change on human health
4. Increasing political awareness of climate change on human health
5. Empowering public health services system for disease prevention and control
6. Generating research and method on epidemiology and medicine to find out the approach in breaking the disease transmission chains
7. Preventing and eradicating climate change vector-related diseases

Based on Bali's ideas and update situations worldwide, the ICCSR 2010 suggested several important recommendations to develop alternative strategy for adaptation to climate change in the health sector as the following:

1. Strengthen the vulnerability and risk assessment methodology in the health sector due to climate change.
2. Develop a framework for policy development supported by needed decrees and regulations.
3. Develop planning and decision-making methodology based on local/regional evidences.
4. Improve inter-sector collaboration and partnerships.
5. Improve community participation, including private and higher education institutions/academics.
6. Strengthen the capability of local governments.
7. Develop networking and sharing of information.
8. Strengthen early warning system and emergency response at the community-level institutions.

13.5 Conclusion

To conclude the health adaptation strategy and efforts undertaken in Indonesia nowadays, it includes the following: (1) to increase awareness of health consequences of climate change, (2) to strengthen the capacity of health systems to provide protection from climate-related risks and substantially reduce health system's greenhouse gas (GHG) emissions, and (3) to ensure that health concerns are addressed in decisions to reduce risks from climate change in other key sectors. The health adaptation program is going through phases as the following:

2010–2014 Preparation Phase: data inventory, existing condition analysis, and choosing appropriate alternative method

2015–2019 Implementation Phase: management, monitoring, maintenance, and evaluation of all programs

2020–2024 Implementation Stability Phase: consistent and stable program implementation management, effective monitoring and evaluation, and sustainable fixing

2024–2029 Implementation Stability Phase: monitoring, evaluation, and development of current program capacity

References

Boer R, Sutardi, Hilman D (2007) Summary for policymakers: climate variability and climate changes, and their Implication in Indonesia [report]. The State Ministry of the Environment and the State Ministry of Public Works, Jakarta

Bucheim J (1998) Coral reef bleaching. Odyssey Expeditions – Tropical Marine Biology Voyages.

Directorate General of Communicable Disease Control and Environment Health (DG CDCEH) (2010) Paper presented on BMKG National Workshop and Seminar on Health Sector 2010. Jakarta

Haryanto B, Franklin P (2011) Air pollution: a tale of two countries. Rev Environ Health 26(1):75–82

Hulme M, Sheard N (1999) Climate change scenarios for the Philippines. Climatic Research Unit at University of East Anglia, Norwich

Indonesian Climate Change Sectoral Roadmap (ICCSR) (2010) Health Sector. Jakarta

Kovats S, Ebi KL, Menne B (2003) Methods of assessing human health vulnerability and public health adaptation to climate change. Series 1. WHO Europe

Ministry of Environment (2005) Indonesian environmental status 2004. Jakarta

Ministry of Environment (2009) Indonesian Environmental Status 2008. Jakarta

Ministry of Health (2008a) Directorate General of Communicable Disease Control and Environment Health's Yearly Report 2008. Jakarta

Ministry of Health (2008b) Indonesian Health Profile 2008. Jakarta

Ministry of Health (2008c) National basic health research report 2007. Institute for Health Research and Development. Jakarta

Ministry of Health (2011) Directorate General of Communicable Disease Control and Environment Health's paper presented on Symposium on Climate Change and Health Impacts, Jakarta

National Agency for Meteorology, Climatology, and Geophysics (BMKG) (2010) Paper presented on BMKG National Workshop and Seminar on Health Sector 2010. Jakarta

National Bureau for Statistics (2005) National Survey on Social and Economic Report 2004. Jakarta

Ratag MA (2001) Model Iklim Global dan Area Terbatas serta Aplikasinya di Indonesia. Paper presented on Seminar Peningkatan Kesiapan Indonesia dalam Implementasi Kebijakan Perubahan Iklim. Bogor, Indonesia, 1 November 2001

World Health Organization for South-East Asian Regional Office (2007) Side event of the COP 13, WHO regional workshop on Protecting Human Health from Climate Change. Bali, Indonesia

Chapter 14
Climate Variability and Human Health in Southeast Asia: A Taiwan Study

Huey-Jen Su, Mu-Jean Chen, and Nai-Tzu Chen

Abstract The local, regional, or global climatic conditions may alter atmospheric compositions and chemical processes and are implicated in increasing frequencies of extreme temperature and precipitation, which can also be intensified through varying urbanization, economic development, and human activities. Adverse health consequences related to such exposures are often of primary concerns for their immediate and sustaining impacts on the general welfares.

In Taiwan, higher cardiovascular or respiratory mortality appears to derive more from low temperature than does from the exposure to high temperature for the general population. Rural residents are less affected than urban dwellers under extreme temperature events if cardiovascular or respiratory mortality is benchmarked. In towns and villages with a high percentage of the elderly living alone, the senior and the disabled people and the aborigines will present a higher mortality associated with extreme temperature.

Extreme rainfall events may also interrupt the chain of food supply and life support and lead to reporting malnutrition of people in affected regions. In addition, contaminated water sources for drinking and recreation purposes, due mostly to flooding after extreme precipitation, are known to be associated with disease outbreak and epidemics. In Taiwan, for waterborne infections, extreme torrential precipitation (>350 mm/day) was found to result in the high relative risk for bacillary dysentery and enterovirus infections when compared to ordinary precipitation (<130 mm/day). Yet, for vector-borne diseases, the relative risk of dengue fever and

H.-J. Su (✉)
Department of Environmental and Occupational Health, College of Medicine,
National Cheng Kung University, Tainan, Taiwan
e-mail: hjsu@mail.ncku.edu.tw

M.-J. Chen • N.-T. Chen
Department of Environmental and Occupational Health, College of Medicine,
National Cheng Kung University, Tainan, Taiwan

National Environmental Health Research Center, National Institute of Environmental
Health Sciences, National Health Research Institutes, Miaoli, Taiwan

© Springer International Publishing Switzerland 2016   237
R. Akhtar (ed.), *Climate Change and Human Health Scenario in South
and Southeast Asia*, Advances in Asian Human-Environmental Research,
DOI 10.1007/978-3-319-23684-1_14

Japanese encephalitis increase with precipitation up to 350 mm/day. Differential lag effects of precipitation appear to be associated with varying risk levels for respective infectious diseases.

Changing patterns of temperature and precipitation have influenced geographical distribution as well as prevalence and incidence rate of climate-related diseases. Availability of medical resources or sanitary capacities is to be taken into account in further analyses before the final adaptation.

Keyword Extreme temperature • Extreme precipitation • Rural • Cardiovascular mortality • Respiratory mortality • Bacillary dysentery • Enterovirus infection • Dengue fever • Japanese encephalitis • Adaptation

14.1 Overview

Climate change has been demonstrated in many regional climate chaos such as frequent extremes including heat wave, temperature drop, precipitation tending to extremes, and uneven distribution (Kerr 2011). These phenomena significantly impact nature ecological system and then further very likely to endanger human survival environment in spreading disease incidences and prevalence even lifting mortality. Epidemic diseases are likely to change their infectious periods and distributions.

Taiwan has been observed the temperature rising 1.4 °C over the past century that is almost twofold of global surface warming about 0.74 °C (Chen 2008). In precipitation, the frequency of droughts and floods was dramatically increased for the past 50 years. Although the average precipitation has no trend to change, the hours of rainfall had significantly decreased trend; therefore, the average rainfall strength (i.e., rainfall amount per hour) has significant increase trend (Chu et al. 2012; Yang et al. 2011). Furthermore, extreme rainfalls regularly brought by typhoons in Taiwan have directly impact this region in hydrology, water resource, agriculture production, public health, ecological environment, etc. These climate extremes become one of the important factors for society and economic development.

Public health effects of global warming were classified as multilevel effects: primary, secondary, and tertiary (Butler and Harley 2010). The primary effect is the direct impact. For example, climate extremes such as heat wave, cold front, drought, flood, tornado, etc., could directly cause human injury and mortality and trigger diseases leading to sudden death. The secondary effect is indirect impact. For instance, climate change could alter disease-spreading routes, biomechanism, or chemistry to impact health indirectly. Global warming could enlarge the epidemic period and distribution of infectious diseases; warming could increase the concentration of air biological and chemical pollutant to enhance risk of human respiration diseases. The tertiary effect is society effect. These disasters dramatically change society, economics, and environment, which brought socioeconomic conflicts and mental stresses on victims.

14.2   The Temperature Variability, Extreme Temperature, and Human Health

Taiwan is located at the border of tropical and subtropical regions with humid and warm climate pattern. In addition, most of the territory of this island is a mountainous region with low population, and most of resident population occupies the lowland area with active commercial business; therefore, urban heating island effect is significant here. In the previous studies, events of extreme high temperatures linked to increased mortality (primary effect) are recognized as one critical challenge to the public health sector.

In Taiwan, higher cardiovascular or respiratory mortality appears to derive more from low temperature than does from the exposure to high temperature for the general population (Fig. 14.1a–d) (Wu et al. 2011). Rural residents are less affected than urban dwellers under extreme temperature events if cardiovascular or respiratory mortality is benchmarked. Significantly increased risk ratios of daily mortality were evident when daily mean heat indices were at and above the 95th percentile, when compared to the lowest percentile, in all cities (Sung et al. 2013). These risks tended to increase similarly among those aged 65 years and older (Wu et al. 2011). Being more vulnerable to heat stress is likely restricted to a shortterm effect, as suggested by lag models which showed that there was dominantly an association during the period of 0–3 days (Sung et al. 2013). In Taiwan, predicting city-specific daily mean heat indices may provide a useful early warning system for increased mortality risk, especially for the elderly. Regional differences in health vulnerabilities should be further examined in relation to the differential social-ecological systems that affect them.

In the other hand, warming is very likely to increase the risk on infectious diseases (secondary effect). Based on health surveillance in Taiwan, extreme high temperature is likely to enhance risks that last for a few weeks on infectious diseases such as dengue fever, scrub typhus, Japanese encephalitis, severe enterovirus, shigellosis, melioidosis, leptospira, etc. Dengue fever risk increases sevenfold under warming period (CEPD 2012). According to spatial analysis, once the average temperature raises 1 °C, 86 townships populated with 7,748,267 in Taiwan will be classified as highrisk level and 203 townships will raise risk to middlerisk level (Fig. 14.1e–f) (Wu et al. 2009; Wu et al. 2007). Temperature was indicated to affect the incidences of Japanese encephalitis, i.e., the incidence increases 19 % as the average temperature increases to 1 °C during the previous 2 months. However, children in Taiwan had vaccine program spouse by government to protect from Japanese encephalitis. As more and more population injected the vaccine, the risk of the diseases could be controlled in the near future. In addition, each 1 °C rise in temperature at previous 6 weeks corresponded to an increase of 1.33-fold risk in the weekly case number of scrub typhus. Consequently, global warming may significantly increase the risks on infectious diseases especially vector-borne in Taiwan; anti-epidemic measures are needed in the future (CEPD 2012).

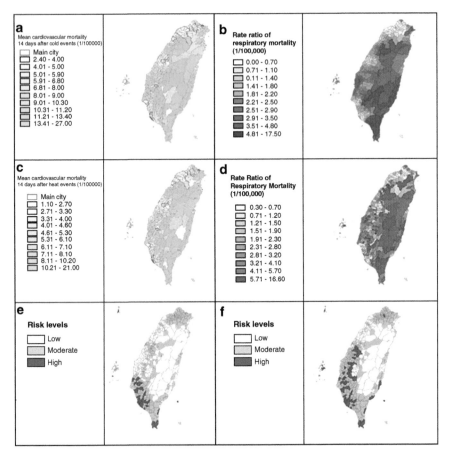

Fig. 14.1 (**a**) The cardiovascular mortality ratio 14 days before and after cold surges, 1993–2004. (**b**) The respiratory mortality ratio 14 days before and after cold surges, 1993–2004. (**c**) The cardiovascular ratio 14 days before and after heat event, 1993–2004. (**d**) The respiratory mortality ratio 14 days before and after heat event, 1993–2004. (**e**) The risk map of dengue in Taiwan, 1998–2006. (**f**) The risk map of dengue in Taiwan with monthly average temperature raises of 1 °C

14.3   Extreme Precipitation and Human Health

Taiwan is located at the typhoon-prone region with mountains crossing the island; torrential rainfalls frequently happen in this region. The torrential precipitation events regularly accompany natural hazards such as flooding and landslides especially in geological vulnerable regions (Su et al. 2011), which may cause more accidental deaths (primary effect).

The incidence of extreme precipitation has increased with the exacerbation of worldwide climate disruption. The torrential rainfalls not only cost huge losses in society but also threaten human health such as lower extremity cellulitis (secondary effect). In 2005, typhoon (Haitang) brought torrential rainfall in southern Taiwan and caused severe flooding. Within the following two weeks, the risks of lower extremity cellulitis incidences on this region were twofold higher than in the regular time.

Daily precipitation levels were significantly correlated with all 8 mandatory-notified infectious diseases in Taiwan (secondary effect). For waterborne infections, extreme torrential precipitation (>350 mm/day) was found to result in the highest relative risk for bacillary dysentery and enterovirus infections when compared to the ordinary rain (<130 mm/day) (Chen et al. 2012). Yet, for vector-borne diseases, the relative risk of dengue fever and Japanese encephalitis increased with greater precipitation only up to 350 mm. Differential lag effects following precipitation were statistically associated with increased risk for contracting individual infectious diseases (Chen et al. 2012).

14.4   Socioeconomic Conflicts and Mental Stress (Tertiary Effects)

The consequences of environmental changes such as reducing air quality, ecological biodiversity, and water resources due to climate change are not only very likely to endanger human's health but also to impact our socioeconomic status even to conflict international relationship. When people face the shortage of human basic needs such as sunlight, air, and water, it does not only enhance mental stress but also possibly damage economics, leading a fragile society. International conflicts become more frequent than before due to fighting for freshwater resources and farmland (Scheffran et al. 2012). For example, India and Pakistan have the issue on water right of Indus River in South Asia. Meanwhile, Bangladesh, sharing 54 rivers with India, often raises regional tension for the water right of Ganges River (Uprety and Salman 2011). In addition, the climate-related disasters would bring mental stress. A survey in Taiwan was conducted with 271 students of junior high schools in mountainous regions of southern Taiwan which were worst affected by Typhoon Morakot. Results indicated 26.9 % students had a diagnosis of posttraumatic stress disorder (PTSD) and 24.0 % had PTSD related to Typhoon Morakot (Yen et al. 2011). As climate change increases the frequency of climate extremes, people has unavoidable lessons to deal with these crises especially integrating with multiple factors such as natural barrier of topography, crop growth, and political power.

14.5   Conclusion

Climate change, especially for extreme weather was acknowledged as a global public health problem. In Taiwan, the increasing temperature and extreme precipitation affected the infectious disease, noncommunicable disease, and mental health via diverse pathway. Although the impacts of climate change have the potential to affect the human health in Taiwan and around the world, the national and local adaptations and actions from the multiple threats of a rapidly changing climate may reduce the health impacts such as increasing public health capacity and improving health literacy, promoting institutional learning, developing tools to facilitate, or embracing adaptive management for certain climate health threats.

References

Butler C, Harley D (2010) Primary, secondary and tertiary effects of ecoclimatic change: the medical response. Postgrad Med J 86:230–234
CEPD (2012) Adaptation strategy to climate change in Taiwan. Coincil for Economic Planning and Development, Taipei
CHEN YL (2008) Observation of climate change in the past century in Taiwan. Sci Dev 424 (in Chinese)
Chen MJ, Lin CY, Wu YT, Wu PC, Lung SC, Su HJ (2012) Effects of extreme precipitation to the distribution of infectious diseases in Taiwan, 1994–2008. PLoS One 7, e34651
Chu HJ, Pan TY, Liou JJ (2012) Changepoint detection of longduration extreme precipitation and the effect on hydrologic design: a case study of south Taiwan. Stoch Env Res Risk A 26:1123–1130
Kerr RA (2011) Climate change. Humans are driving extreme weather; time to prepare. Science 334:1040
Scheffran J, Brzoska M, Kominek J, Link PM, Schilling J (2012) Climate change and violent conflict. Science 336:869–871
Su HJ, Chen MJ, Wang JT (2011) Developing a water literacy. Curr Opin Environ Sustain 3:517–519
Sung TI, Wu PC, Lung SC, Lin CY, Chen MJ, Su HJ (2013) Relationship between heat index and mortality of 6 major cities in Taiwan. Sci Total Environ 442:275–281
Uprety K, Salman MA (2011) Legal aspects of sharing and management of transboundary waters in South Asia: preventing conflicts and promoting cooperation. Hydrol Sci J 56(4):641–661
Wu PC, Guo HR, Lung SC, Lin CY, Su HJ (2007) Weather as an effective predictor for occurrence of dengue fever in Taiwan. Acta Trop 103:50–57
Wu PC, Lay JG, Guo HR, Lin CY, Lung SC, Su HJ (2009) Higher temperature and urbanization affect the spatial patterns of dengue fever transmission in subtropical Taiwan. Sci Total Environ 407:2224–2233
Wu PC, Lin CY, Lung SC, Guo HR, Chou CH, Su HJ (2011) Cardiovascular mortality during heat and cold events: determinants of regional vulnerability in Taiwan. Occup Environ Med 68:525–530
Yang TC, Yu PS, Wei CM, Chen ST (2011) Projection of climate change for daily precipitation: a case study in ShihMen reservoir catchment in Taiwan. Hydrol Process 25:1342–1354
Yen CF, Tang TC, Yang P, Chen CS, Cheng CP, Yang RC, Huang MS, Jong YJ, Yu HS (2011) A multidimensional anxiety assessment of adolescents after Typhoon Morakotassociated mudslides. J Anxiety Disord 25:106–111

Chapter 15
Climate Change and Health: The Malaysia Scenario

Mohammed A. Alhoot, Wen Ting Tong, Wah Yun Low, and Shamala Devi Sekaran

Abstract Malaysia, located in Southeast Asia, has a tropical climate and abundant rainfall (2000–4000 mm annually). Malaysia faces weather-related disasters such as floods, landslides and tropical storm attributed to the cyclical monsoon seasons, which cause heavy and regular downpours. Storms, floods and droughts lead to the rise and emergence of climate-sensitive diseases due to contamination of water, environment and creation of breeding sites for disease-carrying vector mosquitoes. The effect of climate change on health is an area of substantial concern in Malaysia. This chapter examines the trends of six prominent climate-sensitive diseases in Malaysia, namely, cholera, typhoid, hepatitis A, malaria, dengue and chikungunya followed by the environmental and public health policies, programmes and plans to battle the impact on health from the climate change.

Keywords Climate change • Food- and water-borne diseases • Vector-borne diseases • Environmental health • Public health policies • Malaysia

M.A. Alhoot
Medical Microbiology Unit, International Medical School, Management & Science University, Shah Alam 40100, Selangor, Malaysia

Department of Medical Microbiology, Faculty of Medicine, University of Malaya, Lembah Pantai, 50603 Kuala Lumpur, Malaysia

W.T. Tong
Department of Primary Care Medicine, Faculty of Medicine, University of Malaya, Lembah Pantai, 50603 Kuala Lumpur, Malaysia

W.Y. Low (✉)
Dean's Office, Faculty of Medicine, University of Malaya, Lembah Pantai, 50603 Kuala Lumpur, Malaysia
e-mail: lowwy@um.edu.my

S.D. Sekaran
Department of Medical Microbiology, Faculty of Medicine, University of Malaya, Lembah Pantai, 50603 Kuala Lumpur, Malaysia

© Springer International Publishing Switzerland 2016
R. Akhtar (ed.), *Climate Change and Human Health Scenario in South and Southeast Asia*, Advances in Asian Human-Environmental Research, DOI 10.1007/978-3-319-23684-1_15

15.1   Introduction

The world is experiencing climate change. The earth is warming due to greenhouse gasses emission resulting from human activity. Temperature rises contribute to melting of glaciers, increasing average sea levels and changing of precipitation patterns. Extreme weather events are occurring at increasing frequencies and intensities. This can have serious impacts on the environment and human health, as essential necessities for good health such as clean air and water, adequate food and shelter will be affected. Increased temperatures and humidity and shifting rainfall patterns lead to weather-related catastrophes, namely, heat waves, floods, droughts and storms which cause loss of agriculture and homes. These climate events also increase health risks such as climate-sensitive vector-borne and food- and-water-borne diarrhoeal diseases, which currently account for mortality of 1.1 million people and 2.2 million people a year, respectively (WHO 2008). Increases in population numbers and current trends in energy use and development further aggravate the effects of climate change (WHO 2009). The Asia-Pacific region holds the highest number of countries most vulnerable to climate change, namely, Bangladesh, India, Nepal, the Philippines, Afghanistan and Myanmar (Asian Development Bank and International Food Policy Research Institute 2009). Malaysia is also vulnerable to the adverse impacts of climate change and is no exception.

15.2   Demography of Malaysia

Malaysia, located in the Asia-Pacific region, has a total land area of 329,733 km². Malaysia is composed of Peninsular Malaysia and East Malaysia (Sabah and Sarawak) that is separated by the South China Sea (Fig. 15.1). Peninsular Malaysia, in the west, has an area of 131,573 km², while East Malaysia covers about 198,068 km² (Ministry of Science Technology and the Environment Malaysia 2000). Peninsular Malaysia consists of 11 states (Johor, Kedah, Kelantan, Melaka, Negeri Sembilan, Pahang, Penang, Perak, Perlis, Selangor, Terengganu) and 2 federal territories (Wilayah Persekutuan and Putrajaya), while East Malaysia consists of 2 states (Sabah and Sarawak) and Federal Territory of Labuan. Kuala Lumpur is the federal capital of Malaysia.

Malaysia's population comprises of 28.33 million people, including 8.2 % non-citizens. The majority (22.56 million people) reside in Peninsular Malaysia while East Malaysia has a population of 5.77 million people. The population density is 86 inhabitants per square kilometre. The average annual population growth rate (year 2000–2010) is 2.0 % (Department of Statistics Malaysia 2010). The proportion of Malaysia's population by the age of 0–14 years is 27.63 %, 15–64 years is 67.33 % and above 64 years is 5.04 %. While Malaysia has a relatively young population, however, the age structure is moving towards an ageing population (Department of Statistics Malaysia 2010) (Fig. 15.2).

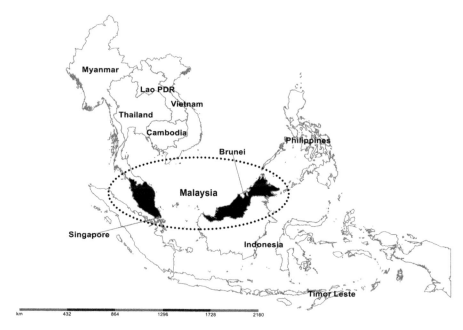

Fig. 15.1 Map of Malaysia

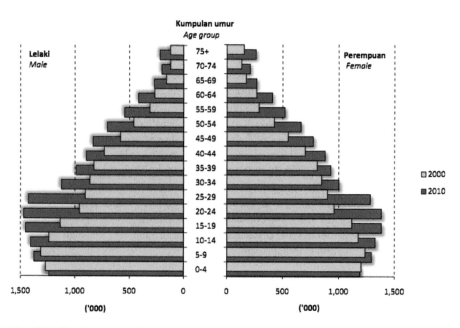

Fig. 15.2 Number of population by sex and age group, Malaysia, 2000 and 2010 (Source: Department of Statistics Malaysia 2010, Population and Housing Census of Malaysia 2010)

Malaysia is a multi-cultural country, with a majority being Bumiputera (67.4 %), followed by 24.6 % Chinese, 7.3 % Indians and 0.7 % other races. East Malaysia is distinct from the Peninsular Malaysia in terms of ethnic composition of its people with a diverse of indigenous residents, which also are part of the Bumiputera group. The largest indigenous group in Sabah is the Kadazan/Dusun ethnic group, which is made up of 24.5 % of the population, while in Sarawak, the largest ethnic group is the Ibans which accounts for 30.3 % of the total citizens in Sarawak (Department of Statistics Malaysia 2010).

Malaysia is a country that is rapidly advancing and has made tremendous progress in socio-economic development since its independence in 1957. Initially, agriculture and mining were the main economic drivers for the country during the post-independence period, but now, Malaysia's economic industry exhibits a multi-sectorial diversification including rubber and oil palm processing and manufacturing, light manufacturing, pharmaceuticals, medical technology, electronics, tin mining and smelting, logging, timber processing, agriculture processing and petroleum production and refining. From 2005 to 2013, Malaysia's gross domestic product (GDP) saw a growth of 4.7 % (Department of Statistics Malaysia 2014a), with a total income value of RM986.7 billion (USD 274.7 billion) in 2013 (Department of Statistics Malaysia 2014b). The World Bank places Malaysia as an upper middle-income country (The World Bank 2014). The level of urbanization is parallel with the country's rapid development whereby all states except Kelantan have an urban population of more than 50 % (Department of Statistics Malaysia 2010).

Malaysia lies at the coordinates of 2°30′N 112°30E which is near the equatorial line. It has a tropical climate with a relatively uniform temperature all year round at 26–28 °C (Ministry of Natural Resources and Environment Malaysia 2011) and abundant rainfall of 2,000–4,000 mm annually (Ministry of Natural Resources and Environment Malaysia 2011). The distribution of rainfall is largely determined by the wind flow patterns and topographic features of the land. The months of November to March experience the Northeast monsoon which carries winds with speeds of 15–50 km/h that bring heavy downpours on some areas along the east coast of Peninsular Malaysia, western Sarawak and the northeast coast of Sabah. From May to August, the west coastal areas of Peninsular Malaysia are affected by the early morning 'Sumatras' during the southwest monsoon. The inland areas covered by mountain ranges are relatively sheltered from monsoon rain influences and receive rainfall of less than 1,780 mm per year. There are also two brief inter-monsoon seasons which bring rain and thunderstorms due to convection currents (Ministry of Science Technology and the Environment Malaysia 2000).

The common weather-related disasters that occur in Malaysia are such as floods, landslides, mudslide and tropical storm, and most of these disasters are attributed to heavy rainfalls (Ibrahim Mohamed Shaluf and Fakhru'l-Razi Ahmadun 2006). Malaysia regularly (almost annually) experiences the occurrence of floods due to the cyclical monsoon seasons, which causes heavy and regular downpours usually in October to March. However, in December 2006, the anomalous event of floods in

Johor Bahru (Southern Peninsular Malaysia), which is located out of the monsoon affected zone, suggests that it could be due to the effects of global warming (Chang 2010).

15.3 Climate Change Trend in Malaysia

Malaysia has recorded a rise in the average temperature since 1951 (Ministry of Science Technology and the Environment Malaysia 2000). Analysis of different regional annual mean temperatures records in various parts of Malaysia indicated warming trends. Generally, the overall correlation analysis of annual mean temperatures over the low land areas has the largest trend over Southern and Central Peninsular Malaysia while the lowest trend is in Sarawak. The rates of increase are about 1.5–2.7 °C per 100 years. The only highland area, Cameron Highland has relatively weak positive correlation as shown in Fig. 15.3. There is strong evidence to link the local warming trends to urbanization process as for Southern and Central Peninsula and also global trends although there is insufficient length of climate data.

The latest data of annual rainfall trend from years of 1951 till year 2013 for different regions in Malaysia shows no overall clear trend of rainfall for the country unless high variability of rainfall probably due to tropical climate. The spatial variability of rainfall lacks the regularity that is generally found with temperature. Therefore, the standardised rainfall anomaly average is used to analyse the rainfall pattern for Malaysia as shown in Fig. 15.4 (Malaysian Meteorological Department (MetMalaysia) 2014). Increase in average temperature can cause rainfall fluctuations that may result in variation of climatic events such as floods, droughts and storms.

These climatic events can have a negative impact on Malaysia's key economic sectors such as agriculture, forestry, water resources, coastal resources, energy and public health. Floods, droughts and storms increase the risk of health diseases due to contamination of water and creation of breeding sites for disease-carrying vector mosquitoes. Environmental hygiene, sanitation and water supply contributed to the spread of these diseases. The effect of climate change on health is an area of substantial concern in Malaysia (Commonwealth Secretariat 2009) particularly with regard to the emergence and rise of climate-sensitive diseases such as food- and waterborne- and vector-borne diseases. These climate-sensitive diseases in Malaysia are much influenced by the climate change variations and Malaysia has developed public health policies of which some were successful while others failed in battling them. This chapter describes the trends of six prominent climate-sensitive diseases in Malaysia, namely, cholera, typhoid, hepatitis A, malaria, dengue and chikungunya followed by the public health policies, programmes and plans to battle the impact on health from the climate change.

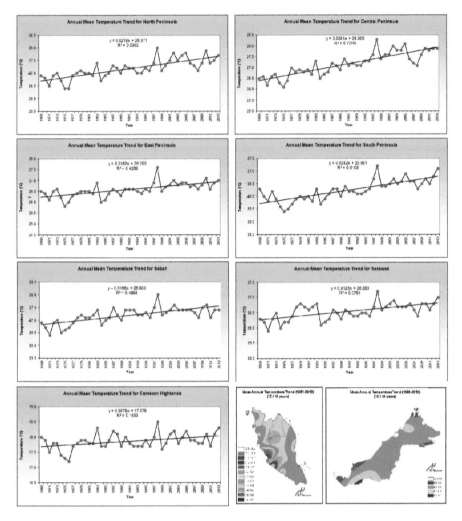

Fig. 15.3 Annual temperature trend for Malaysia 1969 to 2013 (Source: Malaysian Meteorological Department (MetMalaysia) 2014)

15.4 Effects of Climate Changes on Diseases Trends

Before the discovery of the role of infectious agents in the transmission of infectious diseases in the late nineteenth century, human beings have known that climatic conditions affect epidemic diseases. Roman aristocrats retreated to hill resorts each summer to avoid malaria. South Asians learnt early that, in high summer, strongly spiced foods were less likely to cause diarrhoea. Climate change is likely to increase the severity of the weather in which some areas will experience an increase in temperatures, rainfall and flood risk, while other midlatitude areas will experience more

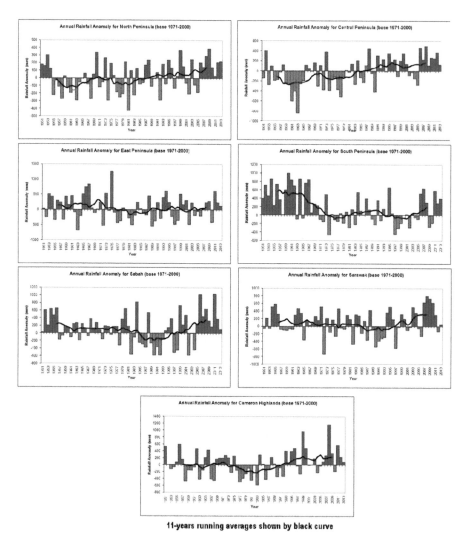

11-years running averages shown by black curve

Fig. 15.4 Standardised rainfall anomaly average for Malaysia 1961 to 2013 (Source: Malaysian Meteorological Department (MetMalaysia) 2014)

droughts. Climate change accelerates the spread of several diseases mainly due to warmer global temperatures resulting in an increase of global average rainfall. Increased precipitation caused by global warming may increase flooding in some areas, which could lead to contamination of drinking water with the possibility of a rise of waterborne disease incidence. Furthermore, rainfall can promote transmission of vector-borne diseases by creating ground pools and other breeding sites for the vectors. Climate-sensitive diseases are among the largest causes of morbidity and mortality globally, including diarrhoea, malaria and other water- and vector-borne infections.

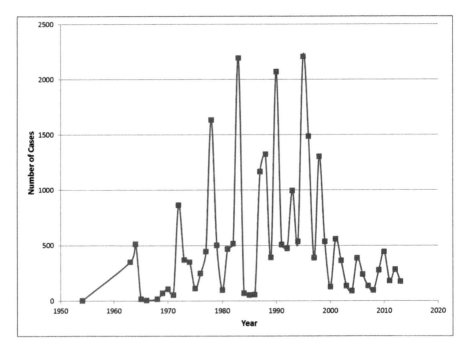

Fig. 15.5 Changes in reported cholera cases in Malaysia from 1954 to 2013 (Sources: WHO (2014). Global Health Observatory (GHO), Ministry of Health Malaysia (2014), Health Facts (2014))

15.5   Food- and Waterborne Diseases

In Malaysia, typhoid fever, cholera, dysentery and hepatitis A are the major waterborne diseases with incidence rates of 0.73, 0.58, 0.28 and 0.41 per 100,000 population (Ministry of Health Malaysia 2014). The spread of these climate-sensitive diseases has been attributed to water crises such as floods and droughts (due to contaminated water) and coupled with improper sewage disposal, personal hygiene and low environmental sanitation standards. In the last century, diarrhoeal diseases such as cholera, typhoid and dysentery were very common in Malaysia due to lack of infrastructure such as inadequate clean water supply and latrines coupled with poor sanitation and hygiene (Engku Azman 2011; Rozlan Ishak 2007) especially in the rural and suburban areas. These infectious diseases are predominantly spread through contaminated water.

   To date, these water-borne diseases remain endemic although there has been a significant decline in the last two decades as shown in Figs. 15.5, 15.6 and 15.7. In 2009, the reported cases showed a low incidence of typhoid, cholera and hepatitis A virus (HAV): typhoid cases, 303; cholera cases, 276; and HAV cases, 40. These diseases usually occur as sporadic outbreaks and the incidence by state is closely associated with the coverage of clean water supply. For example, the incidence of typhoid is highest among the states with the lowest coverage of treated water supply as shown in Fig. 15.8.

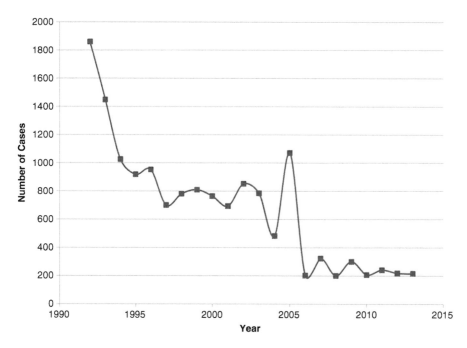

Fig. 15.6 Number of reported cases of typhoid fever in Malaysia from 1992 to 2013 (Sources: Ministry of Health Malaysia (2010). Number of notifiable communicable diseases in Malaysia for 1998–2009, Ministry of Health Malaysia (1992–1999). Annual reports. Ministry of Health Malaysia (2007–2014). Health Facts (2006–2014))

The association between drought and cholera epidemics has been reported in Malaysia (Chen 1970; Khoo 1994; Benjamen 2006; Kin 2007) as the outbreaks have tended to occur in the dry season (May to July) when many are forced to use river water. Thus, the severe drought is considered as one of the main risk factor of cholera epidemics because it affects the water supply. The main affected areas are the coastal and riverine communities especially those with poor environmental sanitation, poor water supply, poor waste disposal and inadequate personal hygiene (Chen 1970; Singh 1972; Yadav and Chee 1990; Khoo 1994; Kin 2007; Benjamen 2006).

15.6 Vector-Borne Diseases

15.6.1 Malaria

Malaria is caused by one of the four species of **Plasmodium** parasite and transmitted by the **Anopheles** mosquito that exhibits definite seasonal prevalence. Although malaria remains the most common vector-borne disease in the world, the number of malaria cases in Malaysia has declined in the last decade. Public health data showed that the number of malaria cases detected in the country has dropped significantly

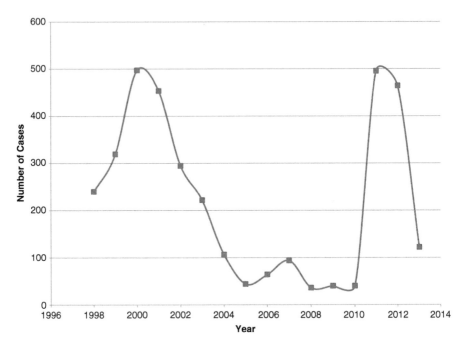

Fig. 15.7 Number of reported cases of hepatitis A in Malaysia from 1998 to 2013 (Sources: Ministry of Health Malaysia (2010). Number of notifiable communicable diseases in Malaysia for 1998–2009, Ministry of Health Malaysia (2007–2014), Health Facts (2006–2014))

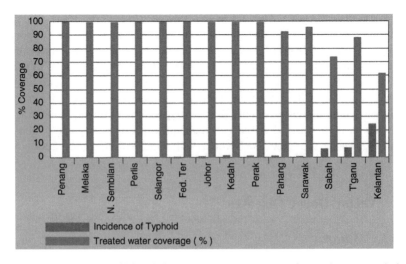

Fig. 15.8 Incidence of typhoid in relation to percentage coverage of treated water supply in the different states in Malaysia (Source: Ministry of Natural Resources and Environment Malaysia (2011). Second National Communication to the United Nations Framework Convention on Climate Change (UNFCCC))

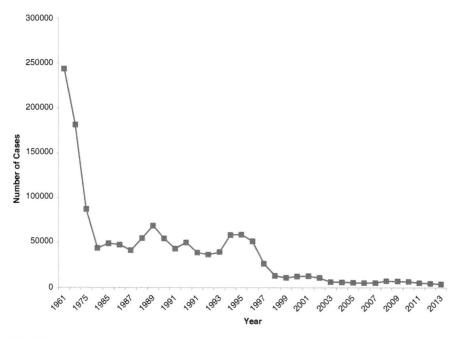

Fig. 15.9 Changes in reported malaria cases in Malaysia from 1961 to 2013 (Sources: Tee (2000). Malaria control in Malaysia, Ministry of Health Malaysia (2010). Number of notifiable communicable diseases in Malaysia for 1998–2009. WHO (2011). World Malaria Report 2011, WHO (2014). Global Health Observatory (GHO), Ministry of Health Malaysia (2007–2014). Health Facts (2006–2014))

from 243,870 in 1961 to 3,851 in 2013 as shown in Fig. 15.9. This decline resulted from effective control measures that applied in the country since 1961 with the opportunities to shift the current control programme into a total elimination programme. Figure 15.10 shows the progressive reduction of malaria incidence in Peninsular Malaysia.

Changes in climate factors such as temperature, rainfall and humidity affect the reproduction, development, behaviour and population dynamics of malaria (Gage et al. 2008). Increased temperature induces the multiplication rate and development of the malaria parasite in the mosquito vector (Ambu et al. 2002). Both, the mosquito vector and malaria parasite are sensitive to changes in temperature and the elevation of ambient temperature of 2 and 4 °C due to climate change will lead to increase in the capacity of the mosquito vectors to transmit malaria by 20 to 30 %, respectively. Thus, the number of malarial cases can be estimated to increase by 15 % if an expected increase of 1.5 °C in ambient temperature in 2050 is observed (Ambu et al. 2002). Furthermore, increased rainfall leads to increase in the number and quality of breeding sites for mosquito's vectors and the increased humidity enhances survival and vectorial capacity of the vectors and hereafter the transmission of the parasites (Ambu et al. 2002). Increases in temperature and rainfall would undoubtedly allow mosquito vectors to survive in areas closely surrounding their current dissemination

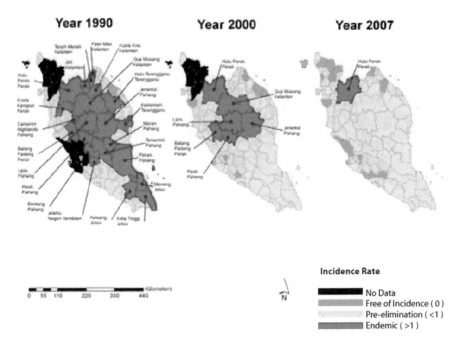

Fig. 15.10 Changes in malaria incidence rate (per 1,000 population) in districts of Peninsular Malaysia according to WHO classification (Source: Ministry of Natural Resources and Environment Malaysia (2011). Second National Communication to the United Nations Framework Convention on Climate Change (UNFCCC))

limits. How far these areas are extended would depend on the magnitude of the change in climate. Also, salt-water intrusion into freshwater of coastal areas can extend breeding sites for malaria vectors and enhance transmission of the disease such as in case of ***Anopheles sundaicus*** which breeds in brackish water in the coastal areas of Peninsular Malaysia (Ambu et al. 2002; Mia et al. 2011). These changes and the different climate factors resulted in slight increase in the recorded cases of malaria in the last 5 years as shown in Fig. 15.9.

There is a variation in the distribution of the infection between the different districts of the country, for example, in 2006, 57.2 % of the cases were from Sabah, 26.7 % from Sarawak and the remaining 16.1 % from Peninsular Malaysia (Ministry of Natural Resources and Environment Malaysia 2011). In areas where malaria is seasonal, a dramatic increase in disease prevalence occurs in all age groups during the annual transmission season because previously acquired immunity in the host is lost in the non-transmission season. Potential extension or curbing of the vector breeding season may lead to shifts in malaria incidence and prevalence. Similarly, climate changes can result in increasing the incidence of malaria infection in endemic areas due to the improvement of the vector-breeding conditions.

15.6.2   *Dengue Fever (DF) and Dengue Haemorrhagic Fever (DHF)*

In terms of human morbidity and mortality, dengue has become an important public health problem and the most important arthropod-borne viral diseases in Malaysia. Dengue is transmitted by the *Aedes* mosquitoes (**A. aegypti** and **A. albopictus**), which are highly sensitive to environmental conditions. Temperature, precipitation and humidity are critical to survival, reproduction and development of the mosquito and these can influence mosquito presence and intensity. Furthermore, warmer temperatures can enhance the disease transmission possibly by allowing the mosquito vector to reach maturity much faster than in colder climate. Also these environmental conditions can then shorten the extrinsic incubation period of dengue virus by reducing the time required for the virus to replicate and disseminate in the mosquito. The extrinsic incubation period is necessary for virus to reach the mosquito salivary glands and transmit to humans (Githeko et al. 2000).

Dengue affects tropical and subtropical regions around the world, mainly in urban and semi urban areas. Climate change might expand the distribution of vector-borne pathogens in both time and space, thereby exposing host populations to longer transmission seasons and immunologically naive populations to newly introduced pathogens (Patz and Reisen 2001). It is possible to accelerate the dissemination of dengue viruses worldwide. Like many vector-borne diseases, dengue fever shows a clear weather-related pattern: rainfall and temperatures affect both the spread of mosquito vectors and the likelihood that they will transmit virus from one human to another. In a cool climate, the virus takes so long to replicate inside the mosquito that most likely would die before it actually has a chance to transmit the virus to another person (Resurgence 2008). Several studies have predicted that global climate change could increase the likelihood of dengue epidemics. One of these studies published an empirical model of worldwide dengue distribution. In this model they use the annual average vapour pressure (a measure of humidity) as a single climate factor to predicted future dengue fever distribution. If humidity were to remain at 1990 levels into the next century, a projected 3.5 billion people would be at risk of dengue infection in 2085, but assuming humidity increases as projected by the Intergovernmental Panel on Climate Change, the authors estimate that in fact 5.2 billion could be at risk (Hales et al. 2002). Other reports have showed correlations between dengue infection and climate variables such as El Niño, temperature, rainfall and cloud cover concluding that climate change could increase the number of people at risk of dengue infection (Haines et al. 2006).

The number of reported dengue fever (DF) and dengue haemorrhagic fever (DHF) cases in Malaysia shows an increasing trend as shown in Fig. 15.7. The recorded number of infected population shows an upward trend from 6,621 cases with incidence rate of 36.4/100,000 population in 1991 to 46,171 case with incidence rate of 148.73/100,000 population in 2010 (Fig. 15.11). This incidence rate exceeds the national target for DF and DHF, which is less than 50 cases/100,000 population. Since 2000, dengue incidence continues to increase unabated. It is

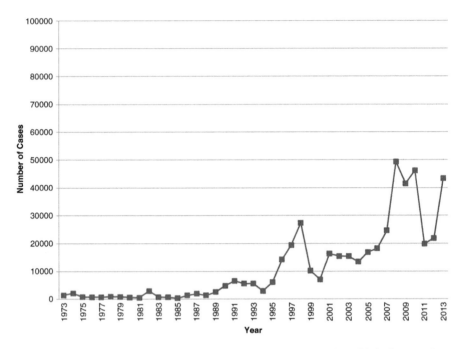

Fig. 15.11 Changes in reported dengue cases in Malaysia from 1973 to 2013 (Sources: Lam (1993). Two Decades of Dengue in Malaysia, Ministry of Health Malaysia (2010). Number of notifiable communicable diseases in Malaysia for 1998–2009. Ang and Satwant (2001). Epidemiology and New Initiatives in the Prevention and Control of Dengue in Malaysia, Director General of Health Malaysia (2011b). Press statement: Current situation of Dengue Fever in Malaysia for week 52/2011(25–31 December 2011), Director General of Health Malaysia (2012a). Current situation of Dengue Fever in Malaysia for week 52/2012, Director General of Health Malaysia (2012b). Press statement: Increased of Mortality of Dengue Cases in Malaysia, Director General of Health Malaysia (2013). Current situation of Dengue Fever in Malaysia for week 52/2013)

prevalent throughout the country with the highest incidence among the most developed and densely populated territories and states. Clinical and laboratory confirmed cases showed that all age groups are affected, with the most vulnerable among school going children and young adults (Mia et al. 2013).

The recorded climate changes in Malaysia showed the increasing rainfall time per year and the ambient temperature provides optimal conditions for mosquitoes to multiply and then spread dengue viruses. *Aedes* mosquito prefers to lay its eggs around human habitation in stagnant clear water in artificial containers rather than natural areas. Even when the containers dry up, the eggs can withstand dryness for about nine months. When exposed to favourable conditions of water and food, the eggs can hatch within a day and emerge as adults within a week. Mosquitoes normally acquire the virus when sucking infected blood from an infected human and it then transmits the virus to others. *Aedes* mosquito has the capacity to sustain the virus in the environment through the transovarial transmission of the virus for up to five generations. Thus, it is generally accepted that changes in precipitation, ambient temperature and humidity may influence the abundance and distribution of the mosquito vectors and by this means increase the ability to spread the infection. A previous study in Malaysia has shown a positive relationship between rainfall and dengue (Li et al. 1985; Hopp and Foley 2003). The study conducted in Jinjang, Selangor, a dengue-prone area found that there was a 120 % increase in the number of dengue cases when the monthly rainfall was 300 mm or more. This may be due to creation of more *A. aegypti* breeding sites due to the rainfall. On the other hand, the extremely heavy rainfall may flush mosquito larvae away or kill them altogether (Promprou et al. 2005). Preliminary findings of an on-going study to develop a climate model for dengue showed that the mean and minimum temperatures were positively associated with the *Aedes* population in Kulim, Kedah. As the minimum temperature increased, the larvae densities also increased leading to increase in the endemicity of the disease (Patz et al. 1998).

15.6.3   Chikungunya

Chikungunya is a reemerging mosquito-borne viral infection. It is caused by a mosquito-borne togavirus belonging to the genus ***Alphavirus***. Many countries neighbouring Malaysia have reported human infections with chikungunya virus. Evidence of presence of chikungunya in Peninsular Malaysia was proved by serological survey of 4384 specimens collected between 1965 and 1969. These results showed that human infection with chikungunya virus appears to be at a low level of activity but is widespread and it is more prevalent in the northern part of the country (Marchette et al. 1980). However, although the serological evidence was present in Malaysia, chikungunya virus has not been known to be associated with clinical illness in the country before 1998. Malaysia experienced the first outbreak of chikungunya when 51 infected subjects were reported in Klang-Selangor between December 1998 and February 1999 (Lam et al. 2001). The majority of the cases were adults, and the clinical presentation was similar to classical chikungunya infections. Chikungunya is among those infectious diseases that extended their occurrence and range as temperatures increase. During the last decade, chikungunya virus has reemerged and caused new epidemics in Malaysia. This emergence may be attributed to the impact of climate change, but also several other factors may be involved such as urbanization, commercial transportation, workers movement, increasing human travel and vector density increase. However, local climatic fluctuations may have exerted a transient impact on chikungunya virus epidemics.

In the last six years, Malaysia has had three outbreaks of chikungunya virus infection. The first two occurred in Perak in 2006, and the third began in Johor in early 2008 until 2009 (Fig. 15.12). Since the reemergence of chikungunya at the end of 2006, Malaysia has been experiencing increasing number of chikungunya cases involving about 10,942 recorded cases until the end of 2010 as shown in Fig. 15.8 (Director General of Health Malaysia 2011a). In the earlier outbreaks, local

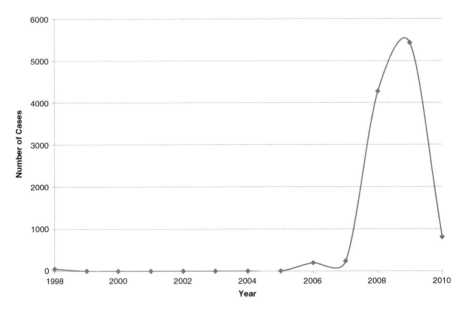

Fig. 15.12 Changes in reported chikungunya cases in Malaysia from 1998 to 2010 (Sources: Lam et al. (2001). Chikungunya infection–an emerging disease in Malaysia, Chua (2010). Epidemiology of Chikungunya in Malaysia: 2006–2009. Director General of Health Malaysia (2011a). Press statement: Current situation of Dengue Fever and Chikungunya in Malaysia for week 52/2010(26/12/2010-31/12/2011))

transmission of the disease has been curtailed, and the disease has not become endemic. Dissimilar to Africa where the primate hosts are the primary reservoirs of the virus, in Malaysia there are no such animal hosts but the necessary mosquito vectors, *A. aegypti* and *A. albopictus*, are present abundantly. That means if a sufficient number of human hosts continue to transmit and perpetuate the virus as is the case for the dengue viruses, chikungunya infection can become endemic. Similar to dengue virus, climate changes in Malaysia such as increasing the ambient temperatures, precipitation and humidity will increase the intensity and activity of the mosquito's vector to transmit the virus (Meason and Paterson 2014; Fischer et al. 2013; Rezza 2008).

In summary, climate changes could play an important role in the endemicity of the common communicable diseases such as dengue, chikungunya, malaria and other diseases that are sensitive to climate and are endemic in Malaysia. The health pattern will be affected due to the climate change such as exposure to high or low temperature, level and duration of raining, contamination of water sources and abundances of the disease-transmitting vectors. Changes in various vector-borne infectious diseases may lead to the expansion of the endemic areas of certain diseases such as malaria and dengue. However, these changes will not affect only the incidence of infectious diseases, but also other risks will arise in both developing and developed countries such as that due to floods and drought. Thus, the risks associated with climate changes will increase over time. Public health policies and programmes will then need to be implemented to handle these climate-sensitive diseases that affect the health.

15.7 Programmes and Public Health Policies on Climate Change and Health

Malaysia's endeavour in addressing climate change is evident since 1970s. The nation's development plans, policies and laws which were formed are related to climate change scenarios so that sustainable development can be achieved. This principle is incorporated in the country development plan, namely, Third Malaysia Plan (1976–1980), the Ninth Malaysia Plan (2001–2005) Vision 2020 and the Third Outline Perspective Plan (OPP-3) (2001–2010). Various environmental-related policies, legislations and plans are formed so that the national developments works are carried out within the environmental standards to ensure the nation's development and environment balance one another. Among them are the Third National Agriculture Policy (1998–2010), National Energy Policy (1979), Fuel Diversification Policy (1981), Environment Quality Act 1974, EQ (Clear Air) Regulation 1978, EQ (Prescribed Activities) (EIA) Order 1987, National Forestry Act 1984, Fisheries Act 1985, Fisheries Maritime Regulations, 1967, Petroleum Mining Act 1986, Petroleum Development Act 1974, Land Conservation Act 1960, National Parks Act 1980 and many others.

Malaysia adopts the 'precautionary principle' or the 'no regret' policy which allows measures to be taken to adapt or mitigate effects of climate change even though it is unsure if the climate change effects will come true (Ministry of Natural Resources and Environment Malaysia 2011). The Malaysian government is receptive towards foreign support on the use of cleaner and environmentally sound technology and encourages other sectors' involvement in such programmes to increase their awareness on global activities in improving the quality of the environment (Lu 2006). Many national programmes and projects were conducted to reduce or mitigate greenhouse gas emissions such as Malaysian Industrial Energy Efficiency Improvement Project (MIEEIP), Clean Development Mechanism (CDM), Malaysia Building Integrated Photovoltaic, etc. and studies have been actively carried out to assess climate and environment interactions and impact of climate change such as National Coastal Vulnerability Index Study and National Study for Effective Implementation of Integrated Water Resources Management (IWRM) in Malaysia, to name a few (WHO 2006). In recent times, two policies were formulated to address climate change more holistically, specifically the National Policy on Climate Change and the National Green Technology Policy to cater for a climate-resilient development, to promote a low carbon economy and to promote green technology (Ministry of Natural Resources and Environment Malaysia 2011).

Malaysia's commitments in addressing climate change also can be seen from its active participation in the national and international arena on multilateral dialogues and collaborations, information sharing and networking. In 1994, the National Steering Committee on Climate Change (NSCCC) was set up and comprised of members from various ministries in the government such as Ministry of Science, Technology and Environment, Ministry of Finance, Ministry of Energy, Communications and Multimedia, Malaysian Meteorological Service and many others which function to formulate and implement policies to address and adapt climate change. Also in the same year, Malaysia ratified a signatory agreement with the United Nations Framework Convention on Climate Change (UNFCC) and its commitment includes preparation of a document which entails the National Greenhouse Gas (GHG) Inventory and reports on actions that has been taken to tackle the issue of climate change. To date, two reports have been submitted to UNFCC, Initial National Communication in 2000 (INC) (Ministry of Science Technology and the Environment Malaysia 2000) and Malaysia Second National Communication to the UNFCC in 2011 (Ministry of Natural Resources and Environment Malaysia 2011) which detailed situation of climate change in Malaysia and strategies that has been carried out and which will be continued and improved to address the causes and impact of the climate impact scenarios. Malaysia is also engaged in various international conventions such as Vienna Convention for the Protection of the Ozone Layer (Montreal Protocol) and ASEAN Agreement on Transboundary Haze Pollution (2002) and even organised the Asia-Pacific Health Ministers' Conference on Climate Change and Health in Kuala Lumpur in 2008.

In 2008, the Cabinet Committee on Climate Change was instituted with the role of determining policy direction and strategies in tackling climate change issues whereby Ministry of Health is one of the members (Commonwealth Secretariat 2009). Provision of proper infrastructures, disease surveillance, investigation and emphasis on research and development were among initiatives undertaken to address the impact of climate change on health. There are already many existing policies and programmes within the country that address climate-sensitive diseases and health conditions such as the vector-borne and food- and water-borne control programmes under the Disease Control Division, Public Health Department, Malaysia (Ministry of Health Malaysia 1999). In terms of provision of healthcare in response to the impact of climate change, comprehensive healthcare service in Malaysia has been successfully extended throughout the country when the goal 'Health for All' was achieved in the year 2000 (Commonwealth Secretariat 2009).

Malaysia has taken a unique approach to climate change adaptation which was coined as 'adaptation through climate change mitigation'. This approach involves action by affected entities, requiring national-, state-, local- and community-level responses. Accordingly, much of Malaysia's adaptation responses come in the form of improved ecosystem management, water resource management and secure agricultural production – each with a backdrop of doing so to improve productivity, efficiency in resource use and optimised economic benefit for the state to the individual. For Malaysia, the health sector also figures prominently into the 'climate change' efforts. Focus is on larval and insecticide controls which are already in

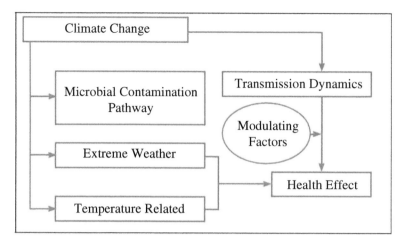

Fig. 15.13 Perception of climate change connection to health-related pathways, dynamics, and modulation in Malaysia. (Source: Solar (2011). Scoping assessment on climate change adaptation in Malaysia Summary October 2011)

place as part of the Health Ministry's Vector-Borne Disease Control Programme. Standard operating procedures for emergency and disaster management are also incorporated at all levels of the national health infrastructure. However, it is still unclear if these actions are adaptive in nature or reactive to current situations (Fig. 15.13).

The Vector-Borne Disease Control Programme is currently in the process of developing the control programme into an elimination programme. Among the key approaches being undertaken include strengthening and improving current strategies by changing the drug regimen to a more effective artemisinin-based combination therapy (ACT) as suggested by the World Health Organization to address the problem of drug-resistant virus, strengthening the surveillance programme particularly in malaria-free but prone areas to prevent reintroduction of the infection and occurrence of outbreaks and improving the case detection mechanism and approaches, including screening of migrant workers. Current and expected programmes and activities for adaptation to both current and projected climate-related health burdens involve the following (Ministry of Natural Resources and Environment Malaysia 2011):

- Emphasis to be placed on entomological surveillance with the recruitment of entomologists at the district level
- Established a Rapid Response Command Centre and incorporated the Centre for Disease Control and Prevention in the 9th Malaysian Plan
- Continued investment in health infrastructures, human resources and research
- Network of public health laboratories and constant vigilance on infectious diseases and strengthening surveillance system and disaster preparedness and response

Next, various policies related to each disease are described.

15.8   Food- and Waterborne Diseases

In controlling food- and waterborne diseases, in 1968, a pilot Rural Environmental Sanitation Programme was conducted in 11 states in Peninsular Malaysia (one per state) with the aim to improve sanitation by provision of clean water supply, building of latrines or sanitary toilets, improvement of environment sanitation by encouraging proper rubbish and sewage disposal and overall cleanliness of rural environment through community efforts *'gotong-royong'*. In 1973, this programme was designed for long term and aimed to cover 80 % of rural population by 1980 (Economic Planning Unit 1976). The target of the programme was to reduce incidences of communicable diseases. Various measures were adopted which involved rural community participation such as promotion of sanitary hygiene practice, building of proper sanitation and usage of sanitation facilities such as clean water supply system and sanitary latrines to ensure proper waste disposal. Identification and priorities were given to water-related diseases areas and encourage involvement of state governmental agencies and voluntary organisation into the programme (Ministry of Health Malaysia 1985). In 1978, the events of drought caused a major cholera outbreak in Peninsular Malaysia. This is due to unavailability of clean water supply to household and people had to obtained water from rivers and wells. At the time, interventions that were carried out to control the epidemic were chlorination of wells, disinfection and constructions of latrines. Also, patients with cholera were treated with tetracycline and immunisation against cholera was carried out (Ministry of Health Malaysia 1980). Since 1983, the oral rehydration salt therapy was introduced nationwide for treatment of cholera. In 1998, the Prevention and Control of Infectious Disease Act (Act 342) was enforced; five food- and waterborne diseases, cholera, typhoid, food poisoning, hepatitis A and dysentery, were made notifiable. Also, the Communicable Disease ControlInformation System (CDCIS), an electronic reporting system, was set up so that data on diseases can be shared at all levels of management for prompting diseases prevention, control and forecast on future outbreaks (Ministry of Health Malaysia 1985). Provision of clean water and environmental sanitation, effective treatment against diseases and mass vaccination for diseases prevention has proven to be effective with evidence of declining trends of the diseases over the years. The Rural Environmental Sanitation and Clean water Supply Programme currently provides 96.36 % coverage of safe water supply to rural community (Ministry of Health Malaysia 2009) and is aiming to give 100 % coverage of sanitary waste disposal and clean water supply in rural areas by 2015 (Commonwealth Secretariat 2009). Currently, cholera, typhoid and other diseases are lowly endemic and only occur as sporadic outbreaks.

In May 2007, the National Crisis Preparedness and Response Centre (CPRC) was established. Equipped with standard operating procedures for emergency and disaster management, the centre is responsible for management of disease outbreak information and coordinating outbreak response activities. This has led to improved registration and reporting system of infectious diseases outbreak with more disease cases being registered (Ministry of Health Malaysia 2009).

In addition, the setting of the Centre for Communicable Disease (CDC) will also improve surveillance, investigation and management of emerging and re-emerging diseases so that there will be instant case notification and prompt actions can be taken in time. Epidemiological investigation will be improved with complete medical staffing and adequate laboratory support for case diagnosis and early confirmation (Commonwealth Secretariat 2009).

15.9 Vector-Borne Diseases

15.9.1 Malaria

Prior to 1967, there were more than 300,000 cases of Malaria in the country annually. At that time, antimalaria drainage and usage of 'antimalaria oil' for larvaciding were measures taken to keep towns in malaria-free Peninsular Malaysia (Ministry of Health Malaysia 1980). In 1967, Malaria Eradication Programme was established with the aim to eradicate malaria by 1982. Among the measures taken in the eradication era were geographical investigation, DDT house spraying, case detection, laboratory diagnosis, registration of malaria cases, chemotherapy, case investigation and follow-up and entomological surveillance (Ministry of Health Malaysia 1980). The success of the programme has led to a significant drop in malaria cases to 44,226 cases in 1980 (Tee 2000). However, because of the difficulty to eradicate Malaria due to land development and migrations of people from one place to another, the vertical programme was changed to 'control' programme and was integrated into health services in the 1980s. In 1986, the malaria control programme was incorporated into the Vector-Borne Disease Control Programme, which is also responsible for 7 other vector-borne diseases, namely, dengue fever and dengue hemorrhagic fever, filariasis, typhus, Japanese encephalitis and scrub, yellow fever and plague. This programme places focus on the control of the diseases with the aim to reduce the impact of vector-borne diseases mortality and morbidity and its recurrence. Among other antimalaria strategies that were also adopted were focal spraying in localities with outbreaks, active case detection and use of insecticide-impregnated bed nets to replace DDT residual spraying (Sirajoon and Yadav 2008). However, the emergence of the resistance to the drug chloroquine has prompted the government to launch the National Anti-Malarial Drug Response Surveillance Programme in 2003 to monitor drug resistance.

Currently, Malaysia has the potential to eliminate malaria completely and is diverting the malaria control programme to an elimination programme (Ministry of Natural Resources and Environment Malaysia 2011) as many areas have been declared as malaria-free. However, there are concerns of the reintroduction and resurgence of malaria due to climate change. Increase in temperature and rainfall will increase the malaria vectors' presence, survival and vectorial capability. In addition, rainfall will cause changes in streamflows and rising of sea levels may lead to creation of more brackish water which will affect the spread of malaria vectors population around the coastal areas. Besides that, the absence of malaria in malaria-free areas is feared to cause lower of natural immunity and reintroduction of malaria. Recognising all factors above, the Vector-Borne Disease Control Programme is currently improving its strategies such as addressing the drug-resistant virus by switching to use of artemisinin-based combination therapy (ACT) and enhancement of the surveillance programme including and improving case detection approaches (Ministry of Natural Resources and Environment Malaysia 2011). In addition to that, research is also being carried out to formulate new insecticide with prolonged residual effects to address the current insecticide, which has a fast diminishing effect. The need for area mapping is important for vulnerability assessment and remote sensing data is currently being investigated to cater to this need (Ministry of Natural Resources and Environment Malaysia 2011).

15.9.2 Dengue Fever (DF) and Dengue Haemorrhagic Fever (DHF)

Following a major DHF outbreak in 1973, a series of actions was implemented to control the outbreak including case detection, case treatments, space spraying of insecticides and regular vector surveillance of the virus host, *Aedes* . In 1975, the Destruction of Disease-Bearing Insects Act (DDBIA) was introduced in which the relevant authority was included in the control of the disease through imposition of penalties to careless and indifferent householders whose compounds were found to have mosquito breeding sites (Tham 2001). This act also necessitates DF and DHF to be reported within 24 h to health authority when it was made a notifiable disease under the act. This is to ensure that dengue control activities will get public support so as to reduce dengue source. The enactment of the Prevention and Control of Infectious Diseases Act 1988 further enhanced the control of dengue under Section 18(d) whereby closure of compound harbouring disease-bearing insects can be done under the legislation. Legislation control of dengue was seen to have a positive impact on the reduction of dengue cases from 10,146 cases in 1999 to 7,118 cases in 2000 (Tham 2001). In 2001, the DDBIA 1975 was amended with heavier penalties.

Human plays the most important role in ensuring our environment is clean and free of dengue vector breeding sites to halt transmission of dengue. Recognising this, disease control through community awareness and participation was intensified through campaigns and programmes such as National Cleanliness and Anti-Mosquito Campaign (1999) which aim to increase awareness of cleanliness in the environment to control the breeding of vector mosquitoes (Ministry of Health Malaysia 2000), the Dengue-Free Schools programme to educate school children and the Communication for Behavioural Impact (COMBI) programme in 2001 (Mohd Raili et al. 2004) and 'Promotion of a Healthy Environment' campaign in 2002 (Abu Bakar and Jegathesan 2001) for prevention and control of dengue.

However, until today, dengue still remains as a major public health problem in Malaysia and myriad of reasons contributed to incessant rise of this indomitable disease. The biological nature of the virus host (Rohani et al. 2008; Lindsay and Mackenzie 1997), increased population growth and disorganised urbanization, poor sanitation and improper waste disposal system and increased rural urban migrations and international travels produce epidemiological environments which can increase viral transmission potential of the mosquito vector, *Aedes Aegypti* (Gubler 1998), thus posing difficulties for vector control. Changes in public health policy in the late 1990s, specifically integration of the vertical organisational structure of the Vector-Borne Disease Control Programme into general health services, caused loss of expertise and funding targeted for vector control. This resulted in the control of vector-borne diseases to be placed under respective local governments in 2000 where there was lack of expertise, resources and political will (Kumarasamy 2006). In December 2010, a drastic measure was undertaken by Malaysia government to curb DF and DHF, that is, the release of 6,000 genetically modified male mosquitoes in an unhabited forested area in Pahang, which is designed to disrupt *Aedes aegypti* mosquito fertility by passing on a gene that would kill the mosquito at its larval stage. This move has raised concerns among environmentalists and experts on possible disruption of the balance ecosystem and emergence of new species of mutated mosquitoes which might introduce new diseases (CBS 2011).

15.9.3 *Chikungunya*

Chikungunya is a virus, which shares the same host as dengue virus, that is the Aedes mosquitoes (especially Aedes aegypti) (Turell et al. 1992; Thonnon et al. 1999). Thus the policies and programmes related to epidemic control of Chikungunya were similar to dengue, which includes public health education, active case detection, fogging and disease source reduction (Lam et al. 2001).

15.10 Conclusions

Patterns of food and water-borne diseases as well as vector-borne diseases are changing due to climate change and these inevitably affect the population and the surrounding environment. Malaysia is proactive in addressing the various health hazards related to climate change. Public health policies and programmes have been instituted and implemented to deal with these infectious diseases that are affected by climate change with evidence of increasing rates and distribution. It is not only the health risks that must be addressed but also the associated social, economic and demographic consequences. The future public health consequences of climate change for people in Malaysia remain uncertain. Thus, climate change adaptation and disaster preparedness for dealing with the climate change adversities should be the utmost priority for human health and well-being.

References

Abu Bakar S, Jegathesan M (2001). Health in Malaysia: achievements and challenges. Planning and Development Division, Ministry of Health Malaysia, Kuala Lumpur

Ambu S, Lim LH, Sahani M, Bakar MA (2002) Climate change-impact on public health in Malaysia. WHO Statement of Support 13

Ang KT, Satwant S (2001) Epidemiology and new initiatives in the prevention and control of Dengue in Malaysia. Dengue Bulletin 25:7–14

Asian Development Bank and International Food Policy Research Institute (2009) Building climate resilience in the agriculture sector of Asia and the Pacific, Philippines

Benjamen PG (2006) Analysis of cholera epidemics in Sarawak from 1994–2003 and molecular characterization of Vibrio cholerae isolated from the outbreaks in Malaysia. Universiti Putra Malaysia, Serdang

CBS (2011) Malaysia fights dengue with mutant mosquitoes [Online]. Available: http://digitaljournal.com/article/303037. Accessed 25 Mar 2012

Chang CH (2010) The impact of global warming on storms and storm preparedness in southeast Asia. Kajian Malaysia 28:53–82

Chen P (1970) Cholera in the Kedah River area. Medical Journal of Malaya 24:247–56

Chua KB (2010) Epidemiology of chikungunya in Malaysia: 2006–2009. The Medical Journal of Malaysia 65:277–282

Commonwealth Secretariat (2009) Commonwealth health ministers' update 2009. Commonwealth Secretariat, London

Department of Statistics Malaysia (2010) Population distribution and basic demographic characteristics 2010. Population and Housing census of Malaysia 2010, Putrajaya, Malaysia

Department of Statistics Malaysia (2014a) GDP by state 2005–2013. In: Department of Statistics Malaysia (ed). Department of Statistics Malaysia, Putrajaya

Department of Statistics Malaysia (2014b) Gross domestic product income approach 2005–2013. In: Department of Statistics Malaysia (ed) Putrajaya. Department of Statistics Malaysia

Director General of Health Malaysia (2011a) Press statement: current situation of dengue fever and chikungunya in Malaysia for week 52/2010(26/12/2010-31/12/2011). In: Ministry of Health Malaysia (ed) Putrajaya. Ministry of Health Malaysia

Director General of Health Malaysia (2011b) Press statement: current situation of dengue fever in Malaysia for week 52/2011(25–31 December 2011). In: Ministry of Health Malaysia (ed) Putrajaya. Ministry of Health Malaysia

Director General of Health Malaysia (2012a) Press statement: current situation of dengue fever in Malaysia for week 52/2012. In: Ministry of Health Malaysia (ed) Putrajaya. Ministry of Health Malaysia

Director General of Health Malaysia (2012b) Press statement: increased of mortality of dengue cases in Malaysia. In: Ministry of Health Malaysia (ed) Putrajaya. Ministry of Health Malaysia

Director General of Health Malaysia (2013) Press statement: current situation of dengue fever in Malaysia for week 52/2013. In: Ministry of Health Malaysia (ed) Putrajaya, Ministry of Health Malaysia

Economic Planning Unit Malaysia (1976) Third Malaysia plan 1976–1980. Chapter 23: Health and family planning, Kuala Lumpur

Engku Azman TM (2011) Environmental challenges in managing food waterborne diseases (FWBD). Malaysian Journal of Public Health Medicine 11

Fischer D, Thomas SM, Suk JE, Sudre B, Hess A, Tjaden NB, Beierkuhnlein C, Semenza JC (2013) Climate change effects on Chikungunya transmission in Europe: Geospatial analysis of vector's climatic suitability and virus' temperature requirements. Int J Health Geogr 12:51

Gage KL, Burkot TR, Eisen RJ, Hayes EB (2008) Climate and vectorborne diseases. Am J Prev Med 35:436–450

Githeko AK, Lindsay SW, Confalonieri UE, Patz JA (2000) Climate change and vector- borne diseases: A regional analysis. Bull World Health Organ 78:1136–1147

Gubler DJ (1998) Dengue and dengue hemorrhagic fever. Clinical Microbiology Review 11:480–496

Haines A, Kovats RS, Campbell-Lendrum D, Corvalán C (2006) Climate change and human health: Impacts, vulnerability and public health. Public Health 120:585–596

Hales S, de Wet N, Maindonald J, Woodward A (2002) Potential effect of population and climate changes on global distribution of dengue fever: An empirical model. Lancet 360:830–834

Hopp J, Foley JA (2003) Worldwide fluctuations in dengue fever case related to climate variability. Clim Res 25:85–94

Ibrahim Mohamed Shaluf, Fakhru'l-Razi Ahmadun (2006) Disaster types in Malaysia: an overview. Disaster Prev Manag 15:286–298

Khoo A (1994) A study of two cholera epidemics in the District of Tawau, Sabah (1989–1991). Southeast Asian Journal of Tropical Medicine & Public Health 25:208–210

Kin F (2007) The role of waterborne diseases in Malaysia. A World of Water: Rain, Rivers and Seas in Southeast Asian Histories 281

Kumarasamy V (2006) Dengue fever in Malaysia: Time for review? Med J Malays 61:1–3

Lam SK (1993) Two decades of dengue in Malaysia. Tropical Medicine 35:195–200

Lam SK, Chua KB, Hooi PS, Rahimah MA, Kumari S, Tharmaratnam M, Chuah SK, Smith DW, Sampson IA (2001) Chikungunya infection–an emerging disease in Malaysia. Southeast Asian J Trop Med Public Health 32:447–51

Li C, Lim T, Han L, Fang R (1985) Rainfall, abundance of Aedes aegypti and dengue infection in Selangor, Malaysia. The Southeast Asian Journal of Tropical Medicine and Public Health 16:560

Lindsay M, Mackenzie J (1997) Vector-borne viral diseases and climate change in the Australian region: Major concerns and the public health response. In: Curson P, Guest C, Jackson E (eds) Climate change and human health in the Asia-Pacific region. Australian Medical Association and Greenpeace International, Canberra

Lu SH (2006) APCT. Chapter 4: Environmental and industrial policies and financing of environmentally sound technologies in Malaysia. Report 2127, Selangor

Malaysian Meteorological Department (MET Malaysia) (2014) *Climate change monitoring* [Online]. Available: http://www.met.gov.my/index.php?option=com_content&task=view&id=1731&Itemid=1595. Accessed 17 Dec 2014

Marchette N, Rudnick A, Garcia R (1980) Alphaviruses in Peninsular Malaysia: II. Serological evidence of human infection. The Southeast Asian Journal of Tropical Medicine and Public Health 11:14–23

Meason B, Paterson R (2014) Chikungunya, climate change, and human rights. Health and Human Rights 16:105–112

Mia MS, Begum RA, Ahchoy E, Raja Datuk Z, Pereira JJ (2011) Malaria and climate change: Discussion on economic impacts. Am J Environ Sci 7:73–82

Mia MS, Begum RA, Er A, Abidin RDZRZ, Pereira JJ (2013) Trends of dengue infections in Malaysia, 2000–2010. Asian Pacific Journal of Tropical Medicine 6:462–466

Ministry of Health Malaysia (1980) Annual report 1980, Kuala Lumpur

Ministry of Health Malaysia (1985) Annual report 1985, Kuala Lumpur

Ministry of Health Malaysia (1992–1999) Annual reports, Kuala Lumpur

Ministry of Health Malaysia (1999) Annual report 1999, Kuala Lumpur

Ministry of Health Malaysia (2000) Annual report 2000, Kuala Lumpur

Ministry of Health Malaysia (2007–2014) Health facts (2006–2014). In Health Informatics Centre Planning and Development Division (ed) Kuala Lumpur

Ministry of Health Malaysia (2009) Annual report 2009, Putrajaya

Ministry of Health Malaysia (2010) Number of notifiable communicable diseases in Malaysia for 1998–2009. In: Disease Control Division (ed), Putrajaya

Ministry of Health Malaysia (2014) Health facts 2014. In: Health Informatics Centre: Planning and Development Division (ed). Ministry of Health Malaysia, Kuala Lumpur

Ministry of Natural Resources and Environment Malaysia (2011) Second national communication to the UNFCCC. Ministry of Natural Resources and Environment Malaysia, Putrajaya

Ministry of Science Technology and the Environment Malaysia (2000) Malaysia initial national communication submitted to the United Nations Framework Convention in Climate Change

Mohd Raili S, Hosein E, Mokhtar Z, Ali N, Palmer K, Marzhuki MI (2004) Applying Communication-for-Behavioural-Impact (COMBI) in the prevention and control of dengue in Johor Bahru, Johore, Malaysia. Dengue Bull 28:39–43

Patz JA, Martens WJ, Focks DA, Jetten TH (1998) Dengue fever epidemic potential as projected by general circulation models of global climate change. Environ Health Perspect 106:147–53

Patz JA, Reisen WK (2001) Immunology, climate change and vector-borne diseases. Trends Immunol 22:171–172

Promprou S, Jaroensutasinee M, Jaroensutasinee K (2005) Climatic factors affecting dengue haemorrhagic fever incidence in Southern Thailand. Dengue Bull 29:41

Resurgence W (2008) Dengue reborn. Environ Health Perspect 116

Rezza G (2008) Re-emergence of Chikungunya and other scourges: the role of globalization and climate change. Ann Ist Super Sanita 44:315–318

Rohani A, Zamree I, Joseph RT, Lee HL (2008) Persistency of transovarial dengue virus in Aedes aegypti (Linn.). Southeast Asian J Trop Med Public Health 39:813–6

Rozlan Ishak (2007) Health effect of climatic change: Malaysian scenarios. Ministry of Health Malaysia

Singh RB (1972) Review of cholera in Malaysia (1900–1970). Med J Malaya 26:149

Sirajoon NG, Yadav H (2008) Health care in Malaysia. University of Malaya Press, Kuala Lumpur

Solar RW (2011) Scoping assessment on climate change adaptation in Malaysia – Summary October 2011. The Regional Climate Change Adaptation Knowledge Platform for Asia, Pathumthani, Thailand

Tee AS (2000) Malaria control in Malaysia. Mekong Malaria Forum 5:6–9

Tham AS (2001) Legislation for dengue control in Malaysia. Dengue Bull 25:109–112

The World Bank (2014) Data – Malaysia [Online]. Available: http://data.worldbank.org/country/malaysia. Accessed 24 Dec 2014

Thonnon J, Spiegel A, Diallo M, Diallo A, Fontenille D (1999) Chikungunya virus outbreak in Senegal in 1996 and 1997. Bull Soc Pathol Exot 92:79–82

Turell MJ, Beaman JR, Tammariello RF (1992) Susceptibility of selected strains of Aedes aegypti and Aedes albopictus (Diptera: Culicidae) to chikungunya virus. J Med Entomol 29:49–53

WHO (2006) Climate change country profile: Malaysia [Online]. Available: www2.wpro.who.int/NR/rdonlyres/3FB0A304-554E.../MAA.pdf. Accessed 26 Feb 2012

WHO (2008) The global burden of disease: 2004 update. WHO, Geneva

WHO (2009) Protecting health from climate change: connecting science, policy and people. WHO Press, Geneva

WHO (2011) World Malaria Report 2011. WHO, Geneva

WHO (2014) Global health observatory data repository [Online]. Available: http://apps.who.int/ghodata/?vid=2250. Accessed 08 Dec 2014

YADAV H, CHEE C (1990) Cholera in Sarawak: a historical perspective (1873–1989). Med J Malays 45:194

Chapter 16
Potential Effects of Climate Changes on Dengue Transmission: A Review of Empirical Evidences from Taiwan

Tzai-Hung Wen, Min-Hau Lin, and Mei-Hui Li

Abstract Effects of climate change on human health become increasingly a growing concern. Especially, the worldwide risk of vector-borne diseases, such as dengue fever, could be strongly influenced by meteorological conditions. Taiwan is unique in terms of its geographical location and abundant ecological environments. It makes Taiwan an unusual epidemic region for dengue disease. This chapter focuses on risk factors related to climate change and geographical characteristics which have been shown to affect dengue transmission in Taiwan. Empirical studies all concluded that relationship between temperature and dengue incidence is highly correlated. Some of studies also concluded that temperature increase has more dramatic influence to dengue incidences in southern Taiwan. However, the effects of other meteorological factors, including precipitation and humidity, on dengue incidence are rather inconsistent. Extreme weather events, such as heavy rainfall, would be possible to damage the habitats of *Aedes* mosquitoes and also beneficial to construct them with lag time of several weeks. The role of extreme weather events in dengue transmission needs to be further investigated. Further studies should also put more focus on the extent of influence of meteorological conditions on human activities and what activity changes may increase or decrease the risk of dengue transmission.

Keywords Climate change • Dengue fever • Medical geography • Taiwan

T.-H. Wen, Ph.D. (✉) • M.-H. Li, Ph.D.
Department of Geography, College of Science, National Taiwan University, Taipei, Taiwan
e-mail: wenthung@ntu.edu.tw; meihuili@ntu.edu.tw

M.-H. Lin, Ph. D
Institute of Health Policy and Management, College of Public Health, National Taiwan University, Taipei, Taiwan
e-mail: greenlibraming@gmail.com

© Springer International Publishing Switzerland 2016   269
R. Akhtar (ed.), *Climate Change and Human Health Scenario in South and Southeast Asia*, Advances in Asian Human-Environmental Research,
DOI 10.1007/978-3-319-23684-1_16

16.1   Introduction

16.1.1   Climate Change, Human Health, and Dengue Transmission

Effects of climate change on human health become increasingly a growing concern and a potential threat for different regions worldwide in past decade. In fact, not all geographical regions and human settlements are equally sensitive and vulnerable to climate changes. South, East, and Southeast Asia are usually considered as more vulnerable than other geographical regions to climate change due to intensive human activities and fast population growth (USAID 2010). Furthermore, areas within the same geographical region can be different in vulnerability to climate change because of social, cultural, and economic conditions.

Climate change in temperature, precipitation, and extreme weather events can both directly and indirectly affect human health. Direct impacts of climate change include higher morbidity/mortality from cold or heat waves and increased injuries and death from extreme weather events. Indirect impacts of climate change on health can occur through changes in waterborne and vector-borne infectious diseases, air and water quality, and food productions. Especially, the worldwide risk of dengue fever is strongly influenced by climate factors. In addition to meteorological characteristics in tropical areas, globalization and population movement also play an important role in escalating risk of dengue transmission (Gubler 2011). Transmission of dengue virus is determined by local meteorological conditions, socioeconomic, and environmental factors.

16.1.2   Dengue Vectors and Natural Environment in Taiwan

Located in the border between tropics and subtropics, with south and north parts split by Tropic of Cancer while east and west separated by Central Mountain Range, Taiwan is unique in terms of its geographical location and abundant ecological environments considered it is a small island with only around 3600 km². However the most serious vector-borne infectious disease in Taiwan, dengue fever, and its vector mosquitoes, *Aedes aegypti* and *Aedes albopictus*, are largely nourished by such geographical features: the lowest monthly mean temperature in northern part of Taiwan is less than 15 °C, whereas lowest average temperature in the South of Tropic of Cancer can be as high as over 20 °C (Wu et al. 2007). This climatic condition pairs with unique geographical characteristics and makes *Aedes aegypti* only distributed in Southeast of Taiwan (as shown in Fig. 16.1) while *Aedes albopictus* is widely located in areas of Taiwan (TaiwanCDC 2011). These two dengue vector mosquitoes are different in temperature sensitivity and behaviors of blood preferences which result in apparent differences in outbreak of indigenous dengue cases between southern and northern in Taiwan and temporal incidences by seasons (Tseng et al. 2009).

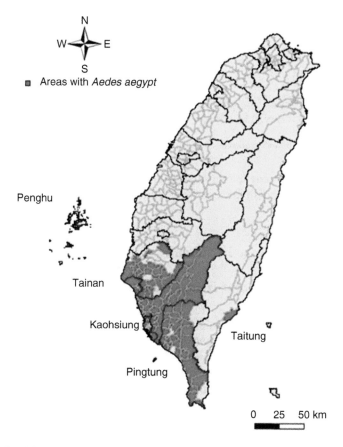

Fig. 16.1 Spatial distribution of major dengue vector – *Aedes aegypti* in Taiwan (Source: TaiwanCDC 2011)

16.1.3   Epidemiology of Dengue Fever in Taiwan

Based on surveillance data by Taiwan CDC, sequences of dengue virus from human cases of local residents and imported cases can indicate that dengue is not an endemic disease but imported cases from other dengue endemic countries and gradually becomes indigenous disease as shown in Fig. 16.2 (Shang et al. 2010). Taiwan is also different from other dengue endemic countries in Southeast Asia in terms of age distribution of dengue hemorrhagic fever (DHF), a fatal hemorrhage symptom after dengue infection in which the reason is still unknown. In Taiwan, dengue hemorrhagic fever usually occurs among adults while children are more prevalent to have such symptom in Southeast Asia (Guzman and Kouri 2002). Special geographical characteristics and patterns of dengue hemorrhagic fever make Taiwan an unusual epidemic region for dengue disease (Wen et al. 2010). This chapter focuses on factors related to climate change and geographical characteristics which have

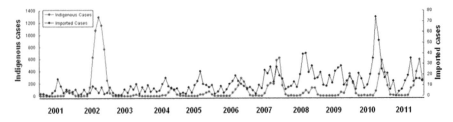

Fig. 16.2 Imported and indigenous dengue cases from 2001 to 2011 (Source: TaiwanCDC 2012)

been shown to affect dengue transmission and also discusses future trends of dengue epidemics under possible scenarios of climate change in Taiwan.

16.2 Climate Change in Taiwan

16.2.1 Temperature and Heat Island Effect

Climate change is a global phenomenon which is inevitable in Taiwan. Chen and Wang (2000) have analyzed the average temperature changes between 1920 and 1997 in Taiwan and found that the average temperatures in all seasons are ascending year after year. The increasing span is 1.3 °C/100 years in spring, 1.6 °C/100 years in summer, and 1.4 °C/100 years and 0.6 °C/100 years for fall and winter, respectively. Comparing with global warming worldwide, besides the average temperature in winter is relatively the same as average global temperature in the northern hemisphere, the temperatures in spring, summer, and fall seasons are all higher than average temperature in the northern hemisphere by two-, four-, and threefolds (Chen and Wang 2000).

It is also found in non-urban setting that the warming trend is 0.63 °C /100 years whereas the data found by weather watch in urban area is as high as 1.12 °C/100 years. Therefore, researchers concluded that the gradual ascending temperature in Taiwan is associated with urban heat island effect (Chen and Wang 2000), which refers to a phenomenon in which the temperature in metropolitan area is significantly higher than surrounding areas. This situation can be interpreted by following four factors in urban settings (Lin 2010):

Roughness:	The atmosphere layer is becoming thicker which causes the pollutant and waste heat difficult to diffuse.
Atmospheric albedo:	The atmospheric albedo can affect solar energy level entering earth surface which indirectly contribute to the amount of suspended particles in the air and have the effects of decreasing and increasing the temperature.
Radiation:	Various surface characteristics may be influential to the energy transmission process. For example, the specific heat

	of concrete buildings is relatively smaller than natural materials; therefore, its temperature is easily risen once it absorbs radiation.
Latent heat flux:	The vegetation and surface water coverage are generally higher in non-urban areas which can transfer sensible heat to latent heat through evapotranspiration.

Empirical studies in Taiwan have shown that urban heat island effect is negatively correlated with vegetation coverage but positively correlated with lack of surface water and concrete structures. According to these studies, it is concluded that the urbanization of Taiwan is correlated with urban heat island effect (Lin et al. 2005). In summary, the climate warming in Taiwan indeed exists and such trend under the influence of heat island effect is more significant in urban areas. According to the prediction made in Taiwan Scientific Report of Climate Change 2011, if global warming effect does not get milder in the future, the average temperature in summer will continue to increase to 27.4 °C by 2100 from 24.5 °C in 1960s as shown in Fig. 16.3a and average temperature in winter will rise to 17.4 °C by 2100 from 14.5 °C in 1960s as shown in Fig. 16.3b.

16.2.2 Precipitation and Typhoons

As a result of climate change, downpours or typhoons could be more frequent in the future. Mean annual rainfall in Taiwan is approximately 2500 mm, but this varies geographically from 1900 mm in southern Taiwan to 2700 mm in northern Taiwan (National Science Council 2011). Taiwan has rain all year round in the north, whereas the south is rainy in summer but dry in winter. Based on available rainfall records, there are no eminent changes in annual precipitations during past 100 years (Hsu and Chen 2002). However, average annual precipitation has showed to be increased in northern Taiwan and declined in southern Taiwan (Hsu and Chen 2002; Yu et al. 2006), especially significant increase in frequency of heavy precipitation around 20 mm/h in Taiwan (Shiu et al. 2009). The overall number of rainy days has decreased in Taiwan with increasing number of heavy precipitation and lower number of light precipitation (Shiu et al. 2009). Furthermore, an increased trend of high amount of rainfall is observed in most rainfall stations of southern Taiwan (Kuo et al. 2011).

Typhoon can bring heavy rains causing floods and landslides in Taiwan. There is an annual average frequency of 3 to 4 typhoons landing Taiwan that strike mostly between July and September. Annual number of typhoons landing over Taiwan has increased with an average of 3.5 per year in the past 30 years comparing with an average of 2.8 per year from 1961 to 1980 (as shown in Fig. 16.4). Moreover, the ratio of typhoon precipitation contributes to annual precipitation that has risen from 15 % in 1970s to 30 % after the year of 2000 (TaiwanEPA 2010). Based on 181 typhoons attacking Taiwan from 1970 to 2009, there are an increasing number of

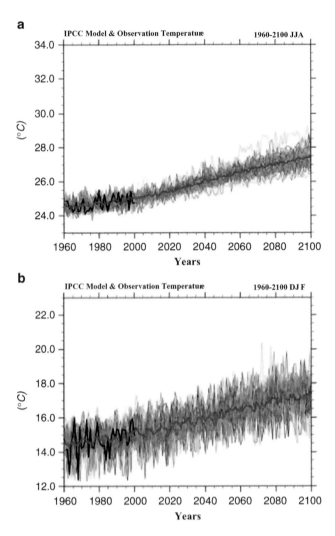

Fig. 16.3 Predicted trend of climatic warming in Taiwan from 2000 to 2100. Panel (**a**) is the trend in summer and panel (**b**) is in winter. (*Black bold line* is real data, *thin lines* with different colors are predictions by various models, and red bold line is the average of these model predictions) (Source: National Science Council 2011)

times typhoons are accompanied by extreme precipitation from once in every 2 year before 2000 to once in every year after 2000 (Taiwan EPA 2010). For example, Typhoon Morakot passed across northern Taiwan between August 6 and August 10 of 2009, but brought a record high rainfall in southern Taiwan causing serious floods in many places. In sum, the frequency of typhoon hits did not markedly increase in Taiwan, but the precipitation followed by typhoons has dramatically increased during past 20 years.

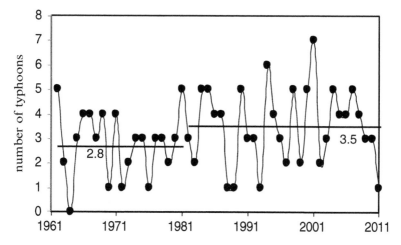

Fig. 16.4 Annual number of typhoon centers landing in Taiwan or only passing through ocean near Taiwan but causing disasters in Taiwan from 1962 to 2011 (Source: Taiwan CWB 2012)

16.3 Climate Factors and Dengue Transmission in Taiwan

16.3.1 Meteorological Factors and Ecology of Dengue Vectors

Climate change could affect the ecology of vectors in Taiwan. Wang and Chen (1997) have proposed five impacts due to climate warming in Taiwan: (1) accelerate growth and bloodfeeding rate of mosquitoes and lead to shorten life spans which also means higher mosquito density and increasing the possibility of mosquito biting and virus infections; (2) extend geographical distribution of vector mosquitoes; (3) facilitate blood digestion within female mosquitoes and the growth of ova which as a result causes nutrition cycle to become shorter during breeding and increases the number and frequency of blood feeding as well as possibility of disease infection; (4) decrease the latent period of pathogen during reproduction; and (5) hasten the activity of pathogens within vectors.

Climate warming is considered as the foremost factor for changing the distribution of mosquito habitats. Guo and Lung (2004) have analyzed the meteorological data and the larva of *Aedes* mosquitoes in Breteau Index (BI) between 1998 and 2002 in Great Kaohsiung Metropolitan and compared previous week and current week to examine the connection between climate factors and mosquito density. The result of the study is shown that the BI in more than half of the monitoring areas is over Level 2; the average temperature, highest temperature, lowest temperature, vapor pressure, weekly precipitation, and frequency are remarkably higher than other weeks. Even though mosquito density indices in Kaohsiung Metropolitan are positively correlated with average temperature, as clearly seen in Fig. 16.5, the areas where BI is over Level 2 in Kaohsiung City are relatively higher than their less populated counterpart, Kaohsiung County. This result is also consistent with the

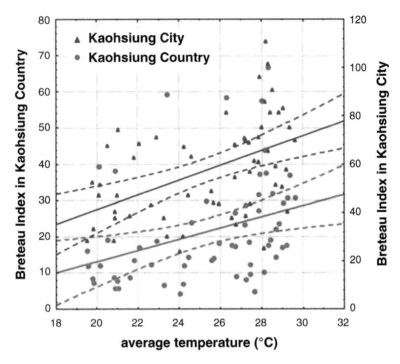

Fig. 16.5 The relationship between Breteau Index (BI) and average temperature in Great Kaohsiung Metropolitan between 1998 and 2002 (Source: Guo and Lung 2004)

description of urban heat island effect as mentioned in previous section, and more populated areas are associated with higher virusborne mosquito density. However, when the study was conducted the same analysis with a time span of month instead of week, the finding did not turn out to be consistent between BI and precipitation. In general, in the areas with the percentage of BI over Level 2, the mosquito density would be significantly positively correlated with the mean temperature and humidity between previous week and current week, when the regression model's R-square of using climatic factors in current week as predictor variables is 0.196 and previous week is 0.204 (Guo and Lung 2004).

Huang has obtained similar results and concluded that temperature, humidity, and precipitation are positively correlated with BI (Huang 2006). Average mosquito index was used as reference sequence and humidity, precipitation, and temperature as comparative serials. The result has shown that the factor values of three climate elements are very close. In other words, mosquito index is not significant affected by any single factor but instead a combination of three climate factors simultaneously. Single factor such as high temperature or rainfall has limited influence to the increase of dengue vectors.

In summary, the relationships between temperature and dengue vector abundance are more consistent in various studies. It is shown in those studies that temperature is indeed an essential factor that causes the increase of vector density, but the factors of humidity and rainfall are still in disagreement. For example, Tseng et al. (2009) have conducted an analysis to all the major cities in Taiwan. They have found that humidity and mosquito density are negatively correlated.

In addition, BI is usually used as an index for monitoring changes of vector abundance. However, BI in Taiwan involves subject investigation and does not include appropriate random sampling process in terms of statistical design which may hinder practical reflection of mosquito abundance. Although ovitrap can be adopted as a surveillance method of monitoring mosquito abundance and detect monthly changes of mosquito population (Pai and Lu 2009), the installation cost of such method is not economical. These possible reasons have become major constraints on understanding the relationships between climate factors and population of dengue vectors.

16.3.2 The Association of Meteorological Factors and Dengue Epidemics

Due to limitation of mosquito indices for dengue vector prediction and inconsistent prediction of dengue outbreak with the adoption of BI (Wu et al. 2007; Chen et al. 2010), researchers have also attempted to study the impact of meteorological factors to dengue epidemics in Taiwan.

Among all the studies involving meteorological factors on dengue epidemics, studies in Taiwan concentrated more on temperature and humidity but turn out with different results. Wu et al. (2007) examined effects of different meteorological factors on dengue incidence in southern Taiwan and found negative association with relative humidity with a lag time of two months. Dengue transmission may be not directly associated with precipitation only, but is affected by combined meteorological conditions, including temperature, rainfall, and humidity, as well as human activities. In other words, higher relative humidity and larger variations of temperature result in lower probability of dengue occurrence with longer lag time.

Temperature variation and dengue incidence are negatively correlated and the major reason is that temperature difference may involve low temperature in the environment which is not beneficial to the growth and breeding of mosquitoes. However, the negative correlation between relative humidity and dengue incidence is unexpected because such finding is in conflict with the growing condition of vectors. The researchers explained that the findings are because data range of relative humidity in the study is wider (63.16–85.29 %). Within this range, the lower the relative humidity is, the more likely dengue-related transmission would occur (Wu et al. 2007). This conclusion is consistent with the Tseng's research findings of related humidity versus vector density studies (Tseng et al. 2009). However, Chen et al. (2010) put more attention on the disparity between Taipei and Kaohsiung. The study result in terms of temperature is the same as other research results. However, relative humidity and precipitation in Taipei have positive effects to dengue incidence, and these two meteorological factors have negative effects in Kaohsiung (Chen et al. 2010). Although both Taipei and Kaohsiung are major cities in Taiwan, Taipei is located at subtropical region while Kaohsiung is at tropical climate region. The two cities are different in terms of natural environmental conditions, with different social development and socioeconomic activities which cause the same climate factors that may result in different epidemic patterns. This further explains that meteorology itself is not the only factor that affects transmission of dengue virus. It is necessary to conduct more empirical studies to figure out relevant mechanisms.

It can be concluded that precipitation changes can affect mosquito-borne disease transmission by altering mosquito breeding habitats. In addition, relative humidity can be an important factor to affect *Aedes* mosquito reproduction and behaviors (Canyon et al. 1999). Besides these studies which emphasize the general meteorological conditions (such as temperature, relative humidity, and precipitation), increases in the occurrence of extreme weather events also could be a possible risk factor of dengue epidemics. Heavy rainfall or flooding may initially flush out mosquito breeding sites, but mosquito populations can come back when the waters recede to leave more standing water in area to provide their breeding sites again. Therefore, an outbreak of dengue can take place with a lag time after an extreme precipitation event. Wang and Chang (2002) reported one surge of dengue cases from October 2001 to April 2002 after Typhoon Nari landed in Taiwan in September of 2001. Hsieh and Chen (2009) analyzed correlation between meteorological factors and twowave dengue outbreak in 2007 and found precipitation bought by two typhoons correlated with dengue incidence cases with a lag time of 7 weeks.

Due to many imported dengue cases in Taiwan every year, whether these imported cases have played a part in indigenous incidences is unknown. Shang et al. (2010) have discovered what the impacts to indigenous dengue incidences are under the effects of imported cases and climate. They have found that high temperature and low humidity can strongly predict the occurrence of imported dengue cases in Taiwan with lag time of 2–14 weeks (Shang et al. 2010). Not only this study addressed the relationships between meteorological factors and dengue incidence, it also turned the focus in explaining the possible mechanisms of dengue outbreaks.

The local studies regarding climate and dengue incidence relations only focus on the discussion of lag time without the consideration of spatial relationships. Yu et al. (2011) use Bayesian maximum entropy to investigate the space-time dependence between climate factors to dengue incidences. This study not only reflected the influence of climates to dengue but also constructed an epidemic model based on spatial perspective and further provided a reference for predicting the dengue incidence.

Despite various research methods, time spans, and regions in investigating the association of climate factors and dengue occurrence in Taiwan, the concluded relationship between high temperature and dengue incidence is consistent. The rise of temperature indeed increases the possibility of dengue occurrence. As for precipitation and humidity, in spite of positive influence of low humidity level to dengue incidence, the results are rather inconsistent and the mechanisms remain unclear and leave room for further exploration.

16.3.2.1 Possible Scenarios of Climate Change and Implications for Dengue Control in Taiwan

From the conclusions obtained in both climate change studies and dengue studies in Taiwan, mosquito density and dengue epidemics would continue to expand in a larger extent in the future if the trend of global warming remains increasing. Tseng's study indicated that if the temperature rises 0.9 °C in the future, then the likeliness of dengue incidences in Pingtung area will increase to 43.81 %. If temperature climbs to 2.7 °C, then the likeliness will increase to 131.43 %, and the probability of dengue incidences will be 21.6 % and 64.7 % in Kaohsiung region (Tseng et al. 2009). The magnitude in Taipei area is smaller in which the possibility of dengue incidence is 1.78 and 5.34 %. It can be concluded that temperature increase has more dramatic influence to dengue incidences in southern Taiwan. Although it has mild impact to northern Taiwan, scientists still need to be cautious of this matter.

For the factors of precipitation and relative humidity, there are still no solid conclusions regardless of relevant climate or dengue epidemics studies. Nevertheless, the issues should be brought up in the future for further understanding such as whether the activity of dengue vectors has a suitable humidity range and heavy rainfall should be regarded as a risk or protective factor to dengue incidences in the long term. It depends on further empirical studies to show whether heavy rainfall would damage or construct the habitats of mosquitoes.

Based on previous literatures, the associations between climatic factors and dengue epidemics in Taiwan can be confirmed. Similar atmosphere conditions could possibly result in different climatic characteristics due to geographical and seasonal variations. In other words, not only the overall impact of climate change is necessary to be considered, but local variations of regional susceptibility of dengue should also be concerned. For example, influence of heat island effect is more significant in localized urban areas than the surrounding urban fringe and rural areas. These localized areas could provide appropriate mosquito habitats and become the potential sources of dengue infection. In addition to natural environmental factors, the effect of human behaviors is also an important factor that makes susceptibility of dengue to be different in urban and rural areas. Wen et al. (2012) analyzed the association between population movement behaviors and spatial-temporal diffusion patterns of a dengue epidemic in Tainan City, Taiwan. The study found that commuting cases diffuse more rapidly across villages than non-commuting cases in the late epidemic period and the role of commuting could be identified as a significant risk factor contributing to epidemic diffusion. Local neighborhood characteristics (number of vacant grounds and empty houses) and high population density are also significant risk factors for dengue diffusion. In summary, geospatial diffusion of dengue epidemics is the result of complex interactions among climate change, mosquito ecology, and human activities.

16.4   Concluding Remarks

Although we have realized there are certain correlations between meteorological conditions and dengue incidences in Taiwan, the mechanisms that cause the dengue epidemics remain unclear. Climate alone cannot be the sole risk factor for the occurrence of dengue. Since the transmission route of dengue involves the complicated interactions between mosquitoes and human behaviors, it would be more helpful to understand the possible mechanisms between climate change and occurrence of epidemics and develop appropriate control measures. Further studies should put more focus on the extent of influence of meteorological conditions on human activities, such as urbanization or population movement behaviors, and what activity changes may increase the risk of dengue transmission. Last but not least, most of current studies lacked consideration of what impacts these interventions may have in response to epidemic outbreak. These are all the issues that are worth to be explored in the future.

Acknowledgments This report was supported by grants from the National Science Council (NSC 982410H002168MY2) and financial support provided by the Infectious Diseases Research and Education Center, Department of Health, and National Taiwan University. Authors also thank Dr. Kun-Hsien Tsai for his constructive suggestions.

Reference

Canyon DV, Hii JLK, Muller R (1999) Adaptation of *Aedes aegypti* (Diptera : Culicidae) oviposition behavior in response to humidity and diet. Journal of Insect Physiology 45(10):959–964

Chen JM, Wang FJ (2000) The longterm warming over Taiwan and its relationship with the Pacific SST variability. Atmos Sci 28(3):221–242. [In Chinese]

Chen SC, Liao CM, Chio CP, Chou HH, You SH, Cheng YH (2010) Lagged temperature effect with mosquito transmission potential explains dengue variability in southern Taiwan: insights from a statistical analysis. Science of the Total Environment 408(19):4069–4075

Gubler DJ (2011) Dengue, urbanization and globalization: the unholy trinity of the 21st century. Tropical Medicine and Health 39:1–9

Guo HR, Lung SC (2004) Effects of climate changes on the risk of occurrence of communicable diseases among Taiwanese residents. (NSC93EPAZ006003). National Science Council, Executive Yuan, Taipei. [In Chinese]

Guzman MG, Kouri G (2002) Dengue: an update. Lancet Infectious Diseases 2(1):33–42

Hsieh YH, Chen C (2009) Turning points, reproduction number, and impact of climatological events for multi wave dengue outbreaks. Tropical Medicine & International Health 14(6):628–638

Hsu HH, Chen CT (2002) Observed and projected climate change in Taiwan. Meteorology and Atmospheric Physics 79(1):87–104

Huang WC (2006) A study on vector index of dengue fever in relation to climate factors and case number. Unpublished master dissertation. China Medical University, Taiwan. [In Chinese]

Kuo YM, Chu HJ, Pan TY, Yu HL (2011) Investigating: common trends of annual maximum rainfalls during heavy rainfall events in southern Taiwan. Journal of Hydrology 409(3–4):749–758

Lin CM (2010) The influence and environmental meaning of urban heat island effect. Journal of Ecology and Environmental Sciences 3(1):1–15 [In Chinese]

Lin W, Tsai HC, Wang C, Wu KY (2005) The subtropical urban heat island effect revealed in eight major cities of Taiwan. WSEAS Transactions on Environment and Development 1:305–311

National Science Council (2011) Taiwan scientific report on climate change. National Science Council, Executive Yuan, Taipei. [In Chinese]

Pai HH, Lu YL (2009) Seasonal abundance of vectors at outdoor environments in endemic and nonendemic districts of dengue in Kaohsiung, south Taiwan. Journal of Environmental Health 71(6):56–60

Shang CS, Fang CT, Liu CM, Wen TH, Tsai KH, King CC (2010) The role of imported cases and favorable meteorological conditions in the onset of dengue epidemics. Plos Neglected Tropical Diseases 4(8), e775. doi:10.1371/journal.pntd.0000775

Shiu CJ, Liu SC, Chen JP (2009) Diurnally asymmetric trends of temperature, humidity, and precipitation in Taiwan. Journal of Climate 22(21):5635–5649

Taiwan Centers for Disease Control (Taiwan CDC) (2011) 'Distribution of *Aedes aegypti* in Taiwan' TaiwanCDC, [Online]. Retrieved April 14, 2012 from: http://dengue.nat.gov.tw/public/Data/167158571.pdf

Taiwan Centers for Disease Control (Taiwan CDC) (2012) 'Notifiable infectious diseases statistics system' Taiwan CDC, [Online]. Retrieved April 14, 2012 from: http://nidss.cdc.gov.tw/

Taiwan Central Weather Bureau (Taiwan CWB) (2012) 'Typhoon database' TaiwanCWB, [Online]. Retrieved March 9, 2012 from: http://rdc28.cwb.gov.tw/data.php

Taiwan Environmental Protection Agency (Taiwan EPA) (2010) Extreme events and disasters are the biggest threat to Taiwan. Typhoon Morakot. Environmental Protection Administration, Executive Yuan, Taipei

Tseng WC, Chen CC, Chang CC, Chu YH (2009) Estimating the economic impacts of climate change on infectious diseases: a case study on dengue fever in Taiwan. Climatic Change 92(1–2):123–140

United States Agency for International Development (USAID) (2010) The Asia-Pacific regional climate change adaptation assessment final report: findings and recommendations. USAID Asia, Washington, DC

Wang CH, Chen HL (1997) Effect of warning climate on the epidemic of dengue fever in Taiwan. Chinese J Public Health 16(6):455–465. [In Chinese]

Wang TL, Chang H (2002) Do the floods have the impacts over vector borne diseases in Taiwan? Annals of Disaster Medicine 1:43–50

Wen TH, Lin MH, Fang CT (2012) Population movement and vectorborne disease transmission: differentiating spatial–temporal diffusion patterns of commuting and noncommuting dengue cases. Annals of the Association of American Geographers 102(5):1026–1037

Wen TH, Lin NH, Chao DY, Hwang KP, Kan CC, Lin KC, Wu JT, Huang SY, Fan IC, King CC (2010) Spatialtemporal patterns of dengue in areas at risk of dengue hemorrhagic fever in Kaohsiung, Taiwan. International Journal of Infectious Diseases 14:e334–e343

Wu PC, Guo HR, Lung SC, Lin CY, Su HJ (2007) Weather as an effective predictor for occurrence of dengue fever in Taiwan. Acta Tropica 103(1):50

Yu HL, Yang SJ, Yen HJ, Christakos G (2011) A spatiotemporal climatebased model of early dengue fever warning in southern Taiwan. Stochastic Environmental Research and Risk Assessment 25(4):485–494

Yu PS, Yang TC, Kue CC (2006) Evaluating longterm trends in annual and seasonal precipitation in Taiwan. Water Resources Management 20:1007–1023

Chapter 17
Summary and Concluding Remarks

Rais Akhtar

It has been known for thousands of years that weather and climate have wide ranging impacts on human health and well-being as described by Hippocrates in his fifth-century B.C. famous treatise: On Airs Waters and Places:

"Whoever wishes to investigate medicine properly, should proceed thus: in the first place to consider the seasons of the year, and what effects each of them produces for they are not at all alike, but differ much from themselves in regard to their changes" (cited in Adams 1891). Increasing recognition and research interest in the process of climate change has led to a growing concern by researchers in assessing the potential mechanisms by which changes in climate could influence population health (Haines et al. 2006). Thus, climate change represents a range of environmental hazards and will affect populations wherever the current burden of climate sensitive disease is high – such as the urban poor in low and middle income countries and landless farming communities. For example, impoverished communities are increasingly experiencing reduced access to safe fresh drinking water as well as food insecurity associated with climate change impacts.

Understanding the current impact of weather and climate variability on the health of populations is the first step towards assessing future health impacts of climate change.

About one fifth of the world's population live in South Asia. There are many reasons to be concerned about the impacts of climate change on this region, many parts of which experience apparently intractable poverty. The health problems caused by climate change in South Asia have been conceptualized as having three tiers of linked effects. In this framework, primary health effects are considered the most causally direct impacts of climate change. They include increased mortality

R. Akhtar (✉)
International Institute of Health Management and Research (IIHMR), New Delhi, India
e-mail: raisakhtar@gmail.com

© Springer International Publishing Switzerland 2016

R. Akhtar (ed.), *Climate Change and Human Health Scenario in South and Southeast Asia*, Advances in Asian Human-Environmental Research,
DOI 10.1007/978-3-319-23684-1_17

and morbidity during heat waves and other "natural" disasters as climatic changes contribute to the increasing frequency and intensity of such disasters. Secondary effects include those resulting from ecological changes that alter the epidemiology of some infectious and chronic diseases. Tertiary effects refer to impacts on health of large-scale events with complex, multidimensional economic and political causation, including migration, famine, and conflict.

Climate change scenario in the South and Southeast Asia has been a major focus in the recent assessment of the IPCC that indicates warming trends and increasing temperature extremes have been observed across most of the Asian region over the past century (high confidence) (IPCC 2014).

The IPCC 4th Assessment Report had earlier stated that climate change, in particular increased risk of floods and droughts, is expected to have severe impact on South Asian countries, since their economies rely heavily on agriculture, natural resources, forestry, and fisheries sectors. About 65 % of the population of South Asia live in rural areas and account for approximately 75 % of the poor who are most vulnerable to the adverse impacts of climate change. Such impacts include increased intensity of rainfall events, the impacts of El Niño on the Monsoon pattern, and increased risk of critical temperature being exceeded. The Southeast Asia too is a hot spot for emerging infectious diseases, including those with pandemic potentials. Emerging infectious diseases have exacted heavy public health and economic tolls. Severe acute respiratory syndrome rapidly decimated the region's tourist industry. Influenza A H5N1 has had a profound effect on the poultry industry (Clark 2003).

Since 2008 the WHO has shown leadership in raising awareness of the threats posed by climate change to health. Specifically, WHO has provided evidence, technical guidance, and piloted approaches to protect health from climate risks. The First International Conference on Health and Climate, which was held at the WHO headquarters in Geneva in August 2014, aimed to (1) enhance resilience and protect health from climate change, (2) identify the health benefits associated with reducing greenhouse gas emissions and other climate pollutants, and (3) support health-promoting climate change policies (Neira 2014).

Thus, if climate change is left unchecked, there is the likelihood of severe, pervasive, and irreversible impacts on people and ecosystems. With these words, the world's scientists have made an impassioned appeal to the world's policymakers to combat what has been described as the challenge of our times (IPCC 2014).

We are aware of the scientific explanations that climate change occurs because excessive amount of greenhouse gases are, and have been, emitted into the atmosphere due to human activity associated with industrialization. Human influence on the climate system is clear, and recent anthropogenic emissions of greenhouse gases are the highest in history. Earth's atmosphere has already warmed by 0.85 °C from 1880 to 2012. Recent climate changes have had widespread impacts on human and natural systems (IPCC 2014).

In view of the above, there is a realization now both among developed and developing countries that climate change makes us all vulnerable. This awareness has resulted in successful negotiations at Lima in December 2014, as well as an agreement between US and China. After Paris Climate Agreement, it is hoped that the world is likely to see increasing efforts to address climate change mitigation.

With a certain amount of climatic change having occurred and with future climatic changes locked in, there is now an urgent need to advance a more concrete and systematic and region-specific approach to implementing policies of health protection. Given the complexity of climate change impacts and adaptive responses, these policies and programs need to be multi-sectoral and to occur in coordination with health, environment, and social services planners.

Moreover, the major challenges in health research in developing countries are inadequate budgetary allocation to health sector and health research, inequitable distribution of health infrastructure within the country, deficiencies in surveillance system, lack of a cross-sectoral and interdisciplinary collaborative approaches in research, and lack of awareness about climate change and associated hazards for population health. Therefore, the need of the hour is to generate relevant information for policymakers on region's vulnerability, impacts, and adaptation in the context of a changing climate and its impact on human health in developing countries.

This book comprises studies from South and Southeast Asian countries with a focus on India, Nepal, Bangladesh, Indonesia, Malaysia, Thailand, and Taiwan. It presents a regional analysis pertaining to climate change and human health, focusing on climate change adaptation strategies in geographically and socioeconomically varied countries of South and Southeast Asia.

South and Southeast Asian countries – e.g., India, Pakistan and Bangladesh, and Thailand – are increasingly suffering from climate change-related heat waves and flooding. More than four thousand people have died of severe heatstrokes in Pakistan and India in two distinct heat wave events during May and June, 2015. A World Bank report warned that "countries in the South East Asia region are particularly vulnerable to the sea-level rise, increases in heat extremes, increased intensity of tropical cyclones, and ocean warming and acidification because many are archipelagos located within a tropical cyclone belt and have relatively high coastal population densities" (The World Bank 2013).

Researchers have identified vulnerable regions in Southeast Asia by analyzing data on spatial distribution of various climate-related hazards (cyclones/typhoons, drought risks, landslides, sea level rise, population density, and adaptive capacity). Water is a key medium through which climate change impacts upon human populations and ecosystems, particularly due to predicted changes in water quality and quantity. Climate change presents a serious obstacle to the realization of the rights to water and sanitation (SIDA n.d). Analysis of recent data from South and Southeast Asia reveals the pattern of access to safe drinking water to the population. Inequalities are not widespread, though improved water access is better in countries such as Bhutan, Maldives, Malaysia, Thailand, India, and Pakistan.

However, there are wide disparities in using improved sanitation, with Maldives, Malaysia, Thailand, and Sri Lanka are better off in comparison with other countries. India, Nepal, and Cambodia fared poorly in the provision of improved sanitation.

Such are the vulnerabilities encompassing the region of South and Southeast Asia, and a microlevel region-specific studies will help understand the climate change impact scenario in the context of human health and support to formulate holistic policies with a focus on spatial development and applicable to such a geographically diversified region.

References

Adams F (1891) The genuine works of Hippocrates. Green and Co; Petersen, New York
Clark E (2003) Sars strikes down Asia tourism. BBC News, Thursday, 15 May 2003, 11:08 GMT 12:08 UK
Haines A, Kovats RS, Campbell-Lendrum D, Corvalan C (2006) Climate change and human health: impacts, vulnerability and mitigation. The Lancet 367:2101, June 24
IPCC (2014) Climate change: impact, adaptation and mitigation, WG II, AR5, Chapter 24 Asia. Cambridge University Press, Cambridge, U.K./New York
Neira M (2014) The 2014 WHO conference on health and climate. Bull World Health Org 92:596
SIDA (n.d.) Climate change and the human rights to water and sanitation: position paper, www.sida.se
The World Bank (2013) Warmer world threatens livelihoods in South East Asia, Press Release, June 19, Washington, DC

Index

Printed by Printforce, the Netherlands